Innovation and Knowledge Creation in an Open Economy
Canadian Industry and International Implications

This study of innovation – its intensity, the sources used for knowledge creation, and its impacts – is based on a comprehensive survey of innovation in Canadian manufacturing firms. The study pays attention to the different actors in the system, who both compete with and complement one another. It investigates how innovation regimes differ across firm sizes and across industries. Owing to the high degree of foreign investment in Canada, special attention is given to the performance of foreign-owned firms. The authors compare the Canadian innovation regime with results of studies of other industrialized countries. The picture of a typical innovator that emerges is a firm that combines internal resources and external contacts to develop a set of complementary innovative competencies and strategies. The study finds that innovating firms not only depend on R&D but also rely on ideas and technology from various other sources, both internal and external to the firm.

John R. Baldwin is Director of Micro-Economic Analysis, Division of Statistics, in Statistics Canada, the Canadian federal government's bureau of statistics. He taught in the Department of Economics at Queen's University from 1971 to 1990, was a senior research director at the Economics Council of Canada and a faculty member of the Canadian Centre for Management Development, and is currently an adjunct professor at Carleton University in Ottawa. Dr. Baldwin is the author of numerous articles and analyses examining industrial organization, structural change and adjustment, and regulation. He is also the author of *The Dynamics of Industrial Competition: A North American Perspective* (Cambridge University Press, 1995; paperback edition, 1998).

Petr Hanel is Professor of Economics at the Université de Sherbrooke and a member of the Centre interuniversitaire de recherche sur la science et la technologie in Montreal, Canada. He is a frequent consultant to the World Bank and other international aid agencies, his principal areas of professional interest ranging from the economics and management of technological change to international economics. Professor Hanel is the author of *The World Trade Organisation and Patterns of International Trade* and *Technology and Canadian Export of Machinery for Paper and Wood Processing Industries*, as well as numerous articles and studies dealing with economic aspects of technological change, industrial and commercial policies, and international trade.

Innovation and Knowledge Creation in an Open Economy

Canadian Industry and International Implications

JOHN R. BALDWIN
Statistics Canada

PETR HANEL
Université de Sherbrooke

PUBLISHED BY THE PRESS SYNDICATE OF THE UNIVERSITY OF CAMBRIDGE
The Pitt Building, Trumpington Street, Cambridge, United Kingdom

CAMBRIDGE UNIVERSITY PRESS
The Edinburgh Building, Cambridge CB2 2RU, UK
40 West 20th Street, New York, NY 10011-4211, USA
477 Williamstown Road, Port Melbourne, VIC 3207, Australia
Ruiz de Alarcón 13, 28014 Madrid, Spain
Dock House, The Waterfront, Cape Town 8001, South Africa

http://www.cambridge.org

© John R. Baldwin and Petr Hanel 2003

This book is in copyright. Subject to statutory exception
and to the provisions of relevant collective licensing agreements,
no reproduction of any part may take place without
the written permission of Cambridge University Press.

First published 2003

Printed in the United Kingdom at the University Press, Cambridge

Typeface Times Ten 10/13 pt.   *System* LATEX 2_ε [TB]

A catalog record for this book is available from the British Library.

Library of Congress Cataloging in Publication Data
Baldwin, John R. (John Russel), 1945–
Innovation and knowledge creation in an open economy : Canadian industry and international implications / John R. Baldwin, Petr Hanel.
p.   cm.
Includes bibliographical references and index.
ISBN 0-521-81086-8
1. Technological innovations – Canada – Management. 2. Research, Industrial – Canada – Management. 3. Knowledge management – Economic aspects – Canada. 4. Manufacturing industries – Technological innovations – Canada. 5. International business enterprises – Technological innovations – Canada – Management. 6. Competition, International. I. Hanel, Petr. II. Title.
HD45 .B26 2003
338′.064′0971 – dc21
                                                                2002067407

ISBN 0 521 81086 8 hardback

To Helena and Adrianne

Contents

List of Tables and Figures	*page* xv
Acknowledgments	xxvii

1	The Economics of Knowledge Creation	1
	1.1 Introduction	1
	1.2 Innovation: Crosscutting Themes	2
	1.2.1 The Nature of Innovation: Core Framework	2
	1.2.2 Heterogeneity of Innovation Regimes and the Environment	4
	1.2.3 Knowledge Externalities, Market Imperfections, and Diffusion	8
	1.3 The Economic Themes	11
	1.3.1 The Nature of Innovation Outputs	12
	1.3.2 The Nature of Innovation Inputs	14
	1.4 The Organization of the Study and Principal Findings	16
	1.4.1 The Innovation Survey	16
	1.4.2 Innovation Intensity	17
	1.4.3 The Sources of Knowledge	18
	1.4.4 Research and Development	19
	1.4.5 Effects of Innovation	20
	1.4.6 Innovation and Research and Development in Small Versus Large Firms	21
	1.4.7 Innovation Regimes and Type of Innovation	22
	1.4.8 The Use of Property Rights	23
	1.4.9 Multinationals and the Canadian Innovation Process	23
	1.4.10 Financing and the Cost of Innovation	24
	1.4.11 Technology, Joint Ventures, and the Transfer of Innovation	26

	1.4.12 Strategic Capabilities in Innovative Businesses		27
	1.4.13 Firm and Industry Characteristics Associated with Innovation		27
2	The Innovation Survey		29
	2.1 Introduction		29
	2.2 Defining Innovation		29
	2.3 Measuring Innovation		34
	2.4 Understanding the Innovation Regime		36
	2.5 Background on the 1993 Canadian Survey of Innovation and Advanced Technology		38
3	Patterns of Innovation: Intensity and Types		43
	3.1 Introduction		43
	3.2 The Incidence of Innovation		45
		3.2.1 Types of Innovations Introduced	51
	3.3 Characteristics of Innovation		52
		3.3.1 Originality and Type of Innovation	52
		3.3.2 The Number of Innovations Introduced	55
		3.3.3 Features of Innovation	57
	3.4 Conclusion		59
4	Sources of Innovations		63
	4.1 Introduction		63
	4.2 Internal Sources of New Ideas		68
		4.2.1 Industry Differences in Internal Sources of Innovative Ideas	70
		4.2.2 Internal Sources of New Ideas by Industry Branch	72
		4.2.3 Combinations of Internal Sources	72
	4.3 External Sources of Ideas and Information for Innovation		74
		4.3.1 Industry Sector Differences in External Sources of Innovative Ideas	78
		4.3.2 Product Versus Process Innovators	81
		4.3.3 Spillovers, Market Transactions, and Infra-technologies	83
	4.4 Sources of New Technology		87
		4.4.1 Industry Sector Differences in Internal Sources of Technology	88
		4.4.2 External Sources of Ideas for New Technology	89
		4.4.3 Relationship Between Internal and External Sources of Technology	90
	4.5 Conclusion		91
5	Research and Development and Innovation		96
	5.1 Introduction		96
	5.2 Definitions: Expenditures on Research and Development		96
	5.3 Expenditures on Innovation		98

	5.4	Research and Development Activity	103
	5.4.1	Frequency of Research and Development	103
	5.4.2	Organization of R&D Facilities	104
	5.4.3	Collaborative Research and Development	105
	5.5	The Valuation of Innovation Strategies by R&D Type	108
	5.6	Industry Differences in Research and Development	109
	5.6.1	Frequency of Research and Development	110
	5.6.2	Organization of Research and Development	112
	5.6.3	Other Research-Related Activities	115
	5.6.4	Patterns of Collaborative Agreements	118
	5.7	Differences in the Research and Development Characteristics of Innovative and Non-innovative Firms	121
	5.7.1	Distinguishing Innovative and Non-innovative Firms	121
	5.7.2	R&D Activity in Innovative and Non-innovative Firms	122
	5.7.3	R&D Differences Across Novelty Types	125
	5.8	Conclusion	127
6		Effects of Innovation	130
	6.1	Introduction	130
	6.2	Changes in Organization of Production Brought About by Innovation in General	133
	6.3	Effect of the Most Profitable Innovation on a Firm's Demand, Share of the Market, Factor Costs, and Profitability	138
	6.3.1	Innovation and Outcomes	138
	6.3.2	Innovation and Sources of Reduction in Production Costs	140
	6.4	Innovation and Government Regulation	142
	6.5	Effect of Innovation on Employment and Skills of Workers	144
	6.5.1	Employment Effect Depends on the Type of Innovation	145
	6.5.2	Discriminant Analysis of Employment Effects	146
	6.6	Innovation and Export Sales	153
	6.7	Conclusion	154
7		Innovation and Research and Development in Small and Large Firms	156
	7.1	Introduction	156
	7.2	Do Small Firms Suffer from an Innovation Gap?	159
	7.3	Sources of Innovations	166
	7.4	Research and Development Activity	170
	7.4.1	Frequency of Research and Development	170
	7.4.2	Organization of R&D Facilities	172
	7.4.3	Collaborative Research and Development	174
	7.5	The Link Between R&D and Innovation	175

	7.6	Impediments Associated with Innovation	179
	7.7	Conclusion	181
8	Innovation Regimes and Type of Innovation	185
	8.1	Introduction	185
	8.2	The Definition of Innovation	187
	8.3	The Data	187
	8.4	How Prevalent Is Innovation?	188
	8.5	Types of Innovation	191
	8.6	Features of Innovation	194
	8.7	The Benefits of Innovation	196
	8.8	Internal Sources of Innovation Ideas	199
	8.9	External Sources of Innovation Ideas	201
	8.10	The Importance of R&D Activity	203
	8.11	Internal Sources for New Technologies	204
	8.12	External Sources of New Technologies	205
	8.13	Protection for Intellectual Property	207
	8.14	Impediments to Innovation	209
	8.15	What Happens to Workers in Innovative Firms?	212
	8.16	Conclusion	214
9	The Use of Intellectual Property Rights	219
	9.1	Introduction	219
	9.2	Canada in an International Context	221
	9.3	Forms of Intellectual Property Protection	224
	9.4	Use of Intellectual Property Rights by Manufacturing Firms Operating in Canada	226
	9.5	Effectiveness of Intellectual Property Protection	229
		9.5.1 Overall Evaluations	229
		9.5.2 Canada/United States Comparisons	233
	9.6	Large Versus Small Firms	234
	9.7	Differences in the Use of Intellectual Property Protection by Innovative and Non-innovative Firms	237
		9.7.1 Distinguishing Innovative and Non-innovative Firms	237
		9.7.2 Intellectual Property Use and Innovativeness	237
		9.7.3 Innovation Differences Across Size Classes	240
		9.7.4 Intellectual Property Protection and the Characteristics of Innovations	244
	9.8	Industry Differences	249
		9.8.1 Industry Use of Intellectual Property Protection	251
		9.8.2 Industry Effectiveness of Intellectual Property Protection	253
	9.9	Multivariate Analysis of Intellectual Property Use and Effectiveness	255
	9.10	Conclusion	260

10	Multinationals and the Canadian Innovation Process	265
	10.1 Introduction	265
	10.2 Characteristics of Canadian and Foreign-Owned Firms: Size and Industry Differences	270
	10.3 Incidence and Organization of R&D	273
	10.3.1 Sector Differences	277
	10.3.2 Probabilistic Models of R&D Organization	279
	10.3.3 R&D Collaboration Partnerships	284
	10.4 Sources of New Ideas and Inspiration for Innovation	287
	10.4.1 How Important as Sources of Innovative Ideas Are Foreign Parents and Sister Companies of Foreign-Owned Firms?	288
	10.4.2 Internal and External Sources of Technology	294
	10.4.3 Transfer of Technology	294
	10.5 Impediments to Innovations	297
	10.6 Do Canadian-Owned Firms Innovate More or Less Than Foreign Affiliates?	299
	10.6.1 Do Foreign-Owned Firms Introduce Process Innovations More Often Than the Canadian-Owned Firms?	302
	10.6.2 Originality of Innovations	303
	10.6.3 The Number of Innovations Introduced	306
	10.7 Use of Intellectual Property Rights	308
	10.8 Effects of Innovation	311
	10.8.1 Do Foreign Affiliates Export More or Less Than Canadian-Owned Firms?	312
	10.8.2 The Impact of Innovation on Employment and Skill Requirements	313
	10.9 Ownership Versus Trade Orientation	316
	10.10 Conclusions	317
11	Financing and the Cost of Innovation	322
	11.1 Introduction	322
	11.2 The Costs of Innovation	323
	11.2.1 Product-Process Innovation Cost Differences	325
	11.3 The Sources of Funds for Innovation	328
	11.4 Government Funding of Innovation	333
	11.4.1 Tax Credits	334
	11.4.2 Use of Tax Credits	335
	11.4.3 R&D Activity, Tax Credits, and Innovation Performance	339
	11.5 Conclusion	345
12	The Diffusion of Innovation	349
	12.1 Introduction	349

	12.2 Technology Transfer Agreements	350
	12.2.1 Characteristics of Technology Transfer Agreements	353
	12.2.2 Rights and Restrictions Associated with Technology Transfer	354
	12.2.3 Restrictions Attached to Technology Transfer Agreements	357
	12.2.4 A Probabilistic Model of Technology Transfer via Licensing	360
	12.2.5 Probability of Continuous Versus One-Time Transfers of Technology	363
	12.3 Participation in Joint Ventures and Strategic Alliances	364
	12.3.1 The Incidence of Joint Ventures and Strategic Alliances	365
	12.3.2 The Relationship Between R&D Collaboration, Joint Ventures, and Innovation	366
	12.3.3 Modelling the Probability of a Firm Using a Joint Venture	368
	12.4 Diffusion of Innovation	371
	12.4.1 The Users of Innovations	371
	12.4.2 Interindustry Flows of Innovations	372
	12.5 Conclusion	374
13	Strategic Capabilities in Innovative Businesses	378
	13.1 Introduction	378
	13.2 Strategic Capabilities and Competencies	380
	13.3 Innovation and Technology Strategies	383
	13.4 Production Strategies	385
	13.5 Marketing and Product-Based Strategies	387
	13.6 Human Resource Strategies	389
	13.7 Specialized Management Strategies	392
	13.8 Conclusion	394
14	Determinants of Innovation	397
	14.1 Introduction	397
	14.2 Empirical Model	400
	14.2.1 The Model	400
	14.2.2 Innovation Variable	401
	14.2.3 Explanatory Variables	402
	14.2.4 Estimation Procedures	409
	14.3 Regression Results	412
	14.3.1 Incidence of Innovation	412
	14.3.2 Patent Use	417
	14.3.3 Type of Innovation	418
	14.3.4 Novelty of Innovation	421
	14.4 Conclusion	423

| 15 | Summary | 427 |

15.1 Innovation Types 428
15.2 The Impact of Innovation 429
15.3 The Innovation Process 431
    15.3.1 Sources of Innovation 431
    15.3.2 The R&D Process 434
    15.3.3 Technology Acquisition 435
    15.3.4 Intellectual Property Rights 436
    15.3.5 Financing 438
    15.3.6 Complementary Strategies for Innovative Firms 439
15.4 Heterogeneity 440
    15.4.1 Differences in Innovation Regimes Across Industries 441
    15.4.2 Differences in Innovation Regimes Across Firm Size Classes 445
    15.4.3 Differences in Innovation Regimes Across Novelty Types 450
    15.4.4 Differences in Innovation Regimes by Nationality 451

Appendix The Innovation and Advanced Technology Survey 455
    A.1 Background 455
    A.2 The 1993 Survey of Innovation and Advanced Technology 456
    A.3 Respondents' Guide: Survey of Innovation and Advanced Technology 484

References 491

Index 507

Tables and Figures

TABLES

2.1	Alternate Measures of Innovation Intensity	*page* 35
2.2	The Types of Sampling Units	39
3.1	The Percentage of Firms That Introduced, or Were in the Process of Introducing, an Innovation During 1989–91, by Size Class	45
3.2	The Percentage of Firms That Introduced, or Were in the Process of Introducing, an Innovation During 1989–91, by Size Class and Individual Industry	47
3.3	Comparison of Technology Patterns in the United Kingdom and United States	49
3.4	The Percentage of Firms That Introduced, or Were in the Process of Introducing, an Innovation During 1989–91, by Industrial Sector and Size Class	50
3.5	The Percentage of Innovating Firms That Introduced, or Were in the Process of Introducing, Product or Process Innovations	51
3.6	Distribution of Product and Process Innovation, by Industrial Sector (% of Firms)	52
3.7	The Distribution of Innovations, by Novelty Type and Industrial Sector (% of Firms)	53
3.8	Distribution of Innovations by Novelty Type and by Individual Industrial Sector (% of Firms)	55
3.9	The Number of Product and Process Innovations Introduced	56

3.10	The Number of Product and Process Innovations Introduced, by Industrial Sector	57
3.11	Features of the Most Profitable Innovation Reported by Size Class, Novelty of Innovation, and Industrial Sector (% of Innovators)	58
3.12	Innovation Features, by Individual Industry (% of Innovators)	60
4.1	Internal Sources of Innovative Ideas, by Novelty Type (% of Innovators Using a Source)	69
4.2	Internal Sources of Innovative Ideas, by Industrial Sector (% of Innovators Using a Source)	70
4.3	Internal Sources of Innovative Ideas, by Individual Industry (% of Innovators Using a Source)	73
4.4	Internal Sources of Innovative Ideas Reported by Firms That Used and That Did Not Use Ideas from Their Own R&D, by Industrial Sector (% of Innovators Using a Source)	74
4.5	Principal External Sources of Ideas for Innovations, by Novelty Type (% of Innovators Using a Source)	76
4.6	Main External Sources of Ideas for Innovations, by Industrial Sector (% of Innovators Using a Source)	78
4.7	Main External Sources of Innovative Ideas, by Individual Industry (% of Innovators Using a Source)	80
4.8	Sources of Internal and External Ideas for Innovations for Product Versus Process Innovations (% of Firms Using a Source)	82
4.9	Correlations Between Inside and Outside Sources of Innovative Ideas	84
4.10	Internal Sources for Technology Used by Innovators, by Novelty Type (% of Innovators Using a Source)	88
4.11	Internal Sources for Technology Required for Innovation, by Industrial Sector (% of Innovators Using a Source)	89
4.12	Correlations Between Internal and External Technology Sources	91
5.1	Percentage of Firms Conducting Research and Development	104
5.2	Percentage of Firms Conducting Research and Development Using Different R&D Delivery Mechanisms	104

5.3	Differences in R&D Delivery Mechanisms Between Ongoing and Occasional Performers (% of Firms)	105
5.4	Percentage of Firms Conducting R&D with R&D Collaborative Agreements	107
5.5	Patterns of Collaboration by Firms (% of Firms Indicating R&D Collaboration)	107
5.6	Average Scores Attached to Innovation Strategies (Mean Score on a Scale of 1–5)	109
5.7	Intensity of Research and Development, by Individual Industry	111
5.8	Delivery Mechanisms for R&D Activity, by Individual Industry	113
5.9	Alternate Measures of the Importance of Research, by Individual Industry	116
5.10	Patterns of Collaboration, by Industrial Sector (% of Firms Reporting Collaborative Agreement)	119
5.11	Frequency of R&D Activity, by Innovator Versus Non-innovator (% of Firms)	122
5.12	Delivery Mechanisms for R&D Activity in Innovators Versus Non-innovators (% of Firms)	123
5.13	Percentage of Innovators Versus Non-innovators Forming R&D Collaborative Agreements	124
5.14	Source of R&D Collaborative Agreements for Innovators Versus Non-innovators (% of Collaborators)	124
5.15	Regional Patterns of R&D Collaborative Agreements in Large Innovative Firms (% of Collaborators)	125
5.16	Regional Patterns of R&D Collaborative Agreements in Large Non-innovative Firms (% of Collaborators)	125
5.17	Regional Patterns of R&D Collaborative Agreements in Small Innovative Firms (% of Collaborators)	126
5.18	Regional Patterns of R&D Collaborative Agreements in Small Non-innovative Firms (% of Collaborators)	126
5.19	Percentage of Firms Performing R&D, by Type of Innovation	127
6.1	Changes in Production Organization Associated with Innovation, by Industrial Sector (% of Innovators)	135
6.2	Changes in Production Organization Associated with Innovation, by Type of Innovation (% of Innovators)	137
6.3	Effects of Innovation on Profit, Factor Costs, and Demand, by Industrial Sector (% of Innovators)	139

6.4	The Percentage of Innovative Firms That Improved Their Market Share, by Size Class	140
6.5	Effects of Innovation on Factor Costs, Working Conditions, and Technical Capabilities, by Industrial Sector (% of Innovators)	141
6.6	Effects of Introducing Innovation in Response to Government Regulations, by Industrial Sector (% of Innovators)	143
6.7	Effects of Innovation on the Number and Skill Requirements of Workers in the Firm (% of Innovators)	144
6.8	Effects of Innovation on the Number and Skill Requirements of Workers, by the Type of Innovation (% of Innovators)	146
6.9	Parameter Estimates of the Function Discriminating Between Firms in Which Innovation Increased or Decreased Employment	150
6.10	Percentage of Observations Classified into Employment-Change Groups	151
6.11	Export Incidence and Export Intensity of Innovative Firms (Employment Weighted)	153
7.1	Percentage of Firms with Innovations, by Size Class	160
7.2	The Percentage of Small and Large Firms That Introduced, or Were in the Process of Introducing, an Innovation in 1989–91, by Individual Industry	161
7.3	Percentage of Innovators with Product Versus Process Innovations, by Size Class	162
7.4	Number of Product and Process Innovations Introduced, by Size Class	163
7.5	Novelty of Innovation, by Size Class (% of Innovations)	164
7.6	Distribution of Sales, by Innovation Category and Size Class (% of Total Sales)	164
7.7	Exports as a Percentage of Sales, by Innovation Type and Size Class	165
7.8	Main Sources of Ideas for Innovations, by Size Class (% of Innovators)	168
7.9	Main Sources of Technologies Associated with Innovations, by Size Class (% of Innovators)	170
7.10	Percentage of Firms Conducting Research and Development, by Size Class	171

7.11	Delivery Mechanisms for Research and Development, by Size Class (as % of Firms Conducting R&D)	172
7.12	Delivery Mechanisms for Research and Development, by Type of R&D Performer and by Size Class	173
7.13	Percentage of Firms Conducting R&D with Collaborative Agreements, by Size Class	174
7.14	Innovative Intensity for Conductors of R&D, by Size Class (% of Firms)	176
7.15	R&D Intensity for Innovators and Non-innovators, by Size Class (% of Firms)	178
7.16	Impediments to Innovation, by Size Class (% of Firms)	181
8.1	Innovation Intensity, by Novelty of Innovation and Size Class (% of Firms, Employment Weighted)	190
8.2	Effects of Innovation, by Novelty of Innovation (% of Innovators, Employment Weighted)	197
8.3	Exports as a Percent of Sales, by Novelty of Innovation (Employment Weighted)	199
8.4	External Sources of Ideas for Innovation, by Novelty of Innovation (% of Innovators, Employment Weighted)	202
8.5	R&D Use, by Novelty of Innovation (% of Innovators, Employment Weighted)	203
8.6	External Sources for Technology, by R&D and Novelty of Innovation (% of Innovators, Employment Weighted)	207
8.7	Impact of Innovation on Workers, by Novelty of Innovation (% of Innovators)	213
9.1	Multiple Use of Statutory Forms of Intellectual Property Protection (% of Firms)	226
9.2	Usage of Individual Forms of Intellectual Property (% of Firms)	227
9.3	Effectiveness of Intellectual Property Protection (Mean Score on a Scale of 1–5)	230
9.4	Effectiveness of Intellectual Property Protection in the United States (Mean Score on a Scale of 1–7)	234
9.5	Multiple Use of Intellectual Property Protection, by Size Class (% of Firms)	235
9.6	Usage of Individual Forms of Intellectual Property Protection, by Size Class (% of Firms)	235
9.7	Effectiveness of Intellectual Property Protection, by Size Class (Mean Score on a Scale of 1–5)	236

9.8	Effectiveness of Alternative Means of Protecting New Products and Processes from Imitation: For Innovators and Non-innovators, Users and Nonusers of Intellectual Property Protection (IPP) (Mean Score on a Scale of 1–5)	241
9.9	A Comparison of Innovation Intensity, R&D, and Intellectual Property Use, by Size Class (% of Firms)	242
9.10	A Comparison of R&D and Intellectual Property Use for Innovators, by Size Class (Indexed to Firms Reporting Sales from Major Product Innovation)	243
9.11	Usage of Intellectual Property Protection for Major Innovation of Innovator, by Region (% of Innovators)	244
9.12	Usage of Intellectual Property Protection for Major Product and Process Innovations (% of Innovators)	246
9.13	Usage of Intellectual Property Protection, by Region and by Product or Process (% of Firms with a Major Innovation That Reported Making Use of Intellectual Property Rights for This Innovation)	247
9.14	World-First/Non-World-First Usage of Intellectual Property Protection for Major Innovation, by Region (% of Innovators)	249
9.15	Industry Patterns of Patent Use	251
9.16	Usage of Individual Forms of Intellectual Property Protection, by Individual Industry (% of Firms)	252
9.17	Effectiveness of Individual Forms of Property Protection, by Industry Sector (Mean Score on a Scale of 1–5)	254
9.18	Regression Coefficients for Utilization of Intellectual Property Protection	257
9.19	Regression Coefficients for Effectiveness Score Attached to Intellectual Property Protection	258
10.1	Composition of the Survey Sample, by Nationality and by Size Class (% of Manufacturing Firms)	270
10.2	Distribution of Firms, by Nationality and Industrial Sector (% of Firms)	272
10.3	Incidence and Delivery Mechanism for R&D, by Nationality (% of Firms)	274
10.4	Delivery Mechanism for R&D, by Industrial Sector and Nationality (% of Firms)	278

10.5	Multivariate Analysis: Performing R&D (All Firms)	281
10.6	Estimated Probability of Performing R&D	282
10.7	Probability Models: Type of Delivery Mechanism for R&D	283
10.8	Internal Sources of Innovative Ideas, by Nationality and Industrial Sector (% of Innovators)	289
10.9	Main External Sources of Innovative Ideas, by Industrial Sector and Nationality (% of Innovators)	291
10.10	Main Sources of Internal and External Innovation Ideas Reported by Foreign Affiliates, Broken Down by Whether Innovation Ideas Came from Related Firms (% of Innovators)	293
10.11	Principal Internal and External Sources of Technology, by Nationality and Industrial Sector (% of Innovators)	295
10.12	Type of Transfer for Technology and Associated License Restrictions (% of Firms)	296
10.13	Impediments to Innovation, by Industrial Sector and Nationality (% of Firms)	298
10.14	Percentage of Firms That Introduced, or Were in the Process of Introducing, an Innovation During the Period 1989–91, by Nationality and Size Class	300
10.15	Innovation Intensity, by Nationality, Size Class and Industrial Sector (% of Firms)	300
10.16	Product Versus Process Innovation Intensity, by Nationality and Size Class (% of Innovators)	302
10.17	Originality of Innovations, by Nationality and Industrial Sector (% of Innovators)	304
10.18	Number of Innovations Introduced and in Progress for Product Versus Process Innovations, by Nationality	307
10.19	Multiple Use of Intellectual Property Protection, by Nationality and Size Group (% of Firms)	309
10.20	Usage of Individual Forms of Intellectual Protection, by Nationality and Size Class (% of Firms)	309
10.21	Effectiveness of Intellectual Property Protection, by Nationality (Mean Score on a Scale of 1–5)	310
10.22	Effects of Innovation, by Nationality (% of Innovators)	311
10.23	Export Incidence and Propensity, by Industrial Sector and Nationality	314

10.24	The Effect of Innovation on the Number and Skill Requirements of Workers, by Nationality (% of Innovators)	316
10.25	Gradations of Innovation Activity, by Degree of Foreign Operations (% of Firms)	318
11.1	Distribution of Total Innovation Cost, by Stages of the Innovation Process (%)	324
11.2	Distribution of Innovation Cost, by Stages of the Innovation Process and by Type of Innovation (%)	326
11.3	Distribution of Innovation Cost, by Stages of the Innovation Process and by Industrial Sector (%)	327
11.4	Firms Financing Innovation Entirely from Internal Sources, by Size Class, Nationality, Novelty, and Sector (% of Innovators)	330
11.5	The Source of Funding of Major Innovations Not Wholly Funded Internally, by Industrial Sector and Novelty Type (% Distribution)	331
11.6	Percentage of Firms Who Conduct R&D and Claim Tax Credits for R&D	336
11.7	Firms Conducting R&D and Claiming Tax Credits, by Firm Category and by Industrial Sector (% of Firms)	337
11.8	R&D Tax Credit Claims, by Innovation Type and by Industrial Sector (% of Innovators)	339
11.9	Use of R&D and Innovation Success (% of Firms)	340
11.10	Use of Tax Credits and Innovation Success (% of Firms)	341
11.11	Industry Differences for R&D, Tax Credit Claims, and Innovation (% of Firms)	343
11.12	Percentage of Innovators Claiming Tax Credits, by Size Class and Industrial Sector	344
12.1	Characteristics of Technology Transfer Agreements for Those Firms Using External Technology Sources, by Size Class (% of Firms with Technology Transfer)	355
12.2	Type of Technology Transfer Agreement, by Sector and Novelty of Innovation (% of Firms with Technology Transfer)	356
12.3	Rights Associated with Technology Transfer Agreements, by Industrial Sector (% of Firms with Technology Transfer)	356

12.4	Territorial Restrictions and Exclusivity Clauses in Licensing Agreements (% of Firms with Licensing Agreement)	358
12.5	Regions to Which Territorial Restrictions Apply (% of Firms with Licensing Agreement)	360
12.6	Probit Regression Coefficients for Licensing and Joint Venture Models	362
12.7	Proportion of Innovating Firms That Formed a Joint Venture or Strategic Alliance to Produce Their Innovation, by Size Class	366
12.8	Proportion of Firms That Formed a Joint Venture or Strategic Alliance to Produce Their Innovation, by Industrial Sector and Novelty of Innovation	366
12.9	The Cross-Classification of Joint Ventures and Collaborative Research (% of Innovators)	367
12.10	Individual Industry Differences in the Importance of Joint Ventures and Collaborative Research	369
12.11	Distribution of Innovation Sales to Types of Purchasers, by Industrial Sector	372
12.12	Method of Transfer of Innovations to Other Firms, by Industrial Sector (% of Innovators)	373
12.13	Production and Use of Innovations, by Industrial Sector (Employment Weighted)	374
13.1	Emphasis Given to Innovation and Technology Strategies (% of Firms Scoring 4 or 5)	384
13.2	Emphasis Given to Innovation and Production Strategies (% of Firms Scoring 4 or 5)	386
13.3	Emphasis Given to Marketing Strategies (% of Firms Scoring 4 or 5)	388
13.4	Emphasis Given to Innovation and Human Resource Strategies (% of Firms Scoring 4 or 5)	391
13.5	Emphasis Given to Innovation and Management Strategies (% of Firms Scoring 4 or 5)	393
14.1	Summary of Dependent and Explanatory Variables Used in Multivariate Analysis	403
14.2	Regression Coefficients for the Determinants of Innovation Activity	413
14.3	Estimated Probability of Introducing an Innovation and Using Patents	414

14.4	Regression Coefficients for the Determinants of Product and Process Innovations	419
14.5	Estimated Probability of Introducing a Product Innovation and Using Patents, and Introducing a Process Innovation and Using Trade Secrets	420
14.6	Regression Coefficients for Determinants of World-First, Canada-First, and Other Types of Innovations	423
14.7	Estimated Probability of Introducing a World-First, Canada-First, or Other Innovation	424

FIGURES

8.1	Innovation Intensity, by Novelty Type	189
8.2	Innovation Intensity, by Individual Industry	190
8.3	Product-Process Innovations, by Novelty Type	193
8.4	Features of Process Innovations, by Novelty Type	195
8.5	Features of Product Innovations, by Novelty Type	195
8.6	Improved Interaction with Customers, by Novelty Type	197
8.7	Market-Share and Profitability Effects, by Novelty Type	198
8.8	Internal Ideas for Innovation, by Novelty Type	200
8.9	Internal Sources of Technology Ideas, by Novelty Type	204
8.10	External Sources of Technology, by Novelty Type	206
8.11	Usage of Intellectual Property Protection, by Novelty Type	208
8.12	Impediments to Innovation, by Novelty Type	211
9.1	National Patent Applications per Capita, 1992	222
9.2	Patent Filings in U.S. Market per R&D Scientists in Home Country, 1992	223
9.3	Distribution of Scores for Users and Nonusers of Patents	232
9.4	Incidence of Use of Intellectual Property in Innovative and Non-innovative Firms	238
9.5	Perceived Effectiveness of Intellectual Property Protection and Other Strategies in Innovative and Non-innovative Firms	239
10.1	Foreign Versus Domestic Firms with R&D Performed on a Regular Basis, by Size Class	275
10.2	Foreign Versus Domestic Firms with R&D Performed Occasionally, by Size Class	276

10.3	Foreign Versus Domestic R&D Performers with R&D Collaborative Agreements, by Size Class	284
10.4	Foreign- Versus Domestic-Firm Use of Different Collaboration Partners	286
10.5	Internal R&D Used for Sources of Ideas by Foreign and Domestic Multinationals, by Industry Sector	290
10.6	Use of External Sources of Ideas, by Foreign and Domestic Firms	292
10.7	Innovation Rate for Foreign and Domestic Multinationals, by Industry Sector	301
10.8	Originality of Innovations for Domestic Versus Foreign Multinationals	305
12.1	Transfers of Intellectual Property Rights, by Type of Innovator	358
12.2	Joint Ventures and Collaboration, by Industrial Sector	368

Acknowledgments

We wish to thank Guy Gellatly, Robert Gibson, and Valerie Gaudreault for their assistance in tabulating answers to the innovation survey and David Sabourin for his assistance with Chapter 14. We are also indebted to numerous colleagues in both North America and Europe who have commented on various chapters, including Richard Nelson, Alfred Kleinknecht, Pierre Mohnen, and John Vardalas. We also owe an intellectual debt to many others whose stimulating conversations have shaped many of the ideas in this book, including Anthony Arundel, David Audretsch, Wes Cohen, Bronwyn Hall, B. A. Lundvall, Jacques Mairesse, Don McFetridge, Luc Soete, and David Wolfe.

Innovation and Knowledge Creation in an Open Economy
Canadian Industry and International Implications

ONE

The Economics of Knowledge Creation

1.1 INTRODUCTION

Innovation is the dynamic force that changes the economy. It provides new products and processes. It generates productivity growth and leads to increases in the standard of living. It is at the heart of entrepreuneurship.

An analysis of innovation is a study in the economics of knowledge creation and application. Studies of innovation have not been as common as other types of studies in industrial organization – of scale economies, scope economies, sunk costs, multiplant economies, competition, and market structure. One of the reasons is that data allowing for broad descriptions of the innovation process have been lacking. Research has had to rely on case studies that are often unrepresentative of the innovation activity that takes place in the entire population. Case studies tend to focus on high-profile new products and processes. By definition, few firms are at the head of the class at any point in time, and focusing on them alone risks giving a distorted view of change.

This study makes use of the first comprehensive innovation survey to cover the Canadian manufacturing sector. The 1993 Innovation and Advanced Technology Survey, carried out by Statistics Canada, was uniquely designed for analytical purposes and differs in key respects from the standardized European Community Innovation Surveys (CIS).[1] Conducted by Statistics Canada in 1993, the innovation survey used here provides an overview of the complex process that produces innovation in Canadian manufacturing. This process is often referred to as the innovation regime

---

[1] See European Commission (1994).

or the innovation system, and it consists of the actors, sources of information, and networks in Canada and abroad, and outcomes associated with the production of innovations.

This book describes the innovation system of Canadian manufacturing firms. In doing so, we build on an emerging, rich survey-based literature that has developed in the economics of innovation. In this chapter, we describe the analytical framework that underpins subsequent chapters.

Innovation takes place via a system of economic actors. It involves a set of activities – ranging from arm's-length transactions between firms, to non-arm's-length transactions that are internal to firms, and finally to transactions with public institutions. As with all economic systems, it consists of a number of interactive parts, sometimes working at arm's length with one another as suppliers and customers and, at other times, working together in collaborative networks. This book describes how these parts fit together.

At the same time, we recognize that the parts fulfil different functions. Actors are different and they both compete with and complement one another. The actors that interact in the innovation system often operate in quite different ways. The participants either act consciously to coordinate decisions or, by acting competitively, influence or determine the overall innovative performance of the economy. There is no single model that serves to explain how an innovation system should or does work. Heterogeneity of purpose and function occurs.

In this system, large firms differ from small firms. Research and development (R&D)-based firms differ from production-based firms. Firms in industries that tend to originate innovations function differently from firms that operate in industries that ingest new materials and new machinery and equipment. Firms also differ in terms of their nationality. About half of all Canadian manufacturing firms are foreign-owned. Cross-border transactions with suppliers, customers, and partners provide them with access to information networks other than those available to domestically owned firms.

The next section presents the methodological hypotheses underlying our approach to the study.

1.2 INNOVATION: CROSSCUTTING THEMES

1.2.1 The Nature of Innovation: Core Framework

The organization of any study of innovation is perforce organized around a set of themes, whose choice depends upon a set of maintained hypotheses about how innovation occurs, or a set of issues whose interest depends upon the validity of a particular set of working hypotheses about how innovation takes place.

The first hypothesis relates to the nature of the business population. On the one hand, the Canadian economy might be described as one where the majority of firms search for innovations and only a minority succeed in the type of short three-year span covered by the survey. On the other hand, the economy may be one where only a minority of firms try to innovate and most of these succeed. If the first description were correct, then it is important to understand what characteristics of a firm lead to successful innovation and what causes a firm to try but to fail. In the second case, we need to understand what distinguishes an innovator from a non-innovator. Or in the case where there is a continuum of innovators, we would ask what distinguishes the more innovative from the less innovative.

Our study is based on the view that the latter description is closer to reality than the first – that only a minority of firms attempts to and successfully introduces major innovations. This view is based on evidence that the number of firms reporting major innovations is small. It leads us throughout this monograph to focus on descriptions of the innovators. As a variant, we also describe the difference between those who produce innovations that differ in terms of their novelty.

A second maintained hypothesis underlying this monograph is that innovation is a result of a process that not only requires firms to search for and create knowledge but also requires a firm to develop a number of complementary competencies.

As a result, a study of innovation needs to examine more than just the R&D intensity of firms. This is partially because innovators require competencies other than just R&D. They need technical competencies on the production side that are often resident in engineering departments.

Therefore, this study goes beyond an examination of the role that R&D plays. In contrast to more traditional studies of innovation that focus almost exclusively on the relationship between R&D and innovation, the present study recognizes that firms pursue a range of strategies, most of which are complementary to R&D.

Innovation requires a set of complementary strategies in many areas of the firm. For example, firms that innovate have a particularly difficult time finding funds for soft knowledge–based assets. This requires the development or acquisition of specific competencies in the area of finance to access highly specialized capital markets. Innovators also need skilled workers, and they need to inculcate them with firm-specific knowledge. This requires the development of human-resource strategies for training and the retention of workers whose training costs are substantial. Innovators also have to penetrate new markets, and this requires special marketing capabilities. In sum, this means that innovators need to develop a range of competencies in addition to the scientific skills that are key to the innovation process.

In pursuing our study of the innovation process in Canada, we are guided by both of the maintained hypotheses outlined above. Our prime interest is the characteristics of innovators. And this interest is wide-ranging. But in pursuing this study, we have organized our facts around a set of themes that reemerge in one chapter after another. These involve, on the one hand, the nature of diversity in the innovation process, and on the other hand, the particular problems that knowledge externalities create.

1.2.2 Heterogeneity of Innovation Regimes and the Environment

1.2.2.1 Sources of Diversity

The competitive and scientific environment of an industry conditions both the nature of innovations that are produced therein and the actors that function in these markets. But there is considerable heterogeneity in both the actors and the nature of innovative activity. As such, it is inappropriate to depict innovation as a process that has unique characteristics and to prescribe a unique, simple route to success. It is difficult to argue that one country spends too little on R&D or that it has the most desirable innovation system until we understand the nature of optimality (Edquist, 1997). And optimality may require heterogeneity, not homogeneity.

An aggregate statistical picture of the average innovator hides the considerable diversity that exists in the population of innovators. New and improved products and processes are responses to challenges and opportunities, which vary both within and across industries. Internal factors that influence innovation are closely related to the size of the firm, as well as the accumulated knowledge and competencies in the firm. External factors are shaped by technological opportunity and market forces.

Two forces are at work that shape the nature of diversity – forces that are purposive and those that are nondeterministic. The progress of creation and accumulation of knowledge creation through regular R&D activity and by alternative means, both inside and outside the firm, by market conditions, changes in organizational structures, and institutional development are all marked by a high degree of uncertainty.

Uncertainty occurs because technological change involves a trial-and-error process. On the one hand, it involves the type of individual and collective experimentation and learning that is stressed in evolutionary economics. On the other hand, it has features of the type of deterministic, rational cause-and-effect process that are stressed by neoclassical economists.

Evolutionary economics has taught us that the creation and diffusion of technological change is multifaceted. Novelty takes on different forms. Innovations of different kinds are created and introduced by different processes in different organizations and systems. However, as in biological evolution, only some innovations survive. This selection process results in the culling of some innovation regimes and the focusing of systems on a reduced set of regimes – for example, the R&D-centric mode of innovation.

Innovation variety occurs partially due to design and partially due to chance. Variety can be found in different motives of economic agents, types of organizations, and institutions that have developed as a result of country-specific cultures. They come from chance happenings in search and learning procedures, especially in relation to scientific discoveries, and finally from unexpected changes in environmental factors (natural as well as economic, social, and political).

The selection process that reduces variety by culling out the less successful in favour of the more successful innovation processes also involves considerable uncertainty. The selection process operates at the level of both the firm and the economy. Firms decide on which innovative ideas will be developed, which internal resources to devote to innovation, and the complementary assets that they must muster or find outside of their organizations. The survival of one technique via selection will depend on the population of techniques that are chosen for the experiment and the institutional structures that exist to support particular modes of innovation. During the selection process, symbiotic relationships develop between firms. Some are based on economies of scale or network externalities. Others involve complementary arrangements with different firms and institutions, such as national research facilities or universities. These relationships are shaped by the type of supporting economic and technological structures – the maturity of financial markets and the type of training programs that exist to help develop a skilled workforce.

Arrayed against this sometimes bewildering complexity associated with evolutionary models of innovation are more traditional neoclassical models that try to organize the array of information into more recognizable segments. These models argue that differences in innovation regimes may reflect not so much random choice as purposive responses to differences in relative prices and opportunities. Small and large firms face different capital costs. They might therefore be expected to choose different capital intensities, both in the production and in the innovation process. When one form of external cooperation is costly, firms are likely to find new forms of cooperation that serve to reduce the costs of investments in knowledge creation. When firms can substitute one type of resource for another more scarce resource in their search for innovation, this involves trade-offs that are handled well within the framework of traditional neoclassical economics.

This book takes the view that there is really no incompatibility between the two schools of thought. Innovation, like any firm strategy, involves choices. Some of these choices are operationalized relatively easily within standard frameworks. Others are not so easily rationalized.

In either case, a picture is required of the innovation process. Developing that picture is the objective of this monograph. Throughout, we focus on a plurality of innovation types. Our study breaks with the traditional or standard way of treating innovation in a firm as dependent only on R&D. We embed innovation more broadly in the firm's set of activities. We argue that ideas for innovations come not only from R&D but also from managers and the production department. Innovations are also triggered by ideas from other firms (from suppliers and customers). We argue that both proprietary information and unpriced spillovers are important. The firm may conduct R&D on its own or it may collaborate with others or it may licence information and technology from other firms (including corporate affiliates).

The study is aimed at understanding how these types and the regimes that support them fit together. We do not treat this diversity as simply an ill-defined nebula. Our objective is to understand differences in types of innovators – small versus large, domestic versus multinational, innovative and less innovative industries – and suggest rationales for the coexistence of different innovation regimes.

1.2.2.2 Types of Diversity
Heterogeneity in the innovation system takes several forms.

First, there are distinct differences in innovation types within industries. Each industry consists of a complex network or system of actors, who often pursue different innovation strategies. Technical progress within an industry takes place on several levels – in the components, in the production process, and in the introduction of improved or new products. Advances are made at different times in different parts of this process, which is coordinated by arm's-length market transactions and via knowledge transfers internal to organizations that may be joined in an interfirm network. Sometimes, such as in the case of multinationals, the latter occur as part of transactions within the same firm. Sometimes, such as with joint ventures, they occur between separate legal entities that combine their resources to share knowledge (Nelson and Rosenberg, 1993).

Second, there are substantial differences in the types of outputs produced by innovative firms. A common distinction that is frequently made is between product and process innovators. Product and process innovation use inputs, such as R&D, in different amounts (Arvanitis and Hollenstein, 1994). We, too, follow this distinction throughout this study in order to examine differences in the development of new products and processes. But we point out that there are few innovations that involve just products or just processes; many involve the simultaneous introduction of new products and new processes. The more complex 'product cum process innovations' have, in general, a greater need for internal competencies, such as skill upgrading, than do the two other innovation types.

Third, there is heterogeneity across size classes. Firm size has received much attention in recent innovation studies (Malerba, 1993; Arvanitis and Hollenstein, 1996; Licht, 1997). The relationship between the size of firm and innovation has been in the forefront of economic studies since J. A. Schumpeter's theory associating successful innovation with larger firm size and monopoly power. More recent theoretical and empirical research (Dasgupta and Stiglitz, 1980a, 1980b; Levin and Reiss, 1988) suggests that size and innovation are mutually dependent. Size may convey an advantage to larger firms when it comes to innovation, but successful innovators grow faster than other firms and become larger than non-innovators (Acs and Audretsch, 1988).

Fourth, there are substantial differences across firms of different nationalities. In today's global economy, the ownership of firms is increasingly international and many firms interact across national borders. About half of Canadian manufacturing firms are foreign-owned. Cross-border transactions with suppliers, customers, and partners provide them with access to information networks other than those available to domestically owned firms. It is important to investigate whether foreign affiliates operating in Canada are integrated into the Canadian innovation system. This study therefore examines whether a firm's conduct and performance are shaped more by ownership or by technological opportunity and market forces.

Fifth, research has shown that innovation systems differ across industries, partially because technological opportunities vary from industry to industry. The incidence and type of innovation is also closely related to the position in the life cycle of a product or a whole new industry. Low rates of innovation are found in traditional industries, such as textiles, wood products, food, and pulp and paper (Evangelista, Sandven, Sirilli, and Smith, 1997).

Several taxonomies of industrial innovation have been constructed with differences in the industry environment as the foundation for their classification. These studies have at their foundation either differences in the technological opportunities of different sectors, some concept of product hierarchy, or the method used to diffuse innovations throughout the economy – issues that relate to spillovers and externalities.

For example, Pavitt (1984) develops a taxonomy based on a classification that divides industries into those that are 1) supplier dominated, 2) production scale intensive – determined by the size and principal lines of activity, and 3) science based. Scherer (1982a, 1982b) chooses to organize his work around a classification that uses the industry where patents are created and where they are used. Robson, Townsend, and Pavitt (1988) extend Scherer's work to develop a stages-model that is based on 1) the intensity of innovation in an industry and 2) the extent to which an industry diffuses products and process innovation to other industries.

In this study, we utilize the Robson et al. (1988) taxonomy that divides the manufacturing sector into those industries that appear to produce a disproportionate percentage of innovations (the core sector) and those that absorb them (the secondary and tertiary 'other' sector). We do so because Robson shows that industries in both the United States and the United Kingdom fit the taxonomy. But in using the Robson taxonomy, we are careful not to refer to the firms in the core sector as innovative and firms in other industries as non-innovative. Both are innovative.

1.2.3 Knowledge Externalities, Market Imperfections, and Diffusion

Generic knowledge is an economic good with unique characteristics. Some new scientific discoveries and new inventions – unless kept secret or protected by a patent – can be used by anybody without diminishing the amount of the knowledge that can be consumed by others.

This ensures the diffusion of innovation by what the economist calls knowledge externalities or spillovers; but it reduces the incentives that private profit-maximizing firms have to produce new knowledge and to innovate.

In markets where firms cannot be sure that they will reap the economic benefit of investments in innovation, firms have less incentive to invest in as much knowledge as would be optimal. Innovation and knowledge creation will be undersupplied. This conventional market-failure analysis (see Arrow, 1962) has been traditionally used to provide an economic rationale for government support of R&D and innovation.

The existence of spillovers presents a delicate trade-off between adequate incentives to innovate and conditions that favour the diffusion of new technology. If intellectual property rights are well protected, investments in innovation will be larger – and, in some cases, more than is socially optimal. Some models even suggest the possibility of oversupply of R&D when private property rights are assured. These arguments are based, among others, on the existence of inefficient patent races that lead to duplicative R&D (Dasgupta and Stiglitz, 1980a; Tisdell, 1995).

Empirical studies have attempted to document the importance of spillovers at the industry and country level (Bernstein, 1997; Hanel, 2000). At issue in this study is not whether there are spillovers, but the extent to which the intellectual property system is used to reduce the effect of these spillovers. We investigate the methods that firms use to mitigate and minimize the problems that arise from having to operate in imperfect knowledge markets. To do so, we examine two related aspects of spillovers. First, we seek to establish the frequency of occurrence of technology spillovers. Second, we investigate the methods that firms use to mitigate and minimize the problems arising from spillovers.

Market imperfections arising from these problems are addressed by government through the creation and enforcement of intellectual property rights – rights that assign ownership to the outcome of ideas that lead to an innovation. While intellectual property rights are meant to stimulate economic activity, there has been little applied research on whether this is the case. There are two major exceptions. Research by Mansfield (1986) and Levin et al. (1987) has challenged the conventional belief that such rights as patents are an effective means of protecting investments in knowledge creation. In this book we also examine why firms make use of the intellectual property system, and whether they perceive intellectual property rights to be as effective in preventing imitation.

While issues of appropriability are seen by some to generate problems, this view is by no means universal. Pavitt (1984, p. 353) argues that most of the knowledge applied by innovating firms is not general purpose, easily transmitted and reproducible, but is applicable only to specific applications and therefore can be adequately protected by innovators. In his study of innovation in the U.K., Geroski (1995, p. 90) concludes that 'spillovers do not always (and perhaps not even often) seriously undermine the incentives to innovate'.

In a related vein, Von Hippel (1988) notes that appropriability problems affect not only the amount of innovation that takes place but also the nexus or location of that innovation. Recognizing imperfections in appropriability, he identifies the stage of a vertically integrated production chain that is most likely to have inherent advantages in appropriating the benefits of an innovation, and postulates that it is this level that will conduct most of the innovative activity. As such, his theory is essentially based on the notion that appropriability exists – but that it is specific to certain stages of the production process.

We recognize that firms manage to internalize externalities of all types, including those associated with knowledge creation. In the case of knowledge creation, firms often do so through the adoption of various strategies other than the use of patents. They make their new product complex; they develop a first-mover advantage; they develop partnerships with other firms. In this study, we examine how important each of these alternatives is – by directly asking firms how they safeguard their innovations and the extent to which they participate in innovation networks.

The nature and extent of these networks has garnered substantial attention – because they provide the means by which the spillover problem can be mitigated. This has implications for the patterns of organization that we might expect to find in innovative firms. For instance, a number of studies have found that firm diversification is related to the science base (percentage of employees that work in R&D) of the industry in which the firm's primary activity is located – Gort (1962), Amey (1964), Gorecki (1975), Grant (1977). This implies that when a firm develops a specialized science-based asset, it often exploits this asset by extending its operations into new industries.

It is for this reason that innovation relies on networks – that actors are tied together in clusters. Suppliers provide customers with new ideas as to how to incorporate new materials or new machinery into the production process. Customers inform suppliers of new machines that are needed in production. Customers and suppliers work with one another. In these commercial transactions, there is room to internalize information leakage. It is with the goal of understanding these networks that we examine the sources of ideas for innovation. An exploration of the sources of innovation also helps us to understand how diffusion occurs and whether it mainly involves unpriced spillovers, or whether it is internalized via market transactions or via alternate methods.

Problems in pricing a highly uncertain good like a process innovation also lead to new forms of organization to reduce the costs of transferring new technologies. The transfer of technology by a firm can be accomplished either through licensing or through the exploitation of its own technology via exports or with production facilities located abroad. The alternative that is chosen will be determined by the relative efficacy in transferring an asset via an arm's-length transaction rather than making a foreign investment or by exporting. We therefore examine the importance of technology transfer as part of the innovation process and the nature of the contracts that are used.

Another issue that is closely related to knowledge spillovers is the role that technological opportunities play in shaping the innovation process. Some industries, it is argued, are more likely to provide greater opportunities for innovation because of their science base. The state of scientific knowledge in some industries makes it more likely that firms therein can take advantage of knowledge advancements to introduce large numbers of new commercial products or new production technologies. An example is the biotechnology sector, where present advances in genome mapping promise rapid advances in new product introduction. This study therefore investigates whether innovators are more likely to succeed when they form partnerships with universities, who are one of the principal creators of scientific knowledge.

1.3   THE ECONOMIC THEMES

While innovation is essentially about disequilibrium and network economics, many other aspects can still be set within the traditional bounds that are used for most economic studies.

A study of innovation requires that attention be devoted to traditional areas involving the delineation of markets and production processes, as well as the nature of transaction costs and how they give rise to market imperfections.

An economics study should perforce define the output being examined and the inputs that are critical to the innovation process. It should also investigate the importance of the industry environment (the technological background and the market forces that shape the dynamics of competition within a sector). It needs to focus on how this environment affects the innovation process and the nature of institutions, both private and public, that facilitate innovation.

The study addresses each of these points in turn.

1.3.1   The Nature of Innovation Outputs

A central finding of this study is that innovative firms are not those that serendipitously stumble across inventions. Innovators differ from non-innovators in that they adopt a purposive stance to find new products and to adopt new processes. The Canadian manufacturing sector is not a world where most firms are engaged in intense innovative activity, where some are rewarded by chance and others are not. It is a world that divides into firms that heavily stress an innovation strategy and those that do not.

Within the innovation group, there are considerable differences in the outputs that are produced and in the strategies to produce them. Innovation studies sometimes focus only on innovations that are paradigm-shifting – new products that are so unique that they transform the whole industrial process.

The development of steam engines transformed industrial processes that relied upon waterpower. In turn, electricity in the late 1800s moved the production process away from steam sources. The modern internal combustion engine and the automobile revolutionized urban areas. The electronic chip and the computer are having a similarly dramatic impact on the production process today – both because of their effects on communications and because of their ability to manage information and to monitor and control production processes.

As critical as the introduction of new, frontier technologies may be, they make up only part of the innovation system. As Nelson and Rosenberg (1993, p. 9) argue, 'most industrial R&D expenditures are on products that have long been in existence'. It is these existing products that serve to define the framework within which improvements can be identified and undertaken.

In this study, we first explore innovation activity in general. By necessity, the issues that can be explored at this level are rather general, since the questions must cover a wide range of types of innovations. At this level, a broad definition of innovation is used. It includes those improvements and/or new products and processes that are new to a firm even though well known and used in other countries in the world or by other firms in Canada.

There is, however, an important methodological difference between our approach and that suggested by the Organization for Economic Co-operation and Development (1997) in their Oslo innovation manual that has been used in the European Community Innovation Surveys. We explicitly recognize that the innovation process aimed at introducing an original world-first innovation is likely to be different in many respects from the imitation of known products and processes. Building on the first innovation survey done by the Economic Council of Canada in the seventies (De Melto, McMullen, and Wills, 1980), we include a separate section of the survey that asks questions about the most profitable innovation introduced by the firm. This innovation-specific section provides information on the differences in the idiosyncracies associated with different degrees of novelty.

In this study, we focus on a wide range of innovation types. On the one extreme are innovations that are, according to the firms that reported them, 'world-firsts'. Scientific progress opens new avenues for technological breakthroughs, and many of these build on previous knowledge. Less original innovations are divided in this study into 'Canada-firsts' – innovations that were introduced abroad first but are for the first time implemented in Canada and 'other' innovations. The former involve technology diffusion from abroad; the latter involve technical diffusion within Canada. While these two categories represent the less spectacular technological innovations, they make a significant economic contribution to overall economic growth.

In examining the different types of innovations, ranging from the more to the less original, we consider how each is produced, who produces them, and where they are produced. Our maintained hypothesis is that the innovation system for each type of innovation has unique features. Differences exist with regards to the type of inputs used (R&D versus production engineering), the use of partners for joint ventures and other collaborative exercises, the extent to which the firm relies on outside technologies, and problems in financing. We cast our net broadly because the activities and investments associated with becoming a leader in the introduction of a new product or process, and those associated with staying near the head of the class, or catching up with the leaders, are each important and probably differ in many respects. Since our interest in innovation often stems from a desire to better understand the determinants of economic performance, we need to better understand the overall innovation system – not just part of it.

While we note that there are wide ranges of innovator types – from leaders to followers, from industries that produce innovations used elsewhere to industries that ingest innovations produced elsewhere – we avoid the mistake of claiming that the successful introduction of innovation depends only or primarily on the producers of the innovations. Some firms are responsible for the production of new machines or new materials. Others adapt these machines and material to their production process. It is difficult to ascribe more importance to one than another. Innovation depends on a web of interactions between the two parties. Technologically progressive users of new innovative processes can have a substantial influence on the machines that are created by innovators, as well as on the extent and speed of adoption of new processes. Users of innovations are often at the forefront of the innovation process, requesting and helping to develop new capital equipment purchased from upstream producers.

1.3.2 The Nature of Innovation Inputs

The innovation process is often defined by the type of inputs that are used in the production process. In particular, since Schumpeter first professed a fascination with the way that large corporations systematize their search for knowledge, the role of research and development laboratories has garnered special attention from economists.

This interest in R&D is not misdirected. The importance of R&D has been confirmed by previous research. Brouwer and Kleinknecht (1996) point out that despite different specifications and different measures of innovative output, innovation surveys have always found that innovation is correlated with R&D – underscoring the importance of the continuous accumulation of R&D-type knowledge for innovation output.

The issue is not whether R&D is important, but rather its degree of importance. Chesnais (1993) argues that a focus on R&D alone is inadequate because 'an R&D system is at best a poor proxy to an innovation system'. In order to understand the system as a whole, we need to evaluate the importance and role of other inputs into the innovation process.

A focus on R&D alone ignores the fact that information in the firm is acquired and developed outside formal R&D systems. Tacit, uncodified knowledge accumulates in the firm through complex interactions that gather, store, and use technical knowledge. An exclusive concentration on R&D ignores the linkages between organizations through which knowledge is transmitted – linkages that transfer information through arm's-length transactions, across subsidiaries within firms, and via alliances or joint ventures.

A narrow focus on R&D using official measures of R&D is problematic for a second reason. Mowery and Rosenberg (1989) have stressed that there is a certain lack of distinctiveness surrounding the definition used for the collection of official R&D statistics. Only a fraction of technological effort is counted as R&D. Not all expenditures on the creation of new and improved products are covered by the OECD (1993a) 'Frascati' definition of R&D. There are important knowledge-creating activities that firms do not consider to be part of R&D. For example, firms without formal R&D departments ascribe a substantial part of their knowledge-creation process to product design teams and not R&D departments (Felder et al. 1996).

Therefore, this study examines the importance of several sources of knowledge that firms use for innovations. We start by noting that innovations are dependent upon the work that goes on in industrial R&D labs, as well as in university or government laboratories. Our interest in the importance of R&D also extends to whether there are economies of scale attached to the R&D process that apparently give large firms an advantage in the pursuit of this activity.

But we also show that ideas originate from other areas of the firm, such as the production or engineering departments, and that universities are an important part of the innovation process. In examining the importance of non-R&D sources of ideas for innovation, we not only outline the other sources of information but also classify them as complements or substitutes for R&D.

The study also investigates the monetary importance of several types of expenditures required in addition to R&D in order to bring an invention to market. These include expenditures on marketing and technology acquisition. Process engineering is always important and rarely considered as R&D. Design activities, solving production problems, and technology-watching all contribute to the innovation process and are rarely considered to be part of R&D. We find that R&D expenditures, such as those defined in the Frascati manual and compiled in international R&D statistics, are often only a small portion of the resources required to support innovation.

Even within a firm, innovation requires more than technical knowledge arising from engineering or R&D. It requires complementary competencies in finance, marketing, and production. This study therefore asks not only what sources of ideas are used for innovation, but also what importance is attached by innovators and non-innovators to competencies in each of several different functional areas of the firm – human resources, financing, and management.

One of the most important complementary capabilities for innovation lies in the area of financing. This study investigates two financing issues that affect innovation – the source of funds for investments in innovation and the extent to which government programs that are used in Canada better support some types of innovation than others. In both cases, we find that the industries that receive innovations from the core sector have to rely mainly on their own internal funds.

Finally, we focus our attention not just on the internal innovation production capabilities of a firm, but also on the extent to which firms reach outside themselves for inputs used to produce innovations. Not all innovative ideas are developed entirely within a firm. We show that innovation arises from a network of firms interacting sometimes at arm's length, sometimes in symbiotic relationships that blur the boundaries of a firm. Often the difficulties in creating and ingesting new knowledge cannot be overcome through arm's-length transactions, and firms expand their boundaries to incorporate other firms into a larger innovative network – they enter into collaborative arrangements to create new technologies and for R&D, either directly through mergers or through joint partners and ventures.

1.4 THE ORGANIZATION OF THE STUDY AND PRINCIPAL FINDINGS

In order to explore the above issues, we organized the chapters of the study as follows.

1.4.1 The Innovation Survey

The second chapter focuses on the data source used in this study – a special survey of innovation in the Canadian manufacturing sector. Until the 1990s, most studies of innovation had to use case studies, which did not permit very comprehensive coverage of the innovation process. Or they used patent statistics or R&D data that were more comprehensive in terms of coverage of a wider range of firms than are covered by case studies, but were restricted in terms of the topics that could be investigated.

This study makes use of the 1993 Canadian Survey of Innovation and Advanced Technology, which is hereafter referred to as SIAT or the Survey of Innovation. It focuses on the population of enterprises operating in the Canadian manufacturing sector and covers a wide set of topics relating to a firm's innovation activities. These topic's range from the nature of innovations being introduced to the relative importance of R&D, to the impact of innovation on firms, to the impediments that firms face, to financing problems, to the use of intellectual property rights, and finally, to the extent to which complementary strategies in marketing, human resources, and management are required for successful innovation. The survey is included as an appendix.

While innovation surveys offer a potentially rich data source, there are a number of difficult measurement issues that have to be resolved if the data are to be useful. This study makes use of a comprehensive innovation survey that was especially designed for analytical purposes. The survey itself was built on the foundation of earlier Canadian work and preliminary versions of the European harmonized innovation survey (based on the OECD (1997) Oslo manual). However, it avoided some of the pitfalls that exist in the latter.

Standardized surveys are useful in providing benchmark data. But standardization involves a compromise among competing underlying concerns, assumptions, and research agendas. These are likely to differ for countries with widely different innovation concerns. The survey used here was designed to provide answers to issues that the standardized OECD methodology can only handle poorly. We believe that the inconvenience that arises because our survey questionnaire is not perfectly compatible with more recent surveys is more than offset by the wealth of data provided that is specific to the Canadian and North American context.

Since the 1993 Canadian Survey of Innovation was specially designed for analytical purposes, we devote the second chapter to an outline of how this was done. The survey was designed to allow differences in innovation regimes across industries and types of innovations to be explored. The chapter focuses on definitional issues, on how the sample frame was chosen, on several operational issues concerning the survey, and on the size of the response rate that was obtained.

1.4.2 Innovation Intensity

The third chapter focuses on the extent to which innovations were being introduced in Canada at the time of the Innovation Survey and measures the importance of innovation by the percentage of firms that report having recently introduced a new product or process. Our objective here is not simply to report innovation rates but to illustrate that the industrial population is heterogeneous and that so too are the types of innovations being brought to market. The chapter examines differences in innovation rates across innovator types – product versus process, small versus large, domestic-controlled versus foreign-controlled firms, and science-based versus consumer goods industries.

Innovation is more frequent in a core set of industries (electrical, machinery, chemicals, pharmaceuticals, computers). Almost half of the firms in these industries introduced an innovation, whereas little more than 25% of firms in the tertiary sectors did so. This may be either the result of greater technological opportunity in these industries or due to the fact that these industries contain more products that are in the early stage of their life cycle. Innovating firms that are found in the core sector feed new technology to the rest of the economy.

The most original 'world-first' innovations are understandably rarer than innovations that introduce to Canada new products or processes created abroad. The larger firms were more likely to report that they created world-first innovations. The core sector was more likely to report a world-first than were the secondary and tertiary other sectors. Only one in six innovations was a world-first, one of three a Canada-first. The most numerous (slightly more than every second innovation) represent the diffusion of technical change, that is, 'innovations' that were new to the reporting firm, but that already existed elsewhere in Canada.

1.4.3 The Sources of Knowledge

Innovation is about knowledge creation, acquisition, and adaptation. The fourth chapter discusses how the economics of knowledge creation affects the organization of firms that are involved in innovation.

It addresses several questions. The first is the extent to which spillovers are important and the source of these spillovers – whether specialized public institutions that provide technical information are seen to be an indispensable element of a national innovation system. It also examines the relative importance of links between affiliated firms and links between firms that arise in the form of normal commercial relations (i.e., between suppliers and customers).

The chapter weighs the relative importance of spillovers compared to market transactions that diffuse innovations. Spillovers are classified into three groups – those associated with commercial transactions, those that fall in the traditional unpriced interfirm group, and those coming from infrastructure facilities like universities. Contrary to the normal emphasis that is placed by neoclassical economic theory on the problem of unpriced spillovers, the survey evidence shows that interfirm transactions dominate the innovation process.

While spillovers are not the most frequent method used to diffuse innovation generally, they are more important in some sectors than others. In particular, they tend to be more important in the downstream technology-using sectors. Here, more use is made of spillovers from competitors. There is also more use made of trade fairs and other avenues for the transfer of information that is more readily codifiable.

The chapter also investigates the extent to which R&D is central to the internal process and how it facilitates the use of external sources. The chapter finds that the R&D-centric model is, by itself, inadequate. Even those firms that use this form of knowledge creation develop other complementary capabilities, especially in management and in their marketing departments. More importantly, there is an alternate mode that focuses on the production department that is essential for process innovation.

1.4.4 Research and Development

Since research and development is seen to have a special and key role in the innovation process, the fifth chapter investigates the importance of this particular factor by examining the extent to which Canadian manufacturing firms incorporate R&D into the innovation process.

This chapter focuses on several dimensions of research and development capacity. It investigates the attitude of firms to the development of innovation and technological capabilities – the stress placed on various business strategies that involve spending on research and innovation. It also examines the commitment of the firm to the phenomenon in question – that is, the existence of and the type of R&D operation.

Differences across industries are examined in order to understand how industry environment conditions the R&D strategy that is adopted. The chapter investigates whether there is a core set of R&D industries that provides a much greater emphasis on R&D activity than elsewhere. It also investigates differences in the extent to which innovation and R&D are closely related.

The chapter reports that contrary to the general impression often left by the official statistics – which in Canada are primarily based on R&D

reported by the minority of firms that claim R&D tax credits – R&D is being performed by about two-thirds of Canadian firms. Knowledge creation and acquisition through R&D in that sense is widespread. What is highly concentrated is a particular form of R&D – that done continuously in separate R&D labs. The latter tends to occur in larger firms and in a small set of core industries that create innovations for transmission to other industries.

1.4.5 Effects of Innovation

The sixth chapter examines the effects of innovation on the organization, activity, and performance of innovating firms. Firms innovate to increase their profitability, which can occur via reductions in costs, improvements in sales, or a combination of both. These general economic objectives of innovative activity are accomplished in many specific ways – by decreasing production costs, by increasing product line diversity, and by improving the quality of the product.

Since not all firms are innovators, it is important to understand the specific effects that Canadian entrepreneurs associate with their innovation – as these delineate both the advantages and impediments to the innovative process. The magnitude of both benefits and impediments determine whether innovation is undertaken.

The chapter finds that while innovation improves the ability of firms to exploit scale economies, these impacts were listed less frequently than improvements in flexibility. In the small Canadian economy, innovation is aimed more at exploiting product-line production economies.

The chapter finds strong evidence that all types of innovation have beneficial effects. Each type of innovator is about equally likely to report benefits of improved profitability. Innovating firms operating in the tertiary 'other' sector reported increased profitability just as often as firms in the secondary and core sectors. This pattern emphasizes the important economic contribution associated with the diffusion of innovation from high- to low-tech sectors and the diffusion of technological change through imitation. Original innovation may not occur as frequently in the downstream sectors, but innovation is just as frequently listed as being profitable in these industries.

The sixth chapter also asks whether innovations that are made to improve regulatory compliance are any less successful in improving the profitability and market share of firms. It finds little evidence to suggest that these types of innovations yield any fewer benefits to the innovators.

Finally, the chapter investigates the degree to which innovation is skill enhancing and the relationship between innovation and employment. While innovations reduced unit labour costs in many firms, they were more frequently associated with increases than decreases in employment. Firms that reported increases in the employment of production workers substantially outnumbered those firms where innovation led to a decline in employment. The employment creation of nonproduction jobs was even more one-sided. Innovation also improved working conditions in almost one-third of innovating firms and increased the need for more skilled workers in almost two-thirds of all innovators.

1.4.6 Innovation and Research and Development in Small Versus Large Firms

At the heart of the Schumpeterian literature is the debate over whether large firms are more innovative than small firms. The seventh chapter focuses on differences in the innovation regime of small and large firms. But rather than trying to assess which group is more innovative, the chapter focuses on how innovation in large firms is different from innovation in small firms. In comparing small and large firms, we argue that it is important not to presume that firms in different size classes should duplicate one another. Small firms are found to possess advantages in some areas and disadvantages in others. Because of their size, small firms may suffer unit cost disadvantages in R&D, but may have offsetting advantages in terms of flexibility and response time to customer needs. We recognize that differences are inherent in a heterogeneous environment.

The seventh chapter therefore asks whether there are different patterns of innovation in small and large firms, specifically whether fewer small than large firms rely on R&D for their innovative ideas.

On average, small firms are less likely to do continuous R&D. They are less likely to innovate. They are less likely to take out patents. However, while they are less likely to innovate, the sheer size of their numbers means that the group accounts for a larger proportion of innovations than large firms do.

The chapter also examines differences in the small-firm sector. It finds that small firms fall into two groups. First, there are those with R&D facilities that conduct R&D continuously and that are only slightly less likely to report innovations than are large firms. Second, there is a set of small firms that operates opportunistically by doing R&D when required, outside of permanent R&D facilities. The latter firms are tied into interfirm networks that allow for the diffusion of new technologies and new products. Small firms in the latter group are numerically more important than the former. The small-firm sector relies heavily on interfirm transactions via commercial networks for the diffusion of knowledge.

The chapter also outlines differences in the problems that small firms and large firms face when they innovate. Small firms perceive that externalities are relatively important in the area of information about technologies, markets, and technical services. They also perceive that there are significant barriers to interfirm cooperation. This is a particularly serious problem because this is the method that they use most frequently for developing new ideas for innovation.

1.4.7 Innovation Regimes and Type of Innovation

Innovations come in many different forms. More importantly, economic growth can derive from innovations of different types. Economic growth is most clearly associated with path-breaking innovations. However, economic growth also results when firms introduce new products or processes that have already been introduced in other countries, but that need to be adapted to special national circumstances.

The eighth chapter examines differences in the innovation process across firms that are producing innovations that differ in terms of their novelty. It examines how innovations divide themselves into world-first, Canada-first, and 'other' purely imitative innovations, and whether these distinctions differ across products and processes. It examines whether small or large firms produce more of one of these types of innovations. It also examines industry differences regarding the novelty of the innovation.

The chapter demonstrates that innovation regimes differ significantly in terms of the sources of inputs, the use that is made of intellectual property, the impediments to innovation that firms must overcome, and the impact of innovation on employment. The more original the innovation, the more likely a firm is to use R&D and outside groups (related firms, industrial research firms, and universities) that can complement their internal research and development facilities.

Problems that innovators face encompass such areas as skill training, market information, regulations and standards, and technical services. The areas that give them the greatest difficulty are lack of skilled personnel, lack of market information, and government standards and regulations. In each of these cases, world-first innovators generally experience these problems more frequently than non-world-first innovators. This suggests that these problems do not block innovation as much as they accompany more intense innovative efforts.

1.4.8 The Use of Property Rights

The problem of appropriability has been repeatedly stressed as an impediment to the knowledge-creation process. The ninth chapter focuses on the use that is made of intellectual property protection in Canada. It examines the extent to which different instruments – patents, trademarks, copyrights, trade secrets, and industrial designs – are used to protect intellectual property. The chapter focuses on the intensity of use and develops a comprehensive picture of the use of different instruments (both statutory and other forms of protection) that protect intellectual property assets. It also measures the degree to which firms perceive different instruments to be more or less effective in protecting intellectual property. Finally, the different forms of intellectual property protection used by a firm are related to its innovation profile.

Intellectual property protection is sought more frequently by larger firms, by foreign-owned firms, and by firms in the core sector. Firms that innovated in the three-year period prior to the survey used intellectual property protection more than those that did not. Being innovative is a primary determinant of the use of intellectual property protection.

Not all forms of statutory protection are sought equally by innovative firms. When the effect of being innovative is separated from the effect of size, nationality, and industry, being innovative has its largest impact on the use of patents and trademarks.

The study also asks how firms evaluate the effectiveness of intellectual property protection. On average, firms score patent protection as being less than 'effective'. But this comes because most firms do not use patents. Those firms that take out patents score them as being effective. This points to a world where some firms choose to be innovators and develop a set of competencies that complement innovation – including a way to protect their intellectual property – and other firms do neither.

1.4.9 Multinationals and the Canadian Innovation Process

Multinational firms are often regarded as having superior abilities that allow them to efficiently transfer the types of technological know-how required for innovation across national boundaries. In the face of market imperfections brought about by difficulties associated with assessing the value of knowledge assets (e.g., due to the difficulty in assessing the value of new technologies), multinationals are seen to be an efficient transfer agent for technological advances from one country to another.

In contrast to this view, some argue that the presence of multinationals in Canada causes a 'truncated' innovation regime. While multinationals may transfer technical know-how, they may not facilitate development of the type of technological infrastructure that is necessary for widespread innovation activity.

The tenth chapter addresses this issue by examining whether the nationality of firm ownership is related to the organization of R&D activity, the sources of innovative ideas, and the incidence and effects of innovation. Throughout, we compare the multinational firm operating in Canada (MNC) to Canadian, domestically owned firms. By comparing the performance of the MNC to domestic corporations, we allow for the fact that technological opportunities and other structural characteristics will condition the nature of the innovation system in Canada.

We compare the multinational firm, first, to Canadian firms in general to see whether there is any indication that foreign firms are disadvantaged relative to those based in Canada. We do not find that the truncated model of the multinational is appropriate in the 1990s. Foreign-owned plants are neither less likely to operate R&D laboratories in Canada nor likely to collaborate less frequently with others.

We also examine differences between foreign multinationals and Canadian firms with a foreign bent – those who have operations abroad or who have export sales. Using this comparison, we find that foreign subsidiaries are not very different from Canadian firms that successfully operate in foreign markets – domestic multinationals. We conclude that it is not nationality as much as the export orientation of the firm that is more closely associated with being innovative. In addition, the survey also presents evidence that foreign-owned firms contribute significantly to technological progress in Canadian industry. They are more likely to perform R&D and introduce innovations in Canada than the locally owned companies.

1.4.10 Financing and the Cost of Innovation

The eleventh chapter focuses on issues relating to innovation costs and financing. Financing costs are sometimes seen to be a particular problem for innovative firms. These firms are risky because of the very nature of their innovation activity. And their assets tend to be concentrated more heavily in soft, knowledge-based investments that offer less secure collateral to lenders. This chapter focuses on three issues relating to the financing of innovation.

The first section examines the composition of innovation costs, with the purpose of ascertaining the importance of research and development costs versus other expenses. Even though basic and applied research expenditures play a key role in debates on innovation, technological change, and industrial policy, they are shown to be neither the only, nor necessarily the most costly, component of the innovation process. Expenditures are required in a number of areas in addition to conventionally measured R&D expenditures. Innovations also require expenditures on the acquisition of technological knowledge (patents and trademarks, licenses, consulting services, and the disclosure of know-how), on development (engineering, design, prototype, and/or pilot plant construction and testing), on manufacturing start-up (engineering, tooling, plant arrangement, construction, and acquisition of equipment), and on marketing start-up activities.

Government policy is often directed at subsidizing expenditures on R&D, since the latter are seen as the focal point of the innovation process. The effectiveness of this policy is dependent, inter alia, on the relative importance of R&D in innovation costs. The chapter indicates that basic and applied research expenditures make up a relatively small percentage of the total expenditures required to bring an innovation to market. Thus, subsidies to this component will have a relatively small impact on total innovation costs. Development is twice as important and, therefore, subsidies in this area will have a larger impact on innovation costs.

The second section of the chapter investigates the evidence that R&D is financed in different ways than are the general assets of the firm. It shows that R&D is much more dependent on internally generated funds than are assets in general and, therefore, is subject to very different financing problems. But surprisingly, external capital markets appear to be able to finance innovation more easily in the more innovative sectors.

Canada subsidizes R&D expenditures with direct subsidies and with tax credits. The third section studies the extent to which the tax-credit subsidy programs offered in Canada for R&D are utilized. This section of the chapter finds that the tax-credit program is not used in the same way by different innovators. In spite of the fact that the criteria of the tax credit program[2] are meant to be sector-neutral, the innovation-producing

---

[2] Federal Income Tax Incentives for Scientific Research and Experimental Development program administered by the Department of Finance and Revenue Canada.

sectors are more heavily subsidized than are the innovation-using sectors.

1.4.11 Technology, Joint Ventures, and the Transfer of Innovation

Economic activities can be organized via arm's-length market transactions or outside of markets within the confines of individual firms. Other chapters develop the theme that innovation involves a substantial interaction between firms that arise during the course of knowledge diffusion. The twelfth chapter examines several examples of transactions that are used to transfer ideas from one firm to another. It investigates the importance of the transfer of technology from one firm to another – the transactions that exchange knowledge on production processes – and asks whether there are particular problems that have to be resolved and how they are overcome.

The chapter finds that technology diffusion is very important. A large proportion of firms acquires technology via a licence when creating an innovation. Technical transfers through licensing arrangements are used more frequently in situations where R&D is not the focus of the innovation process. The chapter also finds that continuous transfer licences for technology tend to be more common among affiliates, thereby confirming that this particularly costly form of contract is used primarily within firms.

An alternate or complementary strategy to technology transfer is the joint development of technology. One such linkage involves collaborative projects with other firms, public and private research institutions, and universities. The second section of the chapter explores the importance of joint ventures – commercial transactions that allow a firm to benefit from scale economies in the innovation process. We investigate whether this particular institution is apparently more suited to some industries, some types of firms, or some types of innovations. Contrary to the technology transfer process that is accomplished with licences, joint ventures are more closely related to R&D intensity. They complement the collaborative R&D process. They are less frequent where the engineering department is the primary source of ideas for innovation. While joint ventures tend to be supported by or are related to collaborative research, they are not more concentrated in the most R&D-intensive core sector. Indeed, they are quite evenly distributed across sectors. The reason is that only the smaller firms in the core sector make very intensive use of joint ventures; large firms in the core sector do not. Once more, this demonstrates substitutability across instruments – this time across firm size classes.

1.4.12 Strategic Capabilities in Innovative Businesses

Firms are collections of competencies. Successful innovation depends upon more than the development of scientific skills. It requires successful human resource strategies to find the skills needed, marketing strategies to reach consumers, financial capabilities to raise capital in a risky environment, and management teams that can put strategies in place across all these functional areas and then monitor them.

The thirteenth chapter is founded on the view that core business strategies in such functional areas as human resources, marketing, finance, and management are an integral part of the capabilities required for innovation.

In this chapter, we examine the differences in emphasis that innovative and non-innovative firms give to human resources, marketing, financing, and management skills and practices. We find not only that innovators and non-innovators differ with respect to the emphasis they give to their scientific competencies, but that they also give considerably more stress to developing their skills in complementary areas like marketing and human resources. Innovators are well-rounded, compleat firms.

1.4.13 Firm and Industry Characteristics Associated with Innovation

While there are benefits from being innovative, not all firms innovate despite the benefits of doing so. The fourteenth chapter is aimed at understanding the conditions that are associated with being innovative. It examines differences between firms that innovate and those that do not innovate using multivariate analysis.

Several issues are addressed. The first is the extent to which the intellectual property regime facing a firm stimulates innovation. The second issue is whether the existence of an R&D unit is likely to lead to innovation. The third is whether there are complementary competencies in human resources, marketing, and finance that, besides the development of a research and development unit, are closely linked to innovation. The fourth is the extent to which a larger average firm size and less competition are associated with a greater tendency to innovate. The fifth deals with the effect of the nationality of a firm on its propensity to innovate. Finally, we examine the importance of the scientific infrastructure.

Throughout, we investigate the extent to which the conditions associated with innovation differ across innovation types – for product in contrast to process innovations and for world-first innovations in contrast to more imitative innovations.

The chapter finds that appropriability conditions do not drive innovation. Moreover, while performing R&D increases the probability of being a successful innovator, so too does giving a stronger emphasis to technological capabilities and to marketing competencies. Size is positively related to innovation. The effects of competition are nonlinear, with intermediate levels of competition being most conducive to innovation. Foreign-controlled firms are not significantly more likely to innovate generally. This result depends upon the inclusion of the competency and size variables. Thus, differences in the raw innovation rates that exist between foreign and domestic firms are accounted for by differences in size and competencies.

TWO

The Innovation Survey

2.1 INTRODUCTION

As productivity growth in Western countries slowed in the post-1973 period, economists and statisticians increasingly turned to understanding the growth process. This interest has led to studies of innovation.

Unfortunately, data on innovation have been difficult to assemble. Data on patents have supported a set of studies. But many innovations are not patented and, therefore, patent data was seen as providing only partial coverage of the innovation process. Data on research and development also existed and could be used to examine differences in the tendencies of small and large firms to innovate; but R&D is only one of the inputs into innovation, and exclusive reliance on R&D data can, therefore, be misleading. Finally, case studies of particular innovations can shed light on the evolutionary process that takes place across the product life cycle. But it is difficult to know how to draw generalizations from case studies that may not be very representative of all firms.

Innovation surveys have evolved in an attempt to provide more detailed data on the process that is behind economic growth. Innovation surveys extend data collection beyond R&D inputs to an examination of some of the other essential ingredients – such as the importance of technology transfer. But their chief claim to originality is the measurement of innovative output.

2.2 DEFINING INNOVATION

Measuring innovative output is difficult. Innovations can be described in many different dimensions. More importantly, economic growth can derive from innovations of different types. Economic growth clearly results from pathbreaking innovations, such as the computer, lasers, or new chemical entities. But innovations that are not at the frontier can also substantially contribute to growth. Economic growth can result when a firm introduces new products or processes that have already been adopted in other countries, but which need to be adapted to special national circumstances. It can occur when firms adopt processes from other industries. Both of these imitative types of innovations involve substantial novelty. Finally, economic growth occurs from more imitative innovations – when firms succeed in improving the products of those who pioneered a new product or process but who could not develop it quickly enough. Competitive pressures that ensure that good ideas become good commercial products or processes also enhance economic progress. Sometimes these changes may simply modify the product to better satisfy consumer demands. At other times, they may involve superior production processes that lower production costs and reduce consumer prices.

Innovation is, therefore, best thought of as a continuous process, whose characteristics often change over the length of the product life cycle. The Gort and Klepper (1982) depiction of the product life cycle provides a foundation that is useful here. Gort and Klepper carefully examined the product life cycle of a number of innovations and delineated four stages. In the first, new products emerge and a small number of firms work to develop a commercial product. In the second stage, there is rapid entry and exit. At this stage, a large number of firms experiment with new product and process designs. Entry is relatively easy, but so too is failure. In the third stage, entry slows, exit continues at high levels, and firms grow larger. It is at this stage that firms begin to move down their cost curve as process innovations allow the successful to exploit scale and scope economies. The final stage of the product life cycle is characterized by a fairly stable market structure with both entry and exit now reduced to lower levels and where oligopolistic competition prevails.

Innovation is important in all stages of the product life cycle. Yet the development of the entirely new product probably only dominates the first stage. Throughout the other stages, improvements occur as firms build on a body of common knowledge or on the work that others have already commenced. In this world, much progress is incremental and few innovations are completely original. Nevertheless, when taken together, the incremental changes have a significant effect.[1]

---

[1] Hollander (1965) details that much of Dupont's productivity growth came from incremental improvements in production technology.

Innovation studies then need to come to grips with the variety of innovations that take place. They need to do so not just because innovations differ but because the purposes to which innovation surveys are put will not be met unless some attempt is made to examine the differences in the types of innovations that are being produced.

Two examples can be given. First, if the objective of innovation surveys is to compare the 'innovativeness' of different countries by describing differences in the extent to which the business populations are innovative, the surveys need to be able to describe what they mean by an innovation. Innovation surveys may find the percentage of firms in two countries that are innovative is the same; but unless they can ensure that the innovations are of a similar 'quality', such a comparison is not very meaningful. Second, if innovation surveys are meant to correlate innovative outputs with innovative inputs (to examine the connection between R&D spending, for instance, and the production of innovations), then the surveys need to be cognizant that these relationships may differ across innovation types. If these relationships vary by type of innovation, policy prescriptions may also differ significantly by type of innovation.

Innovation surveys have generally ignored this problem. The innovation manual used for the European Innovation Survey[2] provides definitions for innovations. Innovation occurs when *a new or changed product is introduced to the market, or when a new or changed process is used in commercial production.*[3] Some attempt is made to distinguish product innovations in terms of importance, but not process innovations. A *significant* product innovation is *a newly marketed product whose intended use, performance characteristics, technical construction, design, or use of materials and components is new or substantially changed*. An *incremental innovation* is an existing product whose *technical characteristics have been enhanced or upgraded.*[4] Changes that are purely aesthetic or product differentiation that does not involve technical changes in production or performance are not included in the definition of innovation. The Oslo Manual (1997) makes similar distinctions.

The European surveys often attempt to distinguish the type of innovation regime that applies to each firm by asking for the percentage of sales accounted for by products that are unchanged, those that are incremental innovations, and those that are significant innovations. But this is

---

[2] European Commission (1994).
[3] European Commission (1997), p. 254.
[4] Ibid.

a difficult question for firms to answer precisely since sales records are not normally kept along these dimensions.[5] In addition, the dividing line between a significant and an 'incremental' innovation is vague and imprecise. Finally, the meaning given to the two terms will differ for an industry leader and an industry follower. A new product for an industry leader will involve a degree of novelty substantially above that of a product that is copied and introduced for the first time by a firm that follows the lead of others. Therefore, it is difficult to use this type of question to examine the extent to which the innovation regime differs across innovation types.

This problem is partly bound up in differences of opinion on the appropriate level at which an innovation survey should be taken – at the level of the firm or at the level of the innovation. The OECD/Eurostat approach (OECD, 1997) has been to develop questions about the sources of innovations, the nature of the technology used, and impediments to innovation at the level of the firm. In large multiproduct firms, the answers to general questions must necessarily cover many different types of innovations – some of which are extremely novel, some of which are more incremental.

Other research on innovation has relied on databases that are painstakingly constructed from lists of innovations that are then linked to firm characteristics, often by way of a survey. The precursor can be found in the U.K. database developed by Science and Technology Policy Research Unit at the University of Sussex (SPRU).[6] This approach has the advantage that it probably ensures the inclusion of only novel innovations. It has the disadvantage that it probably misses the very large number of incremental changes that are so important for overall technical progress.

Subsequent work has used the trade press to estimate the number of innovations that have occurred. One example of this approach is provided by the American Small Business Administration's innovation database as reported in Acs and Audretsch (1990). Examples from other countries are reported in Kleinknecht and Bain (1993). In these cases, both more and less original innovations have been caught – but the coverage of these exercises is difficult to evaluate.

---

[5] Indeed, we test for comparability of the answers attained on a cross-sectional basis to the question on sales and the question on novelty and find that there are substantial differences.

[6] See Townsend et al. (1981); Kleinknecht, Reijnen, and Smits (1992).

Innovation studies, if they are to be comprehensive, need to explore the variety of innovations that take place in the entire population of firms. The 1993 Canadian Innovation and Advanced Technology Survey does so both by using questions that focused on a firm's innovative capacities and by investigating the characteristics of the firm's major innovation.[7] This gives us a broad overview of the firm in terms of information that can be collected at the firm level and more specific information that is best collected at the level of the innovation itself.

The 1993 Canadian survey did so for two reasons. First, preproduction tests indicated that some questions were best answered at the level of the innovation because the answers differed depending upon the type of innovation and because many firms found it difficult to answer some questions at the level of the firm as a whole. Secondly, this approach allowed a relatively precise classification of the novelty involved in the innovation and, therefore, facilitated an investigation of the extent to which the innovation profile differed for firms that produced different types of innovations.

The advantage of the innovation-based approach is that it allows a wealth of data to be compiled on the innovation itself – the same sort of information that is normally collected in an innovation survey that focuses only on the firm. This information is often easier to collect since it is innovation-specific. As indicated, firms find it easier to specify what a particular innovation's costs are than to report their total innovation costs, since they maintain files on particular and important innovations, but they often do not amass data on all 'innovative' activity in the firm.

This approach has a second purpose. It provides more meaningful data than the approach that asks a firm whether it has any innovations – a widespread practice in innovation surveys. The latter will yield relatively high rates of innovation – because most firms engage in some form of innovation or they would not survive. And when almost all firms report themselves as innovators, cross-tabulations of firm characteristics that compare innovators and non-innovators are not likely to provide meaningful information.

The approach that was adopted here was meant to elicit responses only from those firms with an innovation that they considered to be major. In order to accomplish this, firms were asked to fill in a fairly lengthy section on their most important innovation. Those firms with relatively

---

[7] The survey can be found in the Appendix.

minor product improvements were not likely to take the time to do this. In preproduction tests, we discovered that R&D product managers generally had no trouble or hesitation in discussing the characteristics of major innovations. Indeed, in many cases, project files were available on which many questions could be based.

2.3 MEASURING INNOVATION

The 1993 Canadian Survey of Innovation focuses first on developing data on the incidence of innovation. First, the survey asked firms whether they had introduced *a product or a process* innovation in the three years prior to the survey. A product innovation was defined as the commercial adoption of a new product – minor product differentiation was to be excluded. A process innovation was defined as the adoption of new or significantly improved production processes. In both cases, the definition stressed that minor innovations were to be excluded. Second, it asked firms for the percentage of sales in 1993 that came from *a major* product innovation introduced between 1989 and 1991 and the percentage of sales that came either from *a minor or a major* product innovation. The two separate questions were placed in different sections of the questionnaire. The first often went to the R&D or product manager; the second went to the head office.

The first question that asked only for incidence was relatively easy to answer; however, it may suffer from an upward bias if firms did not restrict themselves just to major innovations as they were instructed to do. The second is more difficult to answer because it requires a breakout of sales data that was required of the respondent. Sales data are not always kept by novelty of product. For this reason, the second question should yield a lower innovation rate than the first question. In addition, the percentage of firms answering that they had sales from a major product innovation should yield a lower rate of innovation because it refers only to product innovations and excludes process innovation. On the other hand, the percentage of firms responding that they had sales from a major or minor product innovation could be above or below the number of firms indicating that they simply had a major innovation in the previous three years. It could be below the latter to the extent that the question was inherently more difficult to answer and captured only product innovations. It could be above the latter since it captured minor product innovations, and the question on the overall intensity of innovations was not supposed to do so.

Table 2.1. *Alternate Measures of Innovation Intensity*

	Measure	% of Firms
1)	Firms producing product or process innovations	33.3 (1.0)
2)	Firms reporting sales from major product innovations	21.7 (0.7)
3)	Firms reporting sales from major or minor product innovations	39.5 (0.8)

Note: Standard errors are in parentheses.

The results of the two questions are shown in Table 2.1. The most inclusive question that includes sales from both minor and major product innovations yields the highest innovation rate – some 40%. The question that asks only for sales of major product innovations yields a rate of only 22%. But when firms are asked whether they introduced either a major product or process innovation, the rate is some 33%. In this book, we will generally use the answer to this question to define an innovator, but we also include those firms who responded that part of their sales came from a major product innovation.[8]

Together, the incidence and the sales questions allow us to classify a firm as being innovative or non-innovative. Additional questions on the characteristics of the firm's most important innovation (defined as the one that had the most effect on profits) permitted us to classify the innovator by the characteristics of this innovation. For example, the innovators were asked whether their innovation was a world-first, or first to be introduced in Canada, or just a first for the firm. This information then allows us to classify a firm into one of these categories. Other information on this innovation – such as whether the innovation was patented – is also available.

It should be noted that the information provided depends on the self-evaluation of the firms as to whether they were innovative and not on a panel of experts. The latter has the advantage to some in that it is more objective – although the standards that are employed obviously involve subjective judgement when it comes to classifying a new product as a genuine innovation or only a product improvement.

---

[8] More precisely, an innovator was defined as a firm that answered this question positively, or answered that it had positive sales from an innovation, or that filled out the lengthy series of questions on their most important innovation.

An innovation survey relies on a different type of expert – the manager of a firm. Managers of firms constantly assess themselves against their competitors. They have to keep an eye on the introduction of new products, on the trends in technology use in their industry. They become experts in each of these areas. Therefore, differences between the two sources of data cannot be said to revolve around the extent to which experts provide the source of information.

Instead, they must involve issues, such as bias, that arise from self-evaluation. Self-evaluation bias may occur if managers overstate the frequency with which they innovate because of pressure to appear 'innovative' in a society that values this characteristic highly. While this pressure may exist in a survey done by experts who know the company and may use it as an example, it is less likely in a survey that is done by a statistical agency and where the results of any one company are kept completely confidential. It is also probable that this problem is less important than nonresponse bias. A survey imposes substantial burdens on respondents, and many will be likely to indicate that they are not innovative in order to escape these burdens.

A different problem arises from traditional innovation surveys. There is a certain vagueness to the definition of an innovation. An innovation can range from a minor product improvement to a major new breakthrough. Distinguishing these various outcomes is necessary if we are to evaluate what innovation means. In this survey, we have approached this problem by asking the firm to classify its most important innovation. By doing so, we provide a metric that can be used to evaluate just how innovative a population of innovators actually is.

2.4 UNDERSTANDING THE INNOVATION REGIME

The majority of questions on the survey are aimed at developing a profile both of the innovation process and of the firm in general.

General questions are aimed at establishing whether the firm operates in international markets and is controlled from abroad, as well as the emphasis it gives to a number of strategies – from marketing, technology, management, and innovation to human resources. The answers to these questions allow innovation to be set in context of other activities. They also allow us to investigate which activities are stressed by the most innovative firms.

A section of the questionnaire focuses on R&D since so much attention has been placed on this input to the innovation process. Firms are asked whether they performed R&D, whether this was a continuous process or only done occasionally, the location of R&D, and whether collaborative research was done. The purpose of this section was to provide a brief sketch of the R&D unit; it was not to conduct an intensive analysis of the research and development process.

Another section of the questionnaire focused on the use that was made of intellectual property rights. A set of these questions was put to the entire population of firms – both innovators and non-innovators. The first question asked whether different forms of protection, such as patents or trade secrets, were used. Use is a measure of importance. A second question asked firms to evaluate the efficacy of the different forms of intellectual property protection as a means of protecting their innovation from being copied by their competitors. Finally, the extent to which these property rights were traded was examined. In addition to these questions that were asked at the firm level, several questions were asked about the manner in which innovators protected their most important innovation. The latter allow us to examine whether the use of intellectual property rights varies by type of innovation.

Finally, a large set of questions was posed about the nature of the innovation. These included a set of questions asked at the firm level about the effect of innovations on the firm and about which impediments the firm faced in the innovation process. The survey then delved into a number of issues that we had discovered were best asked not at the firm level, but of the firm's most important innovation. The characteristics were those that were expected to vary by novelty of innovation – such as the source of ideas for the innovation, the source of technology used to produce the innovation, and the effect of the innovation on the demand for labour. This set of questions allows these characteristics to be classified by novelty of the innovation. It also allows firm characteristics – such as the use of R&D, or the use of intellectual property – to be tabulated by the degree of novelty of the firm's most important innovation.

On the input side, the 1993 Canadian Survey of Innovation asks where the major ideas for innovation originate. The research and development unit has long received the lion's share of attention from both statisticians and economists. But other sources of technological capability exist, especially in production engineering. And on the product development side, both marketing and sales personnel potentially play important roles in developing new products.

Inputs for innovation come not only from inside but also from outside the firm. Both suppliers and customers interact to improve product lines and production processes. Consultants, suppliers, research agencies, universities, related firms, and competitors all contribute to innovation. Since innovation is a collaborative affair, the Canadian innovation survey documents the network of contacts that is used to support innovation in Canada.

Several additional topics that touch on the input side to the innovation process are also pursued. The impact of innovation on the demand for labour and on the skill level of the labour force are examined. Finally, the role of new advanced technology as part of the innovation process is explored.

The effectiveness of a country's innovation system depends upon the incentives provided for innovation. Public policy here plays many roles. Sometimes it is interventionist, for example, with regards to offering technical support services. Sometimes it focuses more on establishing broad framework policies like intellectual property laws. The section of the Canadian innovation survey that investigates the extent to which intellectual property laws are used by innovators to protect their innovations from imitators is directly relevant to public policy. But the survey also investigates whether firms experience impediments in other areas where public policies affect innovation – in skill development, in technological support, and in market information services.

2.5 BACKGROUND ON THE 1993 CANADIAN SURVEY OF INNOVATION AND ADVANCED TECHNOLOGY

The data for this book come from the 1993 Survey of Innovation and Advanced Technology (SIAT). The Innovation and Technology Survey was conducted in 1993 using manufacturing firms of all sizes. There were five sections on the innovation part of the questionnaire: section 1 contained general questions; section 2, R&D questions; section 3, innovation questions; section 4, intellectual property questions; and section 5, technology questions (see Table 2.2).

Three types of units were sampled: plants of larger firms whose head office is located elsewhere, the corresponding head offices of these firms, and small firms that have both their management and plant located in the same spot. In large firms, the first four sections were put to management in the head office; the fifth section was addressed to selected plant managers. In small firms, all of the sections were sent to the same location. Together, the head office responses of large firms on general characteristics, R&D innovation, and intellectual property, along with the technology questions

Table 2.2. *The Types of Sampling Units*

	Questions Asked, by Section				
Firm size	General	R&D	Innovation	Intellectual Property	Technology
Head offices (IPs)*	All	All	All	All	
Small firms (NIPs)†	All		Some	All	
Small firms (NIPs)†	All	All			Some
Large plants (IPs)*					All

* IPs = integrated portion of the Business Register of Statistics Canada.
† NIPs = not in integrated portion of the Business Register of Statistics Canada.

answered by their plants, provide a comprehensive overview of the firms' innovative and technological capabilities.

The small firms were handled somewhat differently. In order to reduce response burden, the small firms were separated into two groups. The first group answered sections 1, 3, and 4 – the general, innovation, and intellectual property questions. The second group answered sections 1, 2, and 5 – the general, R&D, and technology questions. In certain sections, small firms were asked only selected questions in order to reduce their response burden.

For the purpose of the survey, larger firms were defined as those that are in the 'integrated portion' of the business register that Statistics Canada maintains, and throughout this study, they are referred to as IPs. These firms are fully profiled by the agency for purposes of maintaining legal and operating organizational structures. These firms account for over 75% of total business revenue. They range in size from 20 employees to over 500 employees, but more than 90% have more than 50 employees. This larger size class, then, should not be identified with the class that the U.S. Small Business Administration defines as large firms – those with more than 500 employees.

Small firms that were included in the survey are not profiled (NIPs – not in integrated portion of the register) in the Business Register. They generally have fewer than 50 employees and tend to have less information on the register. The majority of the latter have fewer than 20 employees.

The group of large firms (IPs) covered here accounts for the majority of production and employment – upwards of 80%. This group received the most detailed questionnaires. By sampling the smaller NIPs as well, the survey attempted to develop a profile of a group of smaller firms who are usually not covered in great depth by innovation surveys. However, in order to reduce the response burden for this group, the numbers of questions posed in the survey were reduced substantially, relative to those sent out to larger firms. In the chapters that follow, we indicate when tables refer to the entire sample, just to larger firms (IPs), and just to the smaller firms (NIPs). While we generally use larger and smaller to refer to the two groups (IPs and NIPs), we also break down our population into more recognizable size classes during the analysis – that is, 0–99, 100–500, 500+. These classifications may be derived from either the entire sample of IPs and NIPs or just the sample of IPs – and are appropriately labelled.

There were 1,595 head offices (answering the first four sections) sampled, 1,954 large plants (answering the last section) sampled, 1,088 of the first group of small firms (answering the first, third, and fourth section) sampled, and 1,092 of the second group of small firms (answering the first, second and fifth section) sampled, for a total of 5,729 units sampled. The sample was taken so as to provide coefficients of variation (CVs) that were in the 5% to 10% range.

The survey was conducted in several steps. Initially, the firm was contacted to determine who within it (both the head office and the plant) should be sent each section. These individuals were contacted by phone to confirm their ability to respond to the survey. Since the questionnaire was complex, initial contacts were made with the company, the purpose of the survey was explained, and contact names for the ideal respondent were solicited. For example, if there was an R&D unit, this part of the questionnaire was directed to the head of this unit. For the intellectual property section, the survey team solicited the name of the individual who was responsible for intellectual property, and then the relevant section of the questionnaire was sent to that person. For the technology sections of the survey, the managers of the plants that were chosen from the sampling process were obtained, and the questionnaire was sent directly to the plant. Finally, the innovation section of the questionnaire was sent to either an R&D manager or a product manager. The general characteristics section went directly to the head office, once again after the name of the individual who might best answer the type of questions contained therein had been solicited. Once the names of contacts were obtained, the relevant parts of the questionnaire were mailed out to the designated individuals. Finally, where necessary, telephone follow-ups were performed. The response rate for the survey as a whole, across all the sections, was 85.5% and ranged from 92.9% in the second group of small firms down to 77.7% in the large plants.

The sample used for the survey consisted of all firms that owned any plants in the manufacturing sector. This population was stratified by region and by size class, and a random sample was chosen by using these stratification variables. The firms' responses that are reported here were probability weighted to provide an accurate representation of the universe of firms from which the survey was taken. We also occasionally report employment-weighted measures. These measures indicate the percentage of employment in firms for a given response. For example, a probability or company-weighted innovation rate of 36% indicates that 36% of the population were innovative. An employment-weighted rate of 72% indicates that firms that innovated accounted for 72% of employment. In subsequent chapters, data that are reported in tables are probability weighted to represent the universe of firms, unless otherwise specified.

For this survey, we did not generally impute nonresponses. We did so where there was additional information from other questions. But often the question was sufficiently complex that imputation was not attempted. For example, information on 'types of licenses' is so firm specific that we judged it to be unwise to fill in missing responses in this case. This means that coverage totals will differ across tables. The percent of innovators, when calculated for all firms, will not be the same as when it is calculated for firms who describe the nature of their technology licenses, because less than 100% of the sample answers the latter question. Readers who compare across tables should remember this. And it is why we use cross-tabulations to indicate differences in categories, and why totals for the tabulations using the cross-tabulations are often omitted. Nevertheless, we only report results for those questions where the respondents are broadly representative of the population.

We should also point out that aggregates reported in tables will differ depending on whether they are being provided for the entire population or just the population of large firms (IPs). The former is the default. If only large firms are used, this is usually indicated in the notes to the table. Population estimates of both small and large firms are used for main aggregates. But many of the differences that are discussed in subsequent chapters involve only large firms because these were the ones requested to provide the largest amount of detail about their innovations. It should be recalled that the large firm sample here includes firms with 20 employees up to more than 2,000 employees.

Finally, it should be noted that estimates from sample surveys are subject to sampling error. Point estimates need to be interpreted in light of the variance of these estimates. We therefore generally include standard errors for individual estimates so that the reader can judge the statistical significance of differences in estimates. Where this would unduly complicate tables, and where the standard errors of the estimates are sufficiently the same, we mention the average size of these estimates in footnotes. But it should be pointed out that the analysis contained in the text relies not just on whether differences are statistically significant; more so, it relies on whether the differences across categories are meaningfully large.

THREE

Patterns of Innovation: Intensity and Types

3.1 INTRODUCTION

This study focuses on the nature of the innovation process in the Canadian manufacturing sector. The innovation process is defined by the intensity and types of innovation, the types of inputs that produce the innovations, the effects of innovation, the importance of research and development, the use of intellectual property in protecting the investment in ideas needed to make innovation work, the financing process, the nature of externalities and networks that are used to diffuse innovations, and the types of skills required to allow innovative firms to succeed.

In this chapter, we begin the process of describing the Canadian innovation process by studying the intensity of different types of innovation. A study of a country's innovation system inevitably faces the questions: How important is innovation? How widespread is it? Therefore, this chapter reports the incidence of innovation, that is, the percentage of firms that introduced a product or process innovation in the three years preceding the survey.

At the foundation of this study is the recognition that the industrial population is heterogeneous and so, too, are the types of innovations being brought to market. Small and large firms may have differential innovation capabilities – with small firms more adept at product development in the early stages of a product life cycle and large firms more capable of the type of process improvements that are critical in the mature stage of the product life cycle (Rothwell and Zegveld, 1982; Acs and Audretsch, 1990). Foreign and domestic firms are seen to have different capabilities of ingesting technology and innovative ideas from abroad, since multinational firms generally provide a more efficacious method of transferring technological knowledge across international boundaries (Caves, 1982; Dunning, 1993). Industries also differ in terms of their contribution to the innovation process, since scientific opportunities are not the same in each sector (Robson et al., 1988). Therefore, this study focuses on the differences that exist in the innovation process. In this chapter, we investigate differences in the intensity of innovation by size class, industry, and nationality of ownership.

Heterogeneity in the innovation process is to be found both in the types of actors and in the types of innovations that are brought to market. Innovations may focus primarily on new product development, or they may focus more on the production process – finding new ways to produce mature products at lower cost. To investigate whether Canadian innovation focuses on the first or second type, this chapter divides innovations into three types – product, process, or a combination of the two. Other major differences occur in the type of innovations produced, since innovations may involve incremental improvements on products and processes that already exist, or they may involve radical change. To investigate this issue, we ask how novel are most innovations that are produced in Canada.

By itself, a measure of incidence yields significant insights into differences that exist across industries or innovation types in Canada. But a note of caution is required at this juncture. The measures that are discussed herein are used to set the scene for subsequent chapters where we investigate how other aspects of the innovation process differ across types of innovators and types of innovations. They are not meant to yield an absolute measure of innovativeness that can be used to compare the intensity of innovation in Canada to that in other countries with any degree of accuracy. The measure that an innovation survey yields for a particular country depends on the survey methodology, on the types of questions asked, on the way in which the survey was conducted, and on the culture of the country. There is such a wide variance in many of these factors that cross-country comparisons of innovation rates are problematic.[1]

---

[1] The 1993 Canadian Survey of Innovation used a long, detailed questionnaire that delved into the characteristics of the major innovation of a firm that indicated it was an innovator. This involved considerable respondent effort. Most of the firms that answered that they had introduced an innovation went on and fully answered the long questionnaire. Non-innovators did not have to do so. Non-respondents who felt burdened by the questionnaire probably included some innovators who did not want to fill in the long-form that was attached to the innovation question. That means the Canadian survey may underestimate the rate of innovation by excluding some of the minor innovations, since those with less

Table 3.1. *The Percentage of Firms That Introduced, or Were in the Process of Introducing, an Innovation During 1989–91, by Size Class*

Type of Weight	Size Class (Employees)					
	All Firms	0–20	21–100	101–500	501–2,000	2,000+
Company weighted	33	30	39	41	58	73
	(1)	(1)	(2)	(3)	(4)	(5)
Employment weighted	52	32	39	47	56	67
	(1)	(1)	(2)	(3)	(4)	(5)

Note: Standard errors are in parentheses.

3.2 THE INCIDENCE OF INNOVATION

We focus our attention first on the percentage of firms that said that they introduced or were in the process of introducing a product or process innovation during the period 1989 to 1991, or that listed a product or process innovation.[2]

One-third of Canadian manufacturing firms introduced, or were in the process of introducing, a product or process innovation during the period 1989 to 1991. The proportion of innovating firms increases from about 30% in the smallest size category to 73% in the largest one. Since firms in the largest size class innovate more frequently than the smaller ones, innovating firms account for more than half (52%) of total manufacturing employment during the 1989–91 period (see Table 3.1).

That the incidence of innovation increases with the size of the firm has been also observed in innovation surveys conducted in France, Italy, and Germany, where the differences between small and large firms appear to be even more significant than in Canada (OECD, 1993b).

In later sections of this book, we examine just the larger firms (IPs), which have generally more than 20 employees. This group has a higher tendency to innovate than the entire population that contains many firms with fewer than 20 employees. Some 39% of the larger firms indicated that

---

important innovations were probably less likely to fill in the long questionnaire on their major innovation.

[2] These are the two questions that defined whether a firm was requested to answer a set of questions on the nature of their major innovation that will be used extensively in later chapters. Firms also were asked to outline the percentage of sales from a major innovation in a different section of the survey. This was not used here since it was not located in the section of the survey containing the characteristics of innovation that are used in subsequent chapters. If it had been used, the overall innovation rate would have increased from 33% to 40% (see Chapter 2).

they had introduced a new product or process in the three years previous to the survey or were introducing one. This increases to 50% if we also consider an innovator to be one that reports sales from major product innovations. And the employment-weighted estimate of innovation using this more extensive definition is 70% – that is, these innovators accounted for 70% of employment in the group of IPs.

Since Schumpeter's (1942) assertion that innovation activity increases more than proportionally with firm size and with industry concentration, economists have debated and investigated the relationship between the size of firm and innovation performance. Cohen and Levinthal (1989) summarize many of the arguments for firm-size differences: 1) that capital market imperfections confer an advantage on large firms in securing finance for risky R&D projects, because size is correlated with the availability and stability of internally generated funds, 2) that there are large, scale economies in the production of R&D or that the returns from R&D are higher where the innovator has a large volume of sales over which to spread the fixed costs of innovation; and 3) that R&D is more productive in large firms as a result of complementary relationship between R&D and other nonmanufacturing activities (e.g., marketing and financial planning) that may be better developed within large firms. Counterarguments revolve around the contention that in large firms, the efficiency of R&D is undermined through loss of managerial control or that the incentives of individual scientists and entrepreneurs in large firms become attenuated as their ability to capture the benefits from their efforts diminishes.

The Canadian evidence that larger firms innovate more frequently than smaller ones that was provided in Table 3.1 may be mainly the result of differences in average firm size across industries. Differences in the minimum efficient size of the firm across industries are related to an industry's age, its technology, and its competitive regime. If less concentrated industries, which have smaller firms, are less innovative than the more concentrated ones, which are dominated by a few large enterprises, we would expect that on average, innovation incidence would increase with the size of the firm, even though most of the relationship could be attributed to industry effects.

In order to determine whether the size effect is pervasive across industries, the percentage of innovating firms is calculated by size class in each major manufacturing industry group (see Table 3.2). The ranking of industries by degree of innovativeness for all firms (column 1) is quite similar to the ranking that would result if the industries were grouped

Table 3.2. *The Percentage of Firms That Introduced, or Were in the Process of Introducing, an Innovation During 1989–91, by Size Class and Individual Industry*

			Size Class (Employees)			
SIC	Industry	All Firms	0–20	21–100	101–500	500+
Core						
31	Machinery	43 (4)	38 (6)	54 (7)	64 (9)	–
33	Electrical and electronic products*	51 (4)	47 (7)	59 (9)	52 (9)	77 (7)
36	Petroleum refining and coal	53 (8)	55 (18)	31 (4)	54 (18)	–
37	Chemicals	49 (5)	36 (10)	44 (12)	50 (18)	76 (10)
3741	Pharmaceuticals	59 (10)	55 (20)	0 NA	73 (19)	81 (14)
Secondary						
15	Rubber and plastic	49 (4)	42 (7)	61 (7)	74 (9)	81 (14)
29	Primary metal	26 (5)	21 (9)	24 (13)	38 (11)	52 (13)
30	Fabricated metal	33 (3)	34 (5)	32 (6)	22 (9)	40 (13)
32	Transportation equipment	35 (4)	24 (6)	52 (8)	50 (9)	70 (9)
35	Nonmetallic mineral products	31 (4)	25 (5)	50 (7)	48 (15)	54 (19)
Other						
10–12	Food, beverage, and tobacco	30 (3)	24 (5)	38 (5)	36 (7)	62 (7)
18&19	Textiles	34 (4)	39 (7)	15 (6)	45 (12)	–
17&24	Leather and clothing	10 (3)	9 (3)	10 (5)	16 (8)	48 (22)
25&26	Wood, furniture, and fixtures	27 (3)	25 (3)	34 (6)	23 (9)	44 (18)
27	Paper	41 (5)	32 (12)	47 (9)	44 (11)	64 (9)
28	Printing and publishing	30 (4)	29 (4)	37 (9)	–	53 (13)
39	Other	42 (5)	40 (5)	58 (11)	70 (14)	–

Note: Standard errors are in parentheses.
* Includes instruments.
NA = not applicable.
– = suppressed for reasons of confidentiality.

by R&D intensity. But what is more significant, the proportion of innovating firms generally increases with the size of firm in each industry. The difference, then, between small and large firms is widespread at the industry level.

There are notable interindustry differences in innovation incidence (Table 3.2). To some extent, the interindustry differences reflect variations in technological opportunity. Technological opportunity in mature industries is lower than elsewhere. Firms in older, mature industries, such as leather and clothing, primary metal, wood products and furniture, food, beverages and tobacco products, as well as those in printing and publishing, innovate less frequently than the 33% average for the entire manufacturing sector. In contrast, scientific progress in biochemistry, electronics, chemistry, and computer science creates technological opportunities that reduce the cost of and increase the demand for innovation. The data on innovation intensity in Table 3.2 reflect this by showing that pharmaceutical firms, producers of electronic equipment (including telecommunication and office equipment), and producers of chemical products, scientific instruments, and nonelectrical machinery are more innovative than the overall average.

Others (Robson, Townsend, and Pavitt, 1988) have observed a consistent pattern across industries in innovation rates for both U.K. and U.S. industries. Robson, Townsend, and Pavitt identify a few core sectors (chemical, mechanical, instruments, and electronics) that are of major importance in the production of technology that gets used over a wide range of sectors, including those outside manufacturing. They derived this taxonomy by examining data from a survey of more than 4,000 important innovations commercialized in the U.K. between 1945 and 1983. American innovation data were derived from a study by Scherer (1982a, 1982b), who developed a large matrix of technology production and use based on the patented inventions of 443 large U.S. corporations in 1974. The U.K./U.S. comparisons are presented in Table 3.3, where an innovation is referred to as a 'technology produced'. Innovations produced in one sector and used in another are called 'product' innovations in the table. Innovations that are produced and used in the same sector are termed 'process' innovations in the table.

A similar pattern of sectoral flows for both the United States and the U.K. is revealed by these innovation data. The core sector is highly innovative in that it produces the majority of innovations (technologies produced – column 1). The core group produces more technologies than it uses, since the ratio of product to process technology is highest

Table 3.3. *Comparison of Technology Patterns in the United Kingdom and United States*

Sector	Percentage of All Technology Produced 1		Ratio of Product to Process Technology Produced 2		Percentage of All Technology Used 3		Percentage of All Product Technology Used 4	
	UK	US	UK	US	UK	US	UK	US
Core	68.3	62.8	3.3	2.6	18.3	18.8	3.3	2.3
Secondary	20.6	23.9	1.3	2.1	16.4	12.7	11.1	7.3
Other manufacturing	8.3	12.0	0.5	1.0	26.0	11.4	30.4	8.0
Nonmanufacturing	2.9	1.3	0.2	0.3	39.4	57.1	55.1	82.4
All groups	100	100	100	100	100	100	100	100

Source: Reprinted from *Research Policy*, vol. 17, M. Robson, J. Townsend, and K. Pavitt, "Sectoral Patterns of Production and Use of Innovations in the UK: 1945–1983," p. 6, copyright 1988, with permission from Elsevier Science.

here – column 2. Thus, a classification of industries based on the intensity of innovations also corresponds closely to one based on the degree of spillovers – since the core sector produces more innovations than other sectors and also has the greatest proportion of its innovations used in other sectors. However, the use to which new technologies are put is relatively even. The distribution of 'technology used' innovations (column 3) is more equal than the distribution of technologies produced (column 1).

These data show that the core sector is highly innovative, producing mainly innovations or 'technologies' used elsewhere. The secondary sector is somewhat less innovative and is more equally balanced between products used in other sectors and processes used in the same sector. The secondary sector uses technology from the core sector but also diffuses technology via new products to the other sector – though with less intensity than the core sector. The remaining industries are the least innovative in the sense that they originate fewer of the new technologies. The other sector absorbs technologies from the core and secondary sectors. Technical progress in the other sector is due in large part to the adoption of innovative products that are produced by the core and secondary sectors – whether these products are material inputs, such as chemicals, or capital inputs, such as machinery and equipment. It should be noted that while this 'other' sector ingests innovations in the way of products

Table 3.4. *The Percentage of Firms That Introduced, or Were in the Process of Introducing, an Innovation During 1989–91, by Industrial Sector and Size Class*

	Size Class (Employees), by Sector					
	0–20	21–100	101–500	501–2,000	>2,000	All
Core	43	54	60	75	80	48
	(4)	(4)	(5)	(6)	(8)	(2)
Secondary	31	43	40	53	85	34
	(3)	(3)	(5)	(7)	(8)	(2)
Other	28	32	33	51	58	29
	(2)	(3)	(4)	(6)	(8)	(1)

Note: Standard errors are in parentheses.

and machinery from the core and secondary sector, this does not make it less innovative. Finding ways to make these products work requires ingenuity – but it probably involves a different set of skills than are required in the core sector.

Since the patterns of production and use of technology are much the same in both the United States and the U.K., the industry taxonomy developed by Robson et al. captures basic differences in the technical characteristics that should be expected to exist in most industrial countries. Therefore, the core, secondary, and 'other' distinction will be used extensively in this study to analyze cross-industry technological patterns in Canada.

The sectoral pattern of innovation intensity in Canada that is presented in Table 3.4 is broadly similar to that found in the United States and the U.K. Industries are grouped into sectors similar to Robson's classification[3] – core, secondary, and other. Firms in the core sector feed technology to the rest of economy and innovate more frequently than those in the secondary and other industrial sectors. When firms of all size categories are considered, half of firms in the core sector (48%) introduced or were in the process of introducing an innovation. Firms in the secondary sector innovated less frequently (34%), but still more than those in other manufacturing industries (29%). Breaking down the innovation incidence within each sector by firm size categories confirms that larger firms innovate more often than smaller ones in all groupings.

---

[3] We adopt the Robson classification, despite some anomalies when applied to Canada, for the sake of international comparisons.

Table 3.5. *The Percentage of Innovating Firms That Introduced, or Were in the Process of Introducing, Product or Process Innovations*

	Product, with No Change in Man. Technology	Product, with Change in Man. Technology	Process, Without Product Change
Introduced	35	45	46
	(2)	(2)	(2)
In progress	20	32	23
	(2)	(2)	(2)

Notes: Only larger firms (IPs) were asked to classify their innovation activity in the three categories. These firms were generally larger than 20 employees. Responses are not mutually exclusive; the row sum therefore may exceed 100%. Standard errors are in parentheses.

3.2.1 Types of Innovations Introduced

Innovation can occur either at the output end of the production process or on the input side. Life-cycle models suggest that product innovation tends to prevail early in the life cycle, while process innovation is more prevalent in the later stages of life. Of course, product and process innovation can also be combined.

To measure the distribution of innovations across these types, innovators are divided into three groups: those introducing product innovations without a change in manufacturing technology (pure product innovation); product innovations that required simultaneous changes in manufacturing technology (product/process innovations); and process changes in manufacturing technology without product change (pure process innovations). The largest group (46%) reported that they introduced pure process innovations (see Table 3.5). Fewer (35%) reported a pure product innovation. But a good 45% reported that they had a product innovation that involved a change in technology.[4]

Whether an industry focuses on product or process innovations is determined by the direction of technological change, competitive pressure, and conditions of appropriability. There are substantial differences in

---

[4] Firms could report more than one category. While 35% reported a product innovation with no changes in technology, only 23% reported *only* a product innovation with no change in manufacturing technology. Some 46% reported a process innovation with no product change, but 15% reported *only* just a process innovation with no product change. The remaining 63% reported either a) both a product and process separately or b) a product with a process innovation.

Table 3.6. *Distribution of Product and Process Innovation, by Industrial Sector (% of Firms)*

Sector	Product, with No Change in Man. Technology	Product, with Change in Man. Technology	Process, with No Product Change
Core	48	37	42
	(4)	(4)	(4)
Secondary	32	50	51
	(4)	(4)	(4)
Other	28	47	46
	(3)	(3)	(3)

Notes: Larger firms (IPs) only. Standard errors are in parentheses.

the product/process focus across industry groups (see Table 3.6). The core group of industries focuses relatively more on pure product than on pure process innovations. The secondary and other sectors downplay pure product innovation relative to the two categories that involve process technology. This accords with the pattern described by Robson et al. (1988). The core sector produces innovations that are diffused as new technology to other manufacturing industries. Therefore, the focus in the core sector is more on products; the focus in the secondary and tertiary (other) sectors tends towards new technologies that reduce production costs and/or improve the quality and flexibility of the production process. Nevertheless, it should be noted that the differences across sectors in the intensity of pure product innovation are greater than the differences in pure process innovation. Process technology is needed everywhere, regardless of whether product innovations are being produced for others or being adopted from elsewhere.

3.3 CHARACTERISTICS OF INNOVATION

3.3.1 Originality and Type of Innovation

Innovations cover a range of improved and new products and processes that differ in terms of novelty. In order to examine differences in the type of innovations being produced, firms were divided into three groups based on the originality of their most important innovation: world-firsts, which are firms whose most important innovation was not previously introduced elsewhere; Canada-firsts, which are firms with innovations that

Table 3.7. *The Distribution of Innovations, by Novelty Type and Industrial Sector (% of Firms)*

Sector	World-First	Canada-First	Other
Core	26	36	38
	(3)	(4)	(4)
Secondary	13	40	47
	(3)	(4)	(4)
Other	11	27	62
	(2)	(3)	(3)
All	16	33	51
	(2)	(2)	(2)

Notes: Larger firms (IPs) only. Standard errors are in parentheses.

were new to Canada; and other, which are those firms whose major innovation was neither a world-first nor a Canada-first. These distinctions were drawn by the innovating firm itself when describing its most important innovation.

Only a small portion of innovators – one in six – was a world-first (see Table 3.7). More than twice as many (33%) described their major innovation as a Canada-first. Some 50% of innovators introduced neither a world-first nor a Canada-first and are classified as other, that is, new to the firm, but introduced previously elsewhere in Canada. This emphasizes the importance of the diffusion process. Most change that is taking place involves innovations that build on ideas already developed. This fact should not be misinterpreted. It does not mean that this type of innovation lacks economic importance, as subsequent chapters will demonstrate.

It is difficult to identify the link in the chain of product development that is the most novel, since small changes in existing processes that do not qualify as making a demonstrably different product or process may nevertheless create a radically improved innovation. Sometimes, small improvements on existing technologies can involve dramatic breakthroughs. For example, the process of developing improved petroleum refineries involved a large number of small improvements whose cumulative effects were substantial (Enos, 1962). The Bessemer furnace increased its efficiency dramatically when superior means were found to improve the pumping of oxygen through the molten pig iron (Mantoux, 1961). Dupont's chemical processes were improved with many small incremental improvements (Hollander, 1965). That does not mean that all incremental innovations have a critical importance – only that this other category should not be just equated with improvements that are merely cosmetic.

A breakdown of the originality of innovation by industry sector (Table 3.7) shows that innovations created in the core industries are more often original than innovations in more traditional fields. As we shall see in a later chapter, the core industries are also the more R&D intensive. Therefore, there is a close association between R&D intensity and the novelty of the innovation output. The proportion of world-first innovations in the core sector is about double that of the secondary and other sectors. The differences across the three sectors with regards to Canada-first innovations are minor. But the pattern is reversed for more imitative 'other' innovations, which are more important in the sectors depending on technology developed elsewhere (Table 3.7). In these sectors, innovation is more likely to involve the purchase of machinery and equipment that are readily available to all, and therefore, progress involves utilizing the innovations of others.

While the results for the core/secondary/other industry groupings show that there are broad intersectoral differences in the novelty of innovations being produced, the results for individual industries within each sector reveal intragroup heterogeneity (see Table 3.8). It is true that the R&D intensive core industries, such as pharmaceuticals, electrical equipment, and chemicals, report the highest proportion of world-first innovations. Nevertheless, there are some non-R&D-intensive industries (rubber and wood products) where world-first innovations also made up a high proportion of the total. On the other end of the scale, printing and publishing, food, beverage and tobacco, leather and clothing, fabricated metals, transport equipment, and machinery were conspicuous by reporting the lowest proportion of world-first innovations. The latter two are normally classified among the more innovative secondary or core sectors.

Industries with firms reporting innovations in the intermediate Canada-first category are those where the transfer of technology to Canada is important. This type of innovation dominates in firms producing machinery, rubber products, primary metals, and transport equipment. It is also very important in such diverse industries as plastics, instruments, fabricated metals, textiles, and leather and clothing.

Companies in printing and publishing, pulp and paper, food/beverages/tobacco, and petroleum refining industries were more likely to concentrate on 'other' innovations, that is, those that were already in use elsewhere in Canada. The latter, while less innovative, nevertheless diffuse technological change within Canada. They are less dramatic than

Table 3.8. *Distribution of Innovations by Novelty Type and by Individual Industrial Sector (% of Firms)*

SIC	Industry	World-First Mean	S.E.	Canada-First Mean	S.E.	Other Mean	S.E.
Core							
31	Machinery	11	(8)	51	(8)	38	(8)
33	Electrical and electronic products	34	(7)	36	(7)	30	(6)
36	Petroleum refining & coal	27	(13)	17	(11)	56	(14)
37	Chemicals	23	(6)	28	(7)	48	(8)
3741	Pharmaceuticals	68	(16)	7	(8)	25	(14)
Secondary							
15&16	Rubber and plastic	19	(7)	40	(8)	41	(8)
29	Primary	24	(11)	42	(12)	34	(12)
30	Fabricated metal	9	(5)	38	(9)	53	(9)
32	Transportation equipment	10	(5)	42	(8)	48	(8)
35	Nonmetallic mineral products	20	(9)	30	(11)	49	(11)
Other							
10–12	Food, beverage, and tobacco	3	(2)	31	(5)	66	(5)
18&19	Textiles	27	(9)	35	(10)	38	(10)
17&24	Leather and clothing	9	(7)	34	(13)	57	(13)
25&26	Wood and furniture	30	(11)	23	(10)	47	(11)
27	Paper	11	(6)	24	(8)	66	(9)
28	Printing and publishing	3	(4)	11	(7)	86	(8)
39	Other	15	(9)	31	(12)	54	(13)

Notes: Larger firms (IPs) only. Standard errors (S.E.) are in parentheses.

the world-first and Canada-first innovations, but their economic impact on the innovating firm in industries selling mostly on local markets is important, as Chapter 6 will demonstrate in more detail.

3.3.2 The Number of Innovations Introduced

The previous sections provide an overview of the *incidence* of innovative activity over a three-year period – where innovation was defined as the introduction of a major new product or process. Equally important is the *intensity* of activity – whether firms are introducing large numbers of innovations.

Table 3.9. *The Number of Product and Process Innovations Introduced*

Number	Product Innovation, with No Change in Man. Technology	Product Innovation, with Change in Man. Technology	Process Innovation, with No Product Change
All firms	3.4 (0.5)	2.4 (0.4)	1.9 (0.1)

Notes: Larger firms (IPs) only. Standard errors are in parentheses.

Intensity is measured here as the number of major innovations that the innovating firm introduced over the 1989–91 period. A breakdown of the intensity is provided in Table 3.9 for innovators that indicated they produced innovations with no change in technology, process innovations with no product change, and product innovations that involved a change in technology. On average, innovative firms with product innovations that did not involve a technology change reported 3.4 major innovations; those firms reporting process innovations with no change in products reported only 1.9 innovations. Innovators reporting innovations that involved the combined product/process category fell between these two extremes with about 2.4 innovations.[5] The lower number of innovations in the process category indicates that introducing major process innovations, which often requires substantial changes in production technology and organization of work on the production floor, is more time-consuming than introducing product innovations.

It is noteworthy that the incidence of pure process innovation (the percentage of innovators falling in this category) is higher than the incidence of pure product innovation, but its intensity (the number of innovations introduced) is lower. On the other hand, the intensity of pure product innovation is lower, but it has a higher incidence. This indicates that the distribution of product innovation is skewed more than is process innovation. There is a smaller percentage of firms engaged in product innovation, but these firms are quite intense innovators. On the other hand, process innovation is spread out across a wider proportion of the population.

The breakdown of the number of innovations by sector shows that both the core and secondary sectors produce a greater number of innovations per firm than does the 'other' sector (see Table 3.10). Therefore,

---

[5] The fact that the mean number of innovations is not large suggests that the approach that asks for characteristics of the most important major innovation captures a significant proportion of the sample of important innovations.

Table 3.10. *The Number of Product and Process Innovations Introduced, by Industrial Sector*

Number	Product Innovation, with No Change in Man. Technology	Product Innovation, with Change in Man. Technology	Process Innovation, with No Product Change
Core	3.5	2.5	2.2
	(0.6)	(0.4)	(0.3)
Secondary	4.9	2.9	2.0
	(1.6)	(1.2)	(0.3)
Other	1.8	2.1	1.7
	(0.2)	(0.3)	(0.2)

Notes: Larger firms (IPs) only. Standard errors are in parentheses.

the sectors with the greatest incidence of innovation are also those with a higher innovation intensity. This is significant because it indicates that the innovations in the other sector are not just cosmetic. Otherwise the numbers of innovations would be significantly higher than are reported here.

3.3.3 Features of Innovation

Information on the features of the major innovations improve our ability to discern the degree of novelty that is imbedded in the innovation. Product innovations may contain new materials, consist of new intermediate products, contain new functional parts, or perform fundamentally new functions. Process innovations can consist of radically new production techniques or just a greater degree of automation of existing processes. Or they may require a new organizational structure.

The importance of each of these characteristics is reported by type of innovation (see Table 3.11). The largest percentage of product innovations involves the use of new materials. Innovators in the 'other' sector are more likely to introduce products with new materials than elsewhere. By way of contrast, firms in the core and secondary sectors are more likely to focus on introducing products with fundamentally new functions. World-firsts, which are more likely to be produced in the core sector, also concentrate more on fundamentally new functions.

Process innovations generally are more likely to involve new production techniques than just a greater degree of automation in existing process techniques. This difference holds across novelty types and across industry sectors. It is, however, the case that new production techniques

Table 3.11. *Features of the Most Profitable Innovation Reported by Size Class, Novelty of Innovation, and Industrial Sector (% of Innovators)*

Feature	All Firms	Smallest Firms (NIPs)	Novelty Type				Industrial Sector		
			World-First	Canada-First	Other	Core	Secondary	Neither
Product innovation only	22	23	20	21	20	36	23	17
Features of product innovation involving									
Use of new materials	40	42	38	33	34	33	31	49
New intermediate products	15	16	16	10	12	16	16	15
New functional parts	29	30	38	28	24	29	34	27
Fundamentally new function	31	31	39	31	23	39	33	26
Other product	12	12	9	8	13	17	11	10
Process innovations only	15	12	18	24	25	12	15	15
Features of product innovation involving									
New production techniques	58	58	55	57	61	42	57	65
Greater degree of automation	39	39	41	33	41	28	38	43
New organization of production	19	18	19	19	21	15	22	18
Other process	6	6	6	6	5	10	3	6
Both product and process innovation	63	65	54	55	65	53	62	68

Notes: Since respondents could indicate more than one feature, most innovations consist of a combination of product and process innovation features. Those that indicated only product or only process innovation features are listed in the respective rows. Those that indicated a combination of both are in the last row (58.1% of all innovations introduced by larger firms). Owing to the fact that smallest firms did not report the originality of their innovations, the cross-tabulations on originality are available for larger firms (IPs) only. Standard errors are generally in the range from 2 to 3.

are listed more frequently by the tertiary other sector that generally introduces innovative products from the core sector – products that have fundamentally new functions. New organizational forms are not introduced very frequently along with process innovations.

A breakdown of innovation characteristics by SIC industry groups (see Table 3.12) demonstrates that the use of new materials is the most frequent feature of innovations in the textile (69%), leather and clothing (75%), and petroleum refining (70%), as well as the chemicals and pharmaceutical industries (48%). In contrast, fundamentally new functions are relatively important in machinery (52%) and chemicals (44%). The introduction of new production techniques is most frequently reported by pulp and paper manufacturers (81%), leather and clothing (78%), rubber and plastics (76%), and wood and furniture (76%).

3.4 CONCLUSION

This chapter has revealed that there is a considerable diversity in the types of innovation taking place. Heterogeneity exists in both the actors and the output of the process.

The intensity and pace of innovation varies considerably across Canadian industry. Every third firm declared that it innovated over a three-year period, but this overall average hides important differences across size classes. More than three-quarters of the largest firms, employing more than two thousand persons, introduced an innovation. Innovating firms accounted for more than half of total manufacturing employment.

Innovation is more frequent in a core set of industries (pharmaceuticals, computers) that have elsewhere been shown to be more innovative. Almost half of the firms in these industries introduced an innovation, whereas fewer than 30% of firms in the tertiary sectors did so. This may be the result of either greater technological opportunity in the former industries or the fact that these industries contain more products that are in the early stage of their life cycle. Innovating firms that are found in the core sector feed new technology to the rest of the economy.

Not all firms focused on the same types of innovations. Even here there is considerable diversity. The largest percentage of innovators reported creating pure process innovations. Just behind are those that reported product innovations with a change in the manufacturing process. The smallest percentage focused on just product innovations. Firms in the core innovative sector are more likely to focus on product than process

Table 3.12. Innovation Features, by Individual Industry (% of Innovators)

SIC	Industry	New Materials	New Inputs	New Functional Parts	New Functions	Other Product Innovations	New Production Techniques	Greater Automation	New Organization of Production	Other Process Innovation
Core										
31	Machinery	29	10	30	52	11	30	19	12	8
33	Electrical and electronic products*	29	17	36	30	27	46	36	16	16
36	Petroleum refining & coal	70	10	5	5	21	52	13	24	17
37	Chemicals & pharmaceuticals	48	24	18	44	3	50	13	19	0
Secondary										
15	Rubber & plastic	53	20	51	35	5	76	41	28	3
29	Primary metal	41	18	5	29	11	66	54	38	10
30	Fabricated metal	20	13	30	29	16	51	40	14	1
32	Transportation equipment	31	9	40	41	3	52	43	37	8
35	Nonmetallic mineral products	34	29	15	31	13	61	18	16	5
Other										
10–12	Food, beverage, and tobacco	49	7	13	27	19	69	41	21	5
18&19	Textiles	69	8	45	25	1	58	42	17	2
17&24	Leather and clothing	75	13	11	33	2	78	54	32	4
25&26	Wood and furniture	36	16	41	21	12	76	40	3	5
27	Paper	48	12	15	16	13	81	45	24	1
28	Printing and publishing	28	25	33	34	2	71	66	36	7
39	Other	72	11	13	23	12	43	25	11	8

Note: Most standard errors range from 5 to 7.
* Includes instruments.

innovations. Firms in the other sectors focus relatively more on process than product innovation.

The most original world-first innovations are understandably rarer than innovations that introduce to Canada new products or processes that were originally created abroad. Only one in six innovations was a world-first, one in three a Canada-first. The most numerous (slightly more than every second innovation) were new to the reporting firm but had already been introduced elsewhere in Canada. Larger firms are more likely to report that they create world-first innovations. The core sector is more likely to report a world-first than the secondary and tertiary other sectors.

The use of new materials is the most frequent feature of product innovation, and the introduction of a new technique is mostly typical of process innovation. It is, thus, not surprising that in the majority of cases, introducing new materials required a change in production techniques. The introduction of radically new processes is the most important characteristic of innovators that focus on the production side.

While this description shows that there is considerable heterogeneity in the types of innovation and the innovators themselves, there are obvious patterns that emerge, patterns that will be investigated further in later chapters. These patterns involve a differentiation of specialization of activity across sectors and actors.

Firms in the core sector create more product and fewer process innovations than do the other two sectors. Firms in the industries belonging to the core sector create product innovations incorporating new technology that are then used as new materials and intermediate products (e.g., chemicals) or as machinery in the other two sectors. The first priority of firms in the core sector is product development. Products in their early stage of development tend to compete more on novelty and technological characteristics than on price (Vernon, 1966), which requires cost-cutting process innovations. As a result, process innovations are less important in the core sector.

The industries in the secondary and in the tertiary other sector receive their new materials and intermediary inputs, as well as new equipment and machinery, from the core sector. They typically produce standard, homogeneous products that compete on price. In order to reduce production costs, they focus on pure process innovation, as well as the combination of product-process innovation, more often than do core-sector firms.

Thus, product innovations that are world-firsts are being introduced more frequently in the core sector, which focuses on new parts and new functions, whereas the other sector incorporates these new products as new materials or new production techniques. As we will show, these differences have a dramatic affect on various parts of the innovation process – from the sources of innovation, to the types of firms involved in the innovation process, to the use of intellectual property rights. We turn to develop these differences further in the chapters that follow.

FOUR

Sources of Innovations

4.1 INTRODUCTION

Innovation results from complex interactions between the impulse of science and the attraction of the market that provide the strategic innovation options facing a firm. Strategic options appear as opportunities, whose profitability depends on such factors as market conditions, the technological environment, the product life cycle, and the skill of a firm's personnel in technology, management, marketing, and other areas. Science provides a stock of available technological knowledge that is the result of accumulated technical expertise, the transfer of new technologies from others, and the firm's own research and development.

Innovation activity depends crucially on the firm's capability to create and acquire knowledge that not only finds inventions but also brings innovations successfully to the marketplace. This capability rests both on a firm's talent for internal problem solving and on its ability to forge productive external linkages via networks, strategic alliances, and user-producer relationships. The process by which firms acquire and generate knowledge is at the heart of innovation activity.

As with any good, innovation results from the combination of inputs that come from both outside and inside the firm. The knowledge-generation process takes outside knowledge and combines it with the firm's own knowledge. But the difference between innovation and the production of existing goods and services is that innovation involves the generation of a good (knowledge) with special characteristics.

Despite the similarities with the normal production process, knowledge generation is unique in several ways. First, imperfections in the market for knowledge may cause firms not to produce knowledge in optimal quantities. Since some forms of knowledge are generic and nonrivalrous, firms will not be able to capture all of the benefits of expenditures on new ideas and will have a tendency to underinvest in the production of this type of knowledge. Secondly, the transfer of knowledge is difficult. By examining the extent to which firms are creating new knowledge and how they are doing so, we can shed light on the extent to which these two problems are being solved.

It should be noted that knowledge externalities do not preclude some knowledge markets from functioning reasonably well. Some of the knowledge required for innovation is transferred through market transactions. And in these instances, acquiring knowledge is little different from the acquisition of other intermediate inputs. For example, suppliers of new machinery and equipment, such as robots or computer-controlled machines, have an incentive to supply information in optimal quantities on how to make their equipment function properly, because their own welfare depends upon their customers fully understanding the advantages of using their products.

But there are other forms of knowledge that, it is sometimes argued, are less likely to be provided in optimal quantities by outside markets – knowledge that has a nonrivalrous nature. This type of knowledge, which cannot be easily priced and is freely available to all, will suffer the problems associated with most public goods, if no solution is found to the pricing problem.

In these situations, new organizations or institutions often emerge to overcome such problems. Intellectual property rights can be tailored to protect the ideas that are generated. Industry R&D institutes can be formed to internalize externalities. Contracts can be signed between the creators of generic knowledge and those that use it so as to reduce the incentive to underproduce knowledge.

Even so, it must be recognized that while new institutions arise to overcome the externality and contractual problems associated with knowledge creation, they do not necessarily eliminate them. Intellectual property rights do not completely protect an idea from being copied. While knowledge can be traded, it is more difficult to transfer and price knowledge than other intermediate goods that are widely traded. This is because the value of knowledge is difficult to ascertain in advance of the transaction. Asymmetries of knowledge about its worth are commonplace. After knowledge has been transferred, it is hard to monitor the contribution of any particular piece of knowledge to the final product.

It is in these situations that alternative mechanisms are created to overcome the problems of the market. For example, a firm's boundaries can be moved outwards to find a mechanism that can be used to evaluate knowledge, ingest it, and use it to develop new products and processes. These arrangements may involve collaborative ventures with other firms, joint working arrangements, or mergers.

An appreciation of the difficulty involved in transferring knowledge assets has led to new insights into different economic institutions. For example, Caves (1982) argues that the geographic extension of the national firm across international boundaries occurs in part to overcome problems in licensing or selling knowledge assets. This theory applies equally well to the development of large multiplant or multiregional firms. The evolution of large firms generally can be attributed not just to the advantages of scale economies in production that large firms possess, but also to the advantages that large multiplant entities have in transferring information via internal mechanisms.

The problems that are perceived to exist in the knowledge-creation process extend beyond just finding mechanisms to transfer external knowledge. Creating knowledge within a firm is perceived to require special capabilities that are scarce. These capabilities involve mastering the complex task of bringing scientific knowledge and principles to bear on the development of new products and processes. For example, the research and development department was developed to formalize the exploitation of scientific knowledge. This R&D facility is seen to be critical for the internal development of new ideas. Because of the need to adapt general knowledge to firm-specific problems, it is also seen to be critical for absorbing new technology from outside the firm (Mowery and Rosenberg, 1989).

While R&D is important to the innovation process, there are other important contributors to innovation located elsewhere within the firm. As Mowery and Rosenberg have stressed, inventions are often the result of discoveries that are made in the production and engineering departments. These discoveries are then turned over to the research department for better understanding of the phenomenon so that it can be commercialized – in particular, so that products resulting from the discovery can be mass-produced. Once the research department has more fully investigated the science behind the invention, production and engineering departments are called upon to turn the inventions into viable commercial products and processes. The contribution of production and engineering departments is critical to the overall success of the innovation process – and in many cases, involves pathbreaking work.

Rosenberg (1976) has also emphasized the importance of engineering departments to innovation that is associated with the evolution of production processes – especially in industries producing standard materials or durable consumer goods. In these industries, operation conditions are difficult, and economies of scale depend on maintenance of capacity in each part of an integrated system of processes. The breakdown of any segment threatens the integrity of the whole. As Rosenberg has demonstrated, production-engineering facilities are used to identify technical imbalances and resolve bottlenecks that allow for improvements in productivity.

The overwhelming attention that has been devoted to R&D has had other consequences. It has led to the widespread adoption of a linear sequential model of technological change that links innovation almost exclusively to structured, permanent R&D activity, in a sequence that flows from R&D to the final product. There are two problems with this simple model. As just argued, the actual process is often likely to be simultaneous and interactive, rather than linear. Second, if the model leads us to neglect other sources of innovation both inside and outside the innovating firm, then we may not come to fully understand the innovation process.

At issue, then, is the type of knowledge-generation process that supports the Canadian innovation system. To develop a picture of the knowledge-creation process, we focus on several questions. To what extent does imperfect appropriation of the results of R&D and innovation by private firms create spillovers that are a potentially important source of ideas and technology for other innovating firms? How important are the specialized public institutions that provide technical information and services – infratechnologies that are seen to be an indispensable element of a national innovation system? Are specialized private firms that provide technical services as important a source of innovative ideas as the public sector? Are interfirm links between affiliated firms the most important source of information that is internalized, or do the links between firms, as a result of the normal commercial relations between suppliers and customers, also serve as an important conduit for the diffusion of ideas that drive innovation? This chapter addresses these issues.

We are also interested in the type of internal sources that are combined with external sources of knowledge in order to produce innovation. In particular, we investigate the extent to which R&D is central to the internal

Introduction 67

process and facilitates the use of external sources. We recognize that ideas and technology sources tapped by innovating firms are to an important degree determined by technological opportunity. Therefore, we explore the extent to which the complexity of the innovation and the industry of the innovator lead to different roles for R&D. We also investigate how different innovations make use of spillovers from external sources like universities. We examine whether the diversity of the sources of ideas used for innovation is related to technological complexity, as well as the relationship of R&D to these other sources. In particular, we ask whether the R&D-centric model of innovation is adequate, or more particularly, where it is most appropriate.

In exploring these issues, we examine several questions:

The first is the extent to which R&D, in contrast to other sources like the production department, was found to be important for the innovation process. We are interested in knowing whether Canadian innovators rely primarily or only partially on R&D departments.

Second, are there differences in the intensity of R&D use? We investigate whether the R&D-centric model is more applicable in some sectors than others.

Third, we ask whether the other internal sources are substitutes for or complements to the R&D process. If they are complements, then the R&D-centric model needs to be expanded. If they are substitutes, a set of alternative models needs to be developed.

Fourth, we ask whether external sources offer substitutes for or provide complements to R&D. In doing so, we divide external sources into those that are presumed to function reasonably well through markets and those where externalities are more important. We outline the situations where externalities are used extensively. We then describe whether external sources act as complements to or substitutes for internal R&D.

Fifth, we investigate how the various types of innovations make use of different information sources.

Once it is recognized that innovation not only involves a technical problem-solving process but also depends on the capability of the firm to access, evaluate, and act on market-related information, it follows that it is useful to separate sources of innovation into two broad categories. The first includes sources of information that led to the idea of introducing a new product or process. The second is more narrowly concerned with the sources of technological solutions that the firm tapped in the process of developing the innovation. Therefore, this chapter looks first at the source of knowledge for innovation in general.

Then it examines sources for technological advances that were implemented during the innovation process.

Throughout, we recognize that different types of innovations (e.g., product versus process) do not respond to the same economic incentives and may rely on different sources of ideas and technology. We, therefore, analyze the survey information on the most important innovation of the firm and on the technology employed for this innovation to determine how characteristics of the innovation vary across industries that differ in terms of their technological opportunities. Next, we classify external sources of innovative ideas into spillovers and market transactions. Then we use these categories to assess their mutual relationships and importance as sources of ideas and technology for innovation. Wherever relevant, the effect of specific aspects of the type of innovation (product-process or differences in novelty) on the knowledge-generation process is examined.

Other aspects, such as the size of firm, the originality of innovation, and the nationality of the firm, are also potentially important determinants of sources of ideas and technology. They are discussed in later chapters.

4.2 INTERNAL SOURCES OF NEW IDEAS

An understanding of the diversity of sources of innovation is derived here from the percentage of innovators[1] who identified various internal and external sources of ideas that were used to generate and develop innovations.[2]

Four main internal sources are identified – management, R&D, sales/marketing, and production.[3] When the idea for innovation originates in the R&D activity of the firm, it is likely that the innovation process is driven by pure scientific and technological progress. Innovations driven by the production department are those that also involve a heavy

---

[1] In this chapter, we deal only with innovators. The next chapter examines the differences between innovators and non-innovators – at least with respect to the nature of their R&D facilities.

[2] Ideally, we would like to know the dollars expended on each source. But data on all these expenditures are not available in most firms. Instead, we rely on a question that requests the sources of ideas used for the most important innovation of the firm. This is a general question. A more specific question could have been put – such as the actual department that did the work or the external source of the partnerships into which the firm actually entered. In fact, the latter will be used in the next chapter to examine R&D collaboration.

[3] An additional 'other' category was included in the list of possible sources provided in the innovation questionnaire but was identified as important by less than 4% of respondents.

Table 4.1. *Internal Sources of Innovative Ideas, by Novelty Type (% of Innovators Using a Source)*

Sector	Management	R&D	Sales & Marketing	Production	Other
All	53 (2)	44 (2)	43 (2)	36 (2)	3 (1)
World-first	43 (5)	66 (5)	40 (5)	26 (4)	6 (2)
Large (500+)	35 (4)	64 (4)	37 (4)	24 (3)	4 (2)

Notes: Larger firms (IPs) only. These generally have more than 20 employees. Standard errors are in parentheses.

problem-solving component. Innovations inspired by the management, or by sales and marketing personnel, are probably introduced as a response to market conditions or opportunities. Of course, the reality of the innovation process is often more complex than the simple technology-market dichotomy would suggest. Managers can also be responsible for watching and tracking new technologies and championing them in the firm. Production departments can develop new, highly original technologies.

When we consider the sources of innovation for larger[4] firms (see Table 4.1), it is management and not R&D that is the most frequently mentioned in-house source of innovative ideas (53% and 44%, respectively). Sales and marketing (43%) has about the same importance as R&D. Production personnel comes last at 36%. Clearly, R&D is important, but it is not the only source of knowledge essential to innovation.

We might expect that the sources of information differ substantially across subgroups. While differences by size class and by degree of originality are dealt with extensively in subsequent chapters, we include here the averages for the most novel group (world-firsts) and the largest firms (greater than 500 employees) to suggest the differences that will be investigated more fully later.

The relative importance of the different sources of ideas and information that are instrumental in the generation and development of an innovation depends significantly on its originality. For the world-first innovations, R&D activity is substantially more important than the average, and the production department or management is substantially less important than the average. Sales and marketing is quite similar to the overall average. Similarly, large firms are more likely to use R&D and less likely to use production than the average.

---

[4] The larger (IPs), which are generally larger than 20 employees.

Clear differences emerge in the pattern of knowledge inputs for the innovation process across different innovation types and different actors. A country's innovation profile, then, will be dictated very much by the internal composition of its innovation activity. In general, the higher in the diffusion chain is the industry, the larger the firm, and the more original the innovation, the more frequent and important is the research and development activity of the firm as a source of innovation.

4.2.1 Industry Differences in Internal Sources of Innovative Ideas

Manufacturing industries are characterized by important differences in the role that they play in the innovation system. As we have demonstrated, a key group of core industries produces more innovations than they use, disseminating many of them to other industries. The latter industries are faced with the task of incorporating new materials as intermediate inputs or adopting new machinery and equipment into the production process.

These industry differences are also associated with fundamental differences in the scientific environment. Advances in science and technology create opportunities that favour creation of new products and processes in some industries rather than in others (Scherer, 1966; Griliches, 1979; Pakes and Griliches, 1984; Levin et al, 1987; Levin and Reiss, 1988; Geroski, 1990).

Taking advantage of the types of technological opportunities that exist in the core sector is likely to require a different knowledge-creation system than in others. More complex technological opportunities are likely to require the formalization of the knowledge-creation process in a research and development department. This is indeed the case. Whereas some 63% of firms indicate that R&D is an important source of ideas for innovation in the core sector, this falls to 45% in the secondary sector and only 29% in the remainder or 'other' sector (see Table 4.2). When these ideas are cross-tabulated by novelty (results not reported here), the importance

Table 4.2. *Internal Sources of Innovative Ideas, by Industrial Sector (% of Innovators Using a Source)*

Sector	Management	R&D	Sales & Marketing	Production	Other
Core	45 (4)	63 (4)	46 (4)	21 (3)	1 (1)
Secondary	55 (4)	45 (4)	42 (4)	48 (4)	3 (1)
Other	57 (3)	29 (3)	41 (3)	38 (3)	5 (2)

Notes: Larger firms (IPs) only. Standard errors are in parentheses.

of R&D increases even further. For example, the majority of firms that introduced a world-first innovation in the core sector report R&D activity (76%) as the principal source of innovation ideas.

By way of contrast, management is less important than R&D in the core sector and more important than R&D in the other sector. This suggests that management and R&D are substitutes for each other and that the choice of one rather than the other depends on the sector within which the firm is located.

While the complexity of the scientific environment is the distinguishing feature of the core sector, technological acquisition is relatively more important in the secondary and other sectors. Consequently, the production department is considerably more important as a source of innovative ideas in both the secondary and other sectors (48% and 38%, respectively) than in the core sector (21%).

The importance of sales and marketing varies less than does the R&D or the production category across sectors, although it is slightly more important in the core sector than elsewhere. Generally, then, customers are important sources of information, and are slightly more important for the industries that supply more innovations to other sectors than they receive from them. But the major differences across sectors are found in the extent to which R&D is utilized to develop new innovative ideas. As we show in Chapter 5, which describes the R&D function, considerable differences across sectors also exist in the way that R&D activity is organized.

In summary, the research-intensive core sector benefits from ample technological opportunities. In these industries, firms conduct R&D to create an in-house capability in order to keep themselves abreast of the latest scientific developments and facilitate the assimilation of new technology that is developed elsewhere. Creating this 'learning' or 'absorptive' capacity enhances a firm's ability to imitate new process and product innovations and to exploit outside knowledge, such as the findings of basic research that provide the basis for subsequent applied research and development (Cohen and Levinthal, 1989). In the secondary and other industries, the contribution of R&D to new knowledge is less important; inspiration for innovation comes more often from management and/or from production, marketing, and sales personnel. Thus, the importance of the sources of innovation in a given industry sector is influenced by the technological opportunities available to firms operating in it, as well as by the position of the sector in the innovation chain. More importantly, the R&D-centric model is more applicable to the industries that make up the core sector than to those in the secondary and other industries. A country's reliance on R&D, then, will depend very much on its industrial structure.

4.2.2 Internal Sources of New Ideas by Industry Branch

The individual industries within each sector broadly follow the sectoral averages described previously. For instance, almost all of the industries in the core sector place greater stress on R&D (see Table 4.3). The exception is machinery, which gives R&D less emphasis than do the other core industries. Outside the core sector, in wood products and in food/beverage/tobacco, management is the dominant source of inspiration for innovation. However, in the secondary and 'other' sectors, there are several industries where innovation ideas come most often from the production department – leather/clothing, nonmetallic minerals. In paper and allied products, the sales department is listed as a primary source of innovative ideas.

There are two conclusions that result from these data. First, the fact that ideas for innovation in firms operating in non-core sectors (secondary and other) originated less frequently in R&D than in core firms suggests that the R&D-centric model, in which research and development is the prime source of innovation, is not in agreement with reality outside the core sector. Second, despite the existence of an overall pattern, there is diversity within sectors. In some industries, where advanced technology is incorporated from elsewhere or where advances are made incrementally, the production department is critical. In other tertiary industries that are primarily consumers of innovations, the sales or marketing department is critical.

4.2.3 Combinations of Internal Sources

Few of the internal sources for innovation ideas are used in isolation. Most firms use more than one source, but not all are combined together. Some are complements with one another. Others act as substitutes. This interaction serves to define differences in the nature of the network upon which innovative firms rely.

Since we are primarily interested in how R&D interacts with the other sources, we divide firms into those obtaining their ideas from this source and those that do not. The percentage of firms that use the other sources is calculated in order to determine the extent to which other sources tend to

Table 4.3. *Internal Sources of Innovative Ideas, by Individual Industry (% of Innovators Using a Source)*

Sector	Management 1	R&D 2	Sales/ Marketing 3	Production 4	S.E.*
Core					
Electrical and electronic products[†]	45	68	45	25	6
Chemicals & pharmaceuticals	44	72	54	17	6
Machinery	47	46	40	22	6
Refined petroleum & coal	58	60	24	21	12
Secondary					
Rubber and plastic	49	49	49	47	8
Primary and fabricated metal	51	40	47	43	7
Transportation equipment	56	50	36	42	8
Nonmetallic mineral products	63	49	33	70	10
Other					
Food, beverage, and tobacco	64	34	46	28	5
Leather & clothing	42	17	37	64	11
Textiles	40	42	36	39	10
Wood and furniture	62	8	16	35	10
Paper	48	32	63	39	9
Printing & publishing	49	9	32	52	10
Other	71	48	52	39	13

Note: Larger firms (IPs) only.
* S.E. (standard errror) is the average for standard errors in the same row.
[†] Includes instruments.

be more or less intensively used when R&D is important. This is done for each of the major industry sectors in order to determine whether these substitute or complementary relationships differ across industries (see Table 4.4).

R&D-using firms tend to use management sources less frequently. Similarly, firms that relied on innovative ideas from the sales and marketing division did not use ideas from management as frequently.[5] Management

---
[5] The connection between the emphasis on sales and other internal sources is, for brevity, not reported here.

Table 4.4. *Internal Sources of Innovative Ideas Reported by Firms That Used and That Did Not Use Ideas from Their Own R&D, by Industrial Sector (% of Innovators Using a Source)*

Sector	R&D	Management	R&D	Sales & Marketing	Production
Core	Did not use	54 (6)	0	41 (6)	17 (5)
	Used	39 (5)	100	49 (5)	24 (4)
Secondary	Did not use	59 (6)	0	38 (6)	46 (6)
	Used	51 (6)	100	47 (6)	51 (6)
Other	Did not use	59 (4)	0	39 (4)	39 (4)
	Used	51 (6)	100	47 (6)	38 (5)

Notes: Larger firms (IPs) only. Standard errors are in parentheses.

acts as a substitute source of ideas for an R&D department. This pattern holds for all three industry sectors, but it is more pronounced for firms in the core sector.

On the other hand, firms using R&D as their source of inspiration for innovation rely more often on specialized divisions, such as sales and production. Moreover, the difference between R&D users and non-R&D users is about the same across all three major sectors. Separating firms according to whether they used or did not use information from the sales and marketing department results in a similar pattern.

4.3 EXTERNAL SOURCES OF IDEAS AND INFORMATION FOR INNOVATION

The innovation process relies heavily on knowledge or ideas from both inside and outside the firm. While the importance of internal sources, such as the R&D group, has received considerable emphasis, external sources also play an important role. Ideas for innovations often originate outside the firm in response to market opportunities and new information, as well as from competition and/or collaboration with other firms and institutions.

These outside sources are diverse. Some are from market partners – either firms that are associated with arm's-length transactions (customers or suppliers) or firms that are part of the same ownership group (related firms). In each of these cases, the transfer of information is accompanied by some other transaction and, therefore, potentially allows for the pricing of the information flow. In the case of customers, the information can be included in the price that is paid for a good or service. In the case of related firms, the relevant contractual relation involves ownership, and the monetary consideration that accompanies a transfer of information may be a direct payment for information, or can be part of the profits remitted to the parent firm. In both cases, the existence of this consideration offers a method for dealing with imperfections in knowledge markets where there are problems in controlling the diffusion of knowledge with less than perfect appropriability mechanisms.

Some ideas for innovation come via other market transactions with third parties – consultants or private research institutes. Once again, these firms operate as part of a market-driven system where knowledge is transferred as part of a commercial transaction.

A third source consists of what are closer to unpriced spillovers. Here, there is less of a well-developed market for the transfer of ideas. These spillovers come closer to the definition of classical externalities, though, in many cases, mechanisms have been developed to provide for some internalization incentives. These sources include public R&D institutions, universities, professional trade literature, patent offices, trade fairs, and exhibitions.[6] This group will be referred to here as the technological infrastructure. Competitors also fall in the group that provide unpriced spillovers. Although some of the knowledge that is derived from competitors may come from joint projects and may be more akin to a market relationship that involves an exchange of resources, most of the spillover here occurs from the observance of the behaviour of competitors and the reverse engineering of competitors' products.[7]

These external sources, taken together, are used by 85% of all innovators. Thus, the innovation process depends very heavily on the diffusion of information from firm to firm.

The most frequently cited external source of new ideas are market-related partners (customers, suppliers, and related firms), which is used by 68% of all innovators (see Table 4.5). Clients are important for 46% of firms – just as many as cite the internal source of R&D or

---

[6] See Hall, Link, and Scott (2000) for a study of the importance of U.S. universities.
[7] As the Chapter 5 will show, joint ventures with competitors are relatively unimportant – certainly relatively less important than competitors are as a source of innovative ideas.

Table 4.5. *Principal External Sources of Ideas for Innovations, by Novelty Type (% of Innovators Using a Source)*

External Source of Ideas	All	Other	Canada-First	World-First	Standard Error
Market-related	68	71	67	67	(4)
Suppliers	28	30	26	30	(4)
Clients/customers	46	55	40	38	(4)
Related firms	15	14	18	12	(3)
Market transactions	15	19	10	19	(3)
Consultants	13	16	9	15	(2)
Private R&D institutions	4	6	1	7	(2)
Spillovers	46	55	38	33	(3)
Competitors	28	41	19	11	(3)
Trade fairs	17	18	17	14	(3)
Professional publications	13	12	14	11	(2)
Public R&D institutions	3	3	2	6	(1)
Patent office	3	1	2	9	(1)
Universities & colleges	3	1	5	4	(1)

Notes: Larger firms (IPs) only. The standard errors are the mean values of the standard errors in each row.

the sales/marketing department. Innovation has, in this sense, important demand-driven components. Many innovations are inspired by suggestions of clients and originate in firms that pay careful attention to the needs of their customers.

W. M. Cohen and R. C. Levin (1989, pp. 1083–90) have also stressed that innovation is determined by the technological opportunities available – by the supply of new ideas. The importance of technological opportunities is confirmed by the fact that the external source with the second highest importance is suppliers, with 28% of firms using this source. Nevertheless, it should be noted that demand sources are considerably more important than supply sources for innovative ideas.

Other market transactions (consultants and private R&D institutions) are less important, being cited by only 15% of innovators. Consultants are important for only about 13% of firms. And as we shall see, these market-driven sources tend to be substitutes for one another, with small firms relying more on consultants and large firms relying more on related businesses. The difference between the importance of market-related partners (68%) and third-party transactions (15%) emphasizes the relative importance of diffusion within well-defined networks that are connected to either the flow of goods or ownership links. There are substantial difficulties in organizing transactions for research activities outside the normal commercial relationships with customers and suppliers.

Spillovers from the technological infrastructure and competitors are also important. Some 46% of firms report that they obtain information from one of these sources. Competitors are the most important group within this category, with 28% indicating that they derived information from this source. This is almost as important as are suppliers in the group of market-related partners and points to the importance of the diffusion of ideas from sources that involve externalities. Trade fairs are second in this group at 17%. Next are professional publications at 13%. Universities, public R&D institutions, and the patent system are each listed as important sources by less than 3% of innovative firms.

Many of the external sources associated with the technological infrastructure provide information and ideas for innovation free of charge, or at a cost lower than the value of information to the innovating firm. The extent to which innovative firms use these 'technological spillovers' (Griliches, 1979) would be expected to vary by the type of innovation. Complex innovations are more likely to require specialized R&D internal facilities and are more likely to use external sources that complement these R&D facilities. World-firsts are more frequent users of public R&D institutions, patent offices, and universities than are more imitative innovations. The public institutions that contribute to the technological infrastructure (Tassey, 1991; Lundvall, 1992) play less of a role in the creation of the less original innovations.

It should be noted that the more novel innovators are no more likely to use other external sources than are the more imitative innovators. Indeed, differences in outside sources across innovators by degree of novelty (Table 4.5) confirm that the more novel innovators are slightly less likely to make use of these sources as a group.

While the importance of suppliers is about the same across the three different types of innovators, the importance of customers is much less for world-first than for 'other' innovators. World-first innovators then substitute internal R&D for customer-driven suggestions.

Finally, world-first innovators are much less likely to draw on competitors and trade fairs. The very novelty of their innovations means that world-first innovators have less to learn from both these sources. Nevertheless, it should be noted that these sources are not ignored completely by the more novel innovators. Even for world-firsts, ideas diffuse from these commonplace sources.

Table 4.6. *Main External Sources of Ideas for Innovations, by Industrial Sector (% of Innovators Using a Source)*

External Source of Ideas	Sectors			Standard Error
	Core	Secondary	Other	
Any of external	80	87	87	(3)
Market-related partners	66	68	70	(4)
Suppliers	24	23	36	(4)
Clients/customers	47	49	44	(4)
Related firms	20	16	12	(2)
Market Transactions	16	18	13	(3)
Consultants	14	18	9	(3)
Private R&D institutions	5	4	5	(2)
Spillovers	41	42	52	(4)
Competitors	24	32	28	(3)
Trade fairs	12	13	25	(3)
Professional publications	14	7	16	(3)
Public R&D institutions	4	2	3	(1)
Patent office	6	1	2	(1)
Universities & colleges	4	2	3	(1)

Notes: Larger firms (IPs) only. The standard errors are the mean values of the standard errors in each row.

4.3.1 Industry Sector Differences in External Sources of Innovative Ideas

Differences in the use of external sources across industry sectors will reflect varying needs that are associated with the chief producers of innovations – the core sector – and the main consumers of innovations in the secondary and the tertiary 'other' sectors.

The evidence on the use of outside sources by industry sectors (see Table 4.6) demonstrates that demand conditions are more uniform across industry sectors than technological opportunity arising from the availability of new inputs and equipment. There are no significant sectoral differences in the importance of clients and customers as an inspiration for innovation. Ideas emanating from clients, that is, users, are the most frequently mentioned external source of inspiration in each sector (44% to 49%). On the other hand, ideas from suppliers are more important in the tertiary other sector.

That the interaction with customers and clients can have an important influence on innovation in supplying sectors is supported not only by innovation studies but also by the relationship between productivity growth of supplying industries and their linkages with downstream user sectors (Wolff and Nadiri, 1993). The evidence provided here shows that demand conditions have relatively equal effects across all sectors – at least as far as providing external ideas.

It should be recalled that the sales/marketing department was slightly more important in the core sector. That and the fact that customers do not play a relatively more important role for external ideas in this sector suggests that the innovation process, while demand driven, is critically stimulated by internal marketing competencies. It is the sales department in the core sector that recognizes innovation opportunities, rather than just customers suggesting new products to firms in the core sector.

While the importance of demand conditions is relatively uniform across sectors, this is not the case for supply conditions. Firms operating in industries belonging to the other sector more frequently use ideas and suggestions coming from their suppliers (36%) than do their counterparts in the core and secondary sectors (23%). This is explained by the fact that firms in the tertiary other sector are the major consumers of new inputs and/or new production machinery, both typically acquired from upstream suppliers. This is an important channel of technological diffusion in an open economy like Canada. Watching competitors provides another valuable source of innovative ideas, especially in the secondary sector.

Although they are used less frequently than the external sources, information and ideas from related firms are particularly important for firms in the core sector, where the largest proportion of foreign-owned firms operate.

Finally, firms in the other sector are significantly more likely to make use of some form of infrastructure spillover. This is primarily because they make greater use of trade fairs – a simple form of information transfer. In contrast, firms in the core sector are significantly more likely to make use of patents and slightly more likely to use public R&D or universities, though the differences here are much smaller than were the differences between world-firsts and more imitative innovations.

The breakdown of main external sources of new ideas for individual industries (see Table 4.7) shows that there are very significant differences among industries belonging to the same sector – much more than was the case for internal sources of new ideas. The similarities in the sectoral averages hide interesting interindustry differences.

Machinery is heavily dependent upon customers, but is much less dependent than other core industries on suppliers (10%), since it is itself the primary supplier of new process technology embodied in machines

Table 4.7. Main External Sources of Innovative Ideas, by Individual Industry (% of Innovators Using a Source)

SIC	Industry	Suppliers	Clients/ Customers	Related Firms	Competitors	Trade Fairs	Tech. Publications	Standard Error
Core								
31	Machinery	10	42	*19*	29	12	11	(6)
36	Refined petroleum and Coal	45	22	*20*	43	25	24	(13)
33	Electrical and electronic products	35	60	*18*	24	18	14	(6)
37	Chemicals & pharmaceuticals	25	39	22	16	3	*18*	(5)
Secondary								
16	Plastic & rubber	32	59	15	18	7	2	(6)
29&30	Primary metal & fabricated metal	20	51	16	39	15	9	(6)
32	Transport equipment	21	46	17	31	12	4	(6)
35	Nonmetallic mineral products	30	37	10	35	13	18	(10)
Other								
10–12	Food, beverage, tobacco	31	45	15	31	21	9	(5)
17&24	Leather and clothing	*46*	13	17	9	*34*	*34*	(8)
18&19	Textiles	40	48	7	*18*	26	14	(8)
25&26	Wood and furniture	34	28	0	0	*20*	*11*	*(10)*
27	Paper	*58*	*61*	12	43	25	15	(8)
28	Printing and publishing	35	*54*	9	*56*	*35*	*41*	*(10)*
39	Other	22	51	*18*	20	16	0	(11)
	All	28	46	15	28	17	13	(2)

Notes: Italicized cells represent the top four industries in each category. Larger firms (IPs) only. The standard errors are the mean values of the individual standard errors in the row.

and equipment. Petroleum refining relies less on clients as an external source than do other core industries, but relies more on suppliers since it is a major customer of new technology. It also relies more heavily upon competitors than do others in the core sector.

In the secondary sector, transport equipment, rubber and plastics, and primary metals rely heavily on clients. Primary metals relies more heavily on competitors. The plastics industry relies more heavily than the others on suppliers and much less on competitors. The nonmetallic mineral products industry relies less on customers for innovative ideas.

In the tertiary 'other' sector, the heaviest user of suppliers is the pulp and paper, as well as the leather and clothing, industry. But pulp and paper also places relatively more emphasis than do other industries on customers. This is also the case for printing and publishing, which is also one of the four top industries stressing the importance of competitors. Leather, clothing, and printing place a very high stress on trade fairs and technical publications.

4.3.2 Product Versus Process Innovators

Differences in the sources of ideas should also be related to whether the innovation involves a product or a process – though the differences here may partially be related to the sector in which the innovator is located.

Process innovations are often taken as a reaction to challenges and opportunities that arise on the production floor. Learning 'by using,' then, would be reflected in a higher percentage of ideas coming from production personnel for process innovation. On the other hand, product innovations should come from a greater variety of sources – from management and from the sales or marketing division – since product innovations often depend on customers for their inspiration.

As expected, product and process innovators feed on different external sources of inspiration (see Table 4.8). The proportion of firms taking into account ideas and suggestions of their clients is greater for those reporting only a product innovation than for those reporting only a process innovation. Moreover, the difference between these two exists across all sectors (data not tabulated here). In the 'other' sector, 50% of product innovators credit clients, while only 37% of process innovators do so. In the core sector, 57% of product innovators and 45% of process innovators do likewise.

Over half of the innovators that introduced a new or improved product *simultaneously* with innovations in the production process utilize ideas and information from clients. Their importance does not vary in a

Table 4.8. *Sources of Internal and External Ideas for Innovations for Product Versus Process Innovations (% of Firms Using a Source)*

Source	Firms with *Only* Product Innovation: No Change in Man. Technology	Firms with Some Combination of Product and Process Innovation	Firms with *Only* Process Innovation: No Product Change
Internal	97 (1)	100	96 (1)
Management	45 (4)	60 (6)	56 (3)
R&D	50 (4)	46 (6)	39 (3)
Sales	57 (3)	46 (6)	33 (3)
Production	27 (3)	37 (6)	42 (3)
External	85 (3)	83 (5)	86 (2)
Market-Related	74 (3)	64 (3)	70 (6)
Suppliers	25 (3)	31 (6)	30 (3)
Clients/customers	55 (4)	54 (6)	39 (3)
Related firms	17 (3)	13 (5)	17 (2)
Market Transactions	16 (3)	12 (4)	16 (2)
Consultants	12 (4)	12 (4)	14 (2)
Private R&D institutions	6 (2)	3 (2)	4 (1)
Spillovers	42 (4)	49 (6)	47 (3)
Competitors	24 (3)	34 (6)	28 (3)
Trade fairs	13 (2)	22 (5)	19 (2)
Professional publications	13 (2)	8 (3)	14 (1)
Public R&D institutions	1 (1)	0	5 (1)
Patent office	4 (1)	2 (1)	2 (1)
Universities & colleges	4 (1)	1 (1)	3 (1)

Notes: Larger firms (IPs) only. Standard errors are in parentheses.

systematic way among sectors. The sales/marketing group is also more important for the two categories that involve product innovations. Moreover, R&D is more important for both the product categories. This corroborates our earlier finding that R&D is more orientated towards product than process innovation.

Suppliers, the second most important external source, are more instrumental in process innovation than in product or combined product and process innovation. This reflects the important role of machinery and equipment producers for technological change in their clients' industries. The production department is also more important as a source of internal ideas for process innovations that do not involve any product innovations.

Related firms are about equally important across all innovation types. This demonstrates the widespread importance of interfirm linkages for the diffusion of new process technologies, as well as the introduction of new products.

Management is particularly important for process innovators and for combined process/product innovation. Technical skills in management are, therefore, of importance for technology diffusion of process innovation.

There are fewer a differences in the importance of spillovers from the technological infrastructure – with the exception that trade fairs are more important for process innovations. These commercial fairs provide an important source of information on new machinery and equipment – especially for smaller firms.

4.3.3 Spillovers, Market Transactions, and Infra-technologies

The literature on endogenous growth and the economics of technological change stresses the importance of spillovers for growth in productivity and economic welfare (Nelson and Winter, 1982; Nelson, 1993; Rohmer, 1994). The importance of the various external information sources has already been demonstrated here. In this section, we ask how they are combined with internal competencies. In particular, we want to know whether they serve as substitutes for or complements to these internal competencies.

For the analysis, internal sources were regrouped into six categories and external sources into four categories.

The three categories used for grouping internal ideas were those firms that used 1) *any R&D*; 2) *R&D with sales*; 3) *R&D combined with sales and production*; 4) *only R&D*; 5) *only production*; and 6) *any production*. These six sources allow us to investigate the R&D versus production alternatives along with several approaches that combine more than one internal source.

The five external sources were: 1) *Market-related partners* – clients, suppliers, related firms; 2) *Market transactions involving technology* – technology transfer agencies, private R&D institutes, consultants, software providers; 3) *Competitors;* and 4) *Spillovers from the infrastructure* – public institutions and information sources providing technological information, government regulations and standards, government industrial development and technology transfer institutions, patent office and literature, public R&D institutions, trade fairs and exhibitions, and professional

Table 4.9. *Correlations Between Inside and Outside Sources of Innovative Ideas*

Outside Sources	Inside Sources					
	Any R&D	R&D with Sales	R&D with Sales and Production	Only R&D	Only Production	Any Production
Market-related partners	−.05 (.27)	.17 (.0001)	.16 (.0001)	−.15 (.0004)	.06 (.16)	.15 (.0004)
Market transactions	.09 (.02)	.07 (.11)	.15 (.0003)	−.02 (.65)	−.02 (.71)	.10 (.01)
Competitors	.03 (.50)	.08 (.07)	.09 (.03)	−.03 (.44)	.04 (.33)	.04 (.41)
Technological infrastructure	.18 (.0001)	.12 (.0005)	.11 (.009)	−.04 (.30)	−.03 (.51)	.11 (.009)
Universities	.004 (.91)	−.009 (.84)	−.013 (.76)	−.04 (.34)	.14 (.001)	.06 (.15)

Notes: Larger firms (IPs) only. Probability values of the significance of the null hypothesis that correlations are zero are in brackets.

and trade literature; and 5) *universities*. In order to examine the connection between the inside and outside sources, we examined the percentage of firms who used both sources and calculated correlations between each. Only the latter are reported here (see Table 4.9).

Internal and external sources in general complement one another. An overwhelming majority of firms (85%) in all sectors use a combination of both internal and external sources. Only one out of seven firms relies exclusively on internal sources. These firms do not present any particular pattern with respect to the originality of innovation, or the industry sector.

Internal sources tend to be used in various combinations with outside sources. In particular, the use of just internal R&D is relatively rare. Only about one in eight firms does not use any internal source of inspiration other than their own R&D. These exclusive 'R&D users' substitute internal R&D for external sources, since the correlations are negative with all the external sources and highly significant with regard to diffusion from external market-related partners.

If firms perform R&D, regardless of their other activities, they are significantly more likely to use outside sources of information from the technological infrastructure. Or alternately, the inside source with the highest correlation to outside sources and the greatest statistical significance is internal R&D. This confirms the contention that internal R&D is important for the internal ingestion of these outside spillovers. But it should be noted that the use of the production division as a source of internal ideas is also correlated with this outside knowledge source, and thus serves an important integrative function as well. More than one internal source makes use of spillovers from the technological infrastructure. However, R&D has the closest connection with the use of information from the outside technological infrastructure. Of those using outside infrastructure spillovers, 56% have internal R&D facilities, while only 37% have R&D facilities if they do not use outside infrastructure spillovers – a difference of some 20 percentage points. Of those using outside infrastructure spillovers, 44% use internal production facilities for ideas, while only 32% use internal production facilities if they do not use outside infrastructure spillovers – a difference of only 12 percentage points.

The connection between R&D and infrastructure support is stronger in the core and tertiary other sectors than in the secondary sector. In the secondary sector, the highest correlation with outside infrastructure use is with the internal production source. The connection between R&D and infrastructure support is also higher for world-firsts and Canada-firsts than for more imitative innovations. The correlation between the use of infrastructure support and the use of the production department is higher for Canada-firsts and for the tertiary other sector than for the core sector.

Since R&D is rarely used alone, it is important to ask what sort of changes occur in the complementarities with various outside sources when it is combined with other internal sources. To examine this, we create two new combinations – those where R&D and the sales department are both used, and those where R&D, sales, and the production department are all used. These combinations occur in 21% and 9% of all cases, respectively. Firms that combine internal R&D with sales or with sales and production are more likely to combine these internal sources with information from market-related partners. The differences here are large. Some 84% of firms that combine R&D and sales use market-related partners, while only 64% of those not doing so make use of this external source. This complementarity is neither significant nor important for those that report just R&D. A similar, though weaker complementarity exists between the R&D/sales/production combinations and other market-based transactions. All of these complementarities are larger and more significant in the core than the secondary and tertiary other sectors – primarily because of the greater importance of related firms in the former sector. This is also the case for world-firsts compared to Canada-firsts and other more imitative innovations.

The R&D/sales/production internal source also is positively related to the derivation of information from competitors and from the public technological infrastructure. Some 41% of this group make use of information from competitors, while only 26% of those not making use of the R&D/sales/production combination derive information from competitors. The respective percentages for those using infrastructure are 51% and 32%. However, there is no close relationship between these internal sources and the use of universities. The strength of the complementarity between this internal combination and competitors does not differ much across sectors, but it is much higher for the infrastructure in the other sector than it is for the core and secondary sectors. Infrastructure complementarities matter most in the innovation-receiving sector.

An alternative strategy for developing sources of inside information is to focus on the production division. Those firms doing so also were significantly more likely to use both market-related partners and other market transactions (Table 4.9). They also were significantly more likely to use infrastructure sources; in particular, they employ more university sources. While the latter difference is not significant for all of manufacturing, it is significant for the other sector. Thus, universities have particular importance for those firms that rely on the production department for their innovations and for the sector that receives most of its innovations from other sectors.

In conclusion, information transfers are an important source of the knowledge-creation process. These transfers do not, however, offer an alternative to firms' own R&D. To use spillovers to their advantage, firms have to create or maintain their expertise by performing internal R&D; but they also combine these outside spillovers with expertise in the production department. This finding extends the observations of Cohen and Levinthal (1989). Some firms use R&D and combine this with outside information from institutes and public sources – but this connection is strongest for the sectors that diffuse innovation to others. In those sectors that have to ingest innovations, outside spillovers from the technological infrastructure are combined with expertise from the production department. Firms in all sectors benefit from spillovers, but they develop different internal mechanisms to do so.

Finally, it should be noted that some firms evolve more comprehensive internal mechanisms where they combine several internal sources. Firms that do so tend to interact more with market-related partners, either customers, suppliers, or related firms, as well as competitors and the technological infrastructure. These firms, then, may be described as the complete innovators, in the sense that they make use of the broadest range of both internal and external sources.

4.4 SOURCES OF NEW TECHNOLOGY

In the process of introducing a new or improved product or a new production process, firms often introduce new technologies. A new technology can involve new machines and equipment such as flexible manufacturing systems, new practices such as concurrent engineering, and new organizational techniques such as work teams.

Since technology acquisition is a special part of the innovation process, the sources of innovative ideas in this domain for large firms[8] are investigated separately. In order to test whether firms saw a difference in the sources of ideas used for a process innovation and for the technology that underlay its innovations in general, the 1993 Canadian Survey of Innovation deliberately investigated sources of ideas for all innovations and then separately investigated the sources of ideas just for the technology going into the innovation. The latter provides a more direct route for understanding the importance of various inputs to technology than simply cross-classifying sources of ideas by whether a firm reports that it is primarily a process innovator, as was done in the section 4.3. Here, as before, we are interested in the extent to which firms used a combination of internal and external sources to find suitable technical solutions to problems arising in the innovation process.

The internal sources of interest are production engineering and R&D. R&D is divided into its two components – research, which is aimed at developing more basic or generic knowledge, and experimental development, which focuses more on solving firm-specific problems with applied research (see Table 4.10).

Most firms solve their innovation-related technical problems by both experimental development and production engineering (52%). Research comes last at 32%. Thus, new technologies come primarily from nonresearch sources. Nevertheless, research is important. The introduction of more radical technology associated with innovations requires research in about one-third of all innovations to solve unusual technical problems or to extend the expertise of the firm to new areas.

---

[8] The larger firms (IPs) are generally larger than 20 employees.

Table 4.10. *Internal Sources for Technology Used by Innovators, by Novelty Type (% of Innovators Using a Source)*

Type of Innovation	Research	Experimental Development	Production Engineering	Standard Error
All	32	52	52	(2)
World-first	47	60	48	(5)
Canada-first	30	53	51	(4)
Other	26	48	54	(4)

Note: The standard errors are the mean values of the individual standard errors in the row.

Once again, the novelty of the innovation determines the importance of research. More original innovations require the type of technology involving research more often than do the imitative ones. Some 47% of innovators that recorded their innovations as being world-first used research to find or create the needed technology, whereas only 26% of those reporting their innovations as neither world- nor Canada-first used research as a source. The importance of experimental development also increases with the originality of innovation. Originality is associated both with more basic research and more experimental development, although the differences in the former area were larger and more significant than in the latter. Solving technical problems encountered in the case of less original 'other' innovations requires R&D less frequently than it requires expertise in production engineering. The latter is the most frequently used internal source of technology needed to solve problems encountered in the introduction of other innovations.

4.4.1 Industry Sector Differences in Internal Sources of Technology

The importance of research also depends on the sector in which the innovation occurs. Firms in the core sector bringing new technology to the rest of the economy resort to research and experimental development more often than do those in the other two sectors, and more frequently than they use production engineering (see Table 4.11). In the core sector, 55% of firms perform research as part of the acquisition of technology in their innovation process; 64% use experimental development. On the other hand, the production-engineering department accounts for less than half of the ideas for new technology in this sector.

Table 4.11. *Internal Sources for Technology Required for Innovation, by Industrial Sector (% of Innovators Using a Source)*

Sector	Research	Experimental Development	Production Engineering	Standard Errors
Core	55	64	44	(4)
Secondary	25	56	66	(4)
Other	18	40	47	(4)
All	32	52	52	(2)

Note: The standard errors are the mean values of the individual standard errors in the row.

The importance of research even for the acquisition of technology is striking. Firms in the core sector resort to R&D, not just because of the importance of product innovations that are produced here and used elsewhere, but also because of the complexity of the knowledge generation process in this sector, even on the technology side.

Firms in the secondary sector resolve technology problems related to their innovations most frequently by using their expertise in production engineering (66%), though experimental development at 56% is second. This is not surprising. Many of the industries included in the secondary sector produce material inputs used in other industries and focus on process innovations. The introduction of process innovations requires production technology expertise, rather than experimentation in research and development.

The same pattern is exhibited in the tertiary 'other' sector, with production engineering and applied experimental development being more important than basic research. The major difference that distinguishes the other sector from the core and secondary groups is that all three internal sources are less important.

4.4.2 External Sources of Ideas for New Technology

In addition to internal resources, firms use an array of external sources of technology to produce new technologies that are required for innovation. As was the case for innovations in general, market-related partners provide the external source most often used. However, contrary to the case for innovations in general, where the most important external source is the customer, information acquisition for new technologies uses suppliers as the most frequent external source of ideas (33%).

The most novel form of innovations generally uses all sources more frequently than less novel innovations – with the exception of publications and trade fairs. Thus, originality is once more marked by a greater diversity of information that supports the knowledge-generation process.

A breakdown of technology sources by industry sector reveals that there are major sector-specific differences with respect to sources of ideas for external technology and with respect to both customers and suppliers, consultants, and 'trade fairs and expositions', whose importance is higher in the tertiary other sector than in the core sector. On the other hand, the core sector tends to rely more on joint ventures, universities, and publications than do the secondary and other sectors.

The downstream secondary and other sectors rely more on market-related partners. Sectors that are diffusers tend to rely more on partners that also specialize in complex R&D-related research or that form part of the backbone of the technological infrastructure – research consortia, universities, and scientific publications.

The fact that suppliers are the most frequently cited external source of new technology for process and complex product-process innovations introduced in the tertiary other sector confirms the hypothesis of E. Von Hippel (1982) that machinery suppliers are likely to play an important role in the introduction of process innovations by numerous downstream users; however, our results show that this occurs most frequently in the sector that ingests new materials and new equipment from other sectors. These are mostly consumer goods–producing industries.

4.4.3 Relationship Between Internal and External Sources of Technology

Most firms (60%) use a combination of internal and external sources of new technology in their innovation process. Only one firm out of four (28%) was able to solve its technical problems by relying exclusively on its own internal resources.

The intensity with which the internal and outside sources are combined differs, or the manner in which they are used varies, as the correlations described in Table 4.12 demonstrate. Firms conducting their own R&D are less frequent users of information from the various external technology sources. The negative relationship is particularly significant for external consultants. Firms that use both research and development internally make significant substitutions against market-related partners, market transactions, and ideas from unrelated firms (competitors). This

Table 4.12. *Correlations Between Internal and External Technology Sources*

	Research	Experimental Development	Production Engineering	Research and Experimental Development	Experimental Development and Production Engineering
Market-related partners	−.06 (.17)	−.11 (.03)	.04 (.34)	−.14 (.003)	−.07 (.13)
Market transactions	−.09 (.06)	−.12 (.01)	.10 (.02)	−.13 (.004)	−.04 (.37)
Unrelated firm	−.07 (.10)	−.08 (.11)	−.03 (.54)	−.12 (.01)	−.13 (.007)
Spillovers	−.05 (.31)	−.03 (.51)	−.03 (.51)	−.02 (.58)	−.015 (.74)

Note: Larger firms (IPs) only.

substitution pattern resembles, though is stronger than, that found for innovation ideas.

On the other hand, there is a positive relationship between the use of production engineering and the use of both market-related partners and market transactions (e.g., consultants), thereby indicating that these sources complement one another.

4.5 CONCLUSION

The innovation process is fed from multiple sources, some internal to the firm, others external to it. Ideas for new and improved products and processes are generated in the course of market transactions with clients and suppliers, with related and unrelated firms, and with other external sources. Ideas for new market opportunities are seized and adapted to a firm's advantage by the management, marketing, and engineering personnel of the firm. Of these, a firm's management is the most frequently reported internal source of ideas for innovation. Managers are not just the passive coordinators of a firm's operations; they are often the sources of inspiration for new ideas and products.

The second most important internal source for innovative ideas (listed by 44% of firms) is R&D activity. This confirms the fact that R&D is, therefore, a key part of the innovation process. The importance of R&D as a source of new ideas for innovation increases with the size of the firm, with the originality of the innovation, and with the innovation intensity of the sector. In smaller firms, which do not have a separate R&D department, its function is often undertaken by management. The high cost of establishing a specialized R&D unit leads many small firms to adopt a less costly, more flexible internal structure where management is the principal source of innovative ideas.

While the results, then, confirm that R&D has a widespread importance, they also show that other sources are used to complement it. Even when R&D is used, it is generally not pursued alone. Many firms list both R&D and the sales department as being important. The sales department brings a customer orientation to the innovation process. The importance of outside stimuli from customers is confirmed by the fact that the most important external source is customers. The degree to which the sales department offers an important complementary resource is quite similar across the main industry sectors.

But there are differences across sectors in the importance of R&D. In general, the higher in the diffusion chain is the industry, the larger the firm, and the more original the innovation, the more important is the research and development activity of the firm as a source of innovation. Since firms in the core sector are more than twice as likely to make use of R&D, the importance of R&D for a country's innovation system will depend upon its industrial structure. Countries that have a greater proportion of secondary and other industries will be less R&D intensive.

All of this suggests that the R&D-centric model is a key part of the innovation process. Of course, we need to add that a key internal complement that is often combined with R&D is an outward sales orientation. But this just adds a marginal facet to the already R&D-centric orientation.

More dramatic changes to this story occur when it is recognized that quite a different model exists side by side with the first. Both management and the production department offer a substitute for R&D and are used extensively in the secondary and other sectors; thus, the two forms of organization tend to be used in quite different places. There is more reliance on the production department in some sectors and for some types of innovations than others. In larger firms operating in the sectors that are innovation consumers, ideas come more frequently from the production than the R&D department. In this way, the production department sometimes substitutes for R&D, and the innovation model at work differs across sectors and innovation types. The relevant innovation model is not simply one where we need note that the marketing/sales department is often an important complement to R&D. There is an alternative model

Conclusion

with the production department and/or management forming the core of the innovation process.

The innovation model also involves important external links. The main external sources of innovative ideas are customers, related firms, and suppliers, in that order. Knowledge diffusion from outside sources is strongly associated with commercial partnerships arising either from trade or ownership connections. The problems of evaluating and ingesting outside information during the innovation process are handled by expanding the boundaries of the firm in two ways. First, the boundaries are extended via mergers, and information essential to the innovation process is exchanged via related firms. Secondly, networks are established with economic partners, and incentive systems are created to overcome asymmetric information problems. These partnership arrangements are far more important than third-party contracts with consultants and research firms. It should also be noted that these arrangements tend to substitute for research and development when firms are not outward oriented, that is, when they do not combine ideas from R&D with the sales department.

On the other hand, when R&D combines with sales to be the internal driving force behind innovation, greater use is made of outside partners like suppliers or consumers or related firms. Those who rely only on R&D use fewer of these sources. There are, therefore, two groups of R&D users. One builds networks with market partners. The other R&D type relies more on its own extensive resources. In both cases, internal expertise in R&D is combined with spillovers from the technological infrastructure. This relationship is stronger in the core sector, because of the greater use of related firms as sources of outside ideas in this sector. A third cluster is made up of those who focus on production expertise internally and rely on spillovers from universities.

While partners provide an important source of ideas, spillovers that entail more traditional externalities are very important, with almost half of the firms receiving information from these sources. Many of these sources are tapped free of charge.

The most frequently mentioned source of spillovers comes from competitors. But some of these spillovers are also acquired at trade fairs and trade conferences, from public and private research institutes, and from trade or patent literature. These spillovers are most important in the other sector, primarily because of the greater role that trade fairs play in the general knowledge-acquisition process for new technologies. R&D facilitates the ingestion of these spillovers. So too does the production department.

The survey results show that the relative importance of some of the sources of internal and external innovative ideas varies by size of the firm, the industry sector, and the type and originality of innovation. Generally, external sources are more important for non-world-firsts. They also tend to be more important for the tertiary other sector. This is primarily because suppliers are more important as an external source for this sector. Spillovers from all sources that we have classified as coming from the technological infrastructure are more important in the innovation-receiving sector – because of the importance of trade fairs for this sector. In contrast, related firms are more important for world-firsts. Moreover, innovators employ the R&D-oriented external sources (universities, industrial research labs) more frequently in the core sector and when they are creating world-firsts.

The finding that the intensity of recourse to market-related external sources of technology is inversely proportional to the technological complexity of the sector supports Mowery and Rosenberg (1989), who argue that internal and external sources of knowledge are used in quite different situations. Their analysis of independent research laboratories in the United States indicated that these facilities specialize in activities that do not require a high degree of interdependence between specific research activities and other manufacturing and nonmanufacturing functions within the firm. These are routine research tasks that often utilize specialized equipment and facilities. They are likely to be utilized by firms introducing less complex innovations. Projects that focus on the improvement of production processes employed by many firms or the analysis of materials, rather than projects developing new products, are likely to be supplied by independent research organizations. On the other hand, innovations in technologically advanced industries require specific technical knowledge and often a close interaction with various corporate functions. According to these authors:

Much of the necessary knowledge for such research and development projects may be highly specific to a given firm and cannot be produced by an organisation not engaged in both the production process and research. The specificity of such technical knowledge stems ultimately from the fact that production and acquisition of detailed technical knowledge, or the use and acquisition of such a knowledge, are frequently joint activities. (Mowery and Rosenberg, 1989, pp. 80–83)

Finally, it is important to note that the introduction of innovation marshals internal as well as external sources of technical expertise. Internal expertise in production engineering is the chief source used to

Conclusion

master the technology that supports product and process innovation. Production engineering is an important internal source of technology, especially for the less original, Canada-first, and 'other' innovations. A firm's experimental development and research departments are more frequently reported as sources of technology solutions than is production engineering in the case of product innovations – especially in the core sector. The reverse is true for new and improved production processes.

Overall, the majority of firms (60%) combine both internal and external technology sources. The outside ideas for new technologies needed for process innovation come most frequently from suppliers of machinery and/or inputs. Most of these external sources are available to innovating firms free of charge and constitute an important complementary source of innovative ideas, especially in the other sector. In contrast, world-first innovations make greater use of university and government labs. Internal R&D tends to be combined more with external R&D for world-first innovations and in the core sector.

In summary, the input knowledge-production process relies heavily but not exclusively on R&D. Alternate models that rely on production and management exist. These alternatives tend to be found in quite different sectors. Finally, outside sources of information are important to most innovators. Classical spillovers can be found in about half of the cases – with competitors providing most of the stimulus here. But the largest proportion of outside knowledge comes through market partnerships between customers and suppliers, as well as between firms and related parties.

FIVE

Research and Development and Innovation

5.1 INTRODUCTION

Innovation requires a conscious effort to develop new ideas for both products and processes. During the nineteenth century, scientific progress was less structured and typically arose from ideas developed in many different areas of the firm. While sources for ideas still flow from different areas, the twentieth century has seen increasing emphasis placed on formal research and development (R&D) facilities as a source of innovation. The renowned research laboratories of AT&T, General Electric, Dupont, and IBM epitomize the organized pursuit of scientific knowledge and its application to production problems. It is true that science played an important role in the early industrial revolution in the United Kingdom; nevertheless, in the twentieth century, the modern corporate enterprise has harnessed science and technology in a new and more extensive fashion within its organizational bounds. Systematic management rules are used to pursue scientific knowledge in the interest of economic well-being.

Since research and development is seen to have a special and key role in the innovation process, this chapter investigates its importance by examining the extent to which Canadian manufacturing firms incorporate R&D into the innovation process.

5.2 DEFINITIONS: EXPENDITURES ON RESEARCH AND DEVELOPMENT

Because of the importance that is today attached to research and development, considerable effort has been devoted to measuring the inputs to this process. New data on the importance of research and development expenditures within different systems of national innovation have emerged. In order to develop internationally comparable data, definitions of research and development expenditures have been prepared by the OECD (1992) – the *Frascati Manual* – to guide international measurement practice.

Research and development in the manufacturing sector, as defined by Statistics Canada (1989, p. 51), is the

> systematic investigation carried out in the natural and engineering sciences by means of experiment and analysis to achieve a scientific or commercial advance. Research is original investigation undertaken on a systematic basis to gain new knowledge. Development is the application of research findings or other scientific knowledge for the creation of new or significantly improved products or processes. If successful, development will result in devices or processes ... likely to be patentable.

While this definition covers many areas of expenditure that result in improvements to products or processes, it is not exhaustive. The line between research and development and other product or process–improving expenditures is difficult to draw in a precise fashion. This is because the definition of research and development is not congruent with the work done by all scientists and engineers in a firm. As Statistics Canada (1989, p. 52) notes:

> Research and development is intended to result in an invention that may subsequently become a technological innovation ... Much of the work done by scientists and engineers is not R&D, since they are primarily engaged in 'routine' production, engineering, quality control, and testing.

Innovation is often conceived of as a linear process – going from basic research through to engineering and commercial production. When the linear model is appropriate, the main difficulty in measuring research and development involves the definition of the dividing point between the research process and the stage of commercialization. Statistic Canada's definition draws the line on the basis of the disappearance of uncertainty. An essential requirement of the definition of research and development is that 'the outcome of the work is uncertain' (Statistics Canada, 1989, p. 52). Once the form of the new product or process has been resolved, then the expenditures are no longer considered to be R&D, although they would still be part of the innovation process. A pilot plant could be considered as part of R&D, but once testing has been completed, the boundary is reached. 'Hence, the costs of tooling, construction drawings and manufacturing blueprints, and production start-up are not included in development costs' (Statistics Canada, 1989, p. 52).

5.3 EXPENDITURES ON INNOVATION

Innovation involves a complex interactive process between science and technology and market forces. The traditional view regards technological progress as dependent upon the application of previously developed basic scientific knowledge. In this world, innovation is the result of the application of 'upstream' knowledge produced in the research and development division to 'downstream' activities involving production or design and engineering.

While this model is appropriate to some innovative activities, it is by no means universal. Many innovations have been made as scientists and engineers struggle to solve downstream problems on the factory floor (Von Hippel, 1988). These changes make use of scientific knowledge but not necessarily frontier research being done upstream in the company.

The development of the type of innovation discovered outside of the R&D lab often has a complex feedback effect on research and development as companies try to refine the solution or understand the process that has produced a new product or process. In this case, technological knowledge precedes scientific knowledge and fosters research in turn.

In the case of nineteen-century innovations in metallurgy, technology frequently drove science. New alloys were often developed by a trial-and-error method. Only after their development was science used to understand the forces that shaped the new compounds. In these instances, the role of R&D was to improve or to monitor the consistency of the properties of the new product that resulted from variations in input quality or variations in the production process. Research played a similar role in the milling industry where quality control and measurement allowed the economies of large-scale production to be exploited, all the while providing the consistency of flour required by millers. The task of ascertaining the appropriate characteristics of inputs was important to developments in construction, where understanding the properties of both concrete and steel allowed for a greatly expanded range of uses of both these materials. Food processing improvements have relied heavily on research that delineated the properties required of various food inputs for successful preservation. In all cases, improvements in the performance of outputs necessitated the development of precise standards for inputs, and this required the systematic grading, classifying, and chemical testing of components.[1]

The role of shop-floor discovery is still important today. It is still the case that technological knowledge often precedes scientific knowledge. When firms were queried about whether technological or R&D capabilities were more important in a recent Statistics Canada survey (Baldwin, Chandler, and Papailiades, 1994), they stressed that technological capabilities were more important factors contributing to their success than research and development factors. This emphasis on technology confirms that 'much of the work of the scientist today involves systematizing and restructuring in an internally consistent way the knowledge and practical solution and methods previously developed by the technologist' (Mowery and Rosenberg, 1989, p. 33).

In summary, firms have an extensive innovation system; only one part of that system is the research and development division, and even that division can be much more inclusive than the Frascati manual allows. Instead of trying to fit the innovation system into a narrowly defined box, the 1993 Innovation Survey focused instead on understanding the nature of the system in its entirety.

The survey recognized that the innovation process is often not linear. Innovations can originate in a number of different departments, each of which has responsibilities outside the scientific development area. Many iterations occur in the development of a new product or process, and interaction between the R&D group and other departments occurs at many levels. As such, some firms have difficulty in subdividing their innovation system into components that are amenable for expenditure measurement. Those that face this problem will be unable to provide a very precise measurement of research and development. Even when estimates of research and development can be provided, they are not (nor were they meant to be) exhaustive of all innovation expenditures.

Traditional approaches that equate important innovative activity with just upstream activities or basic research ignore the important contribution made by expenditures on problem solving or the development of downstream knowledge. They omit important expenditures that produce unintended innovations and that are by-products of other activities. Much

---

[1] For a description of how scientific knowledge and basic research followed technological discovery in metallurgy, construction, and food processing, see Mowery and Rosenberg (1989, pp. 37–55).

of the work normally described as routine leads to product improvements or process improvements – substantial or otherwise.

Even if research was equated with just upstream functions, there are difficulties in measuring the amount of research being performed because the dividing line between upstream and downstream activities is fuzzy and imprecise. The line between research and development and the rest of innovation expenditures is difficult to draw in practice. Hollander (1965) points out in his study of Dupont that major progress was made through incremental improvements in the production process. Mowery and Rosenberg (1989) note that many improvements have been made in food processing essentially by scientists who worked in the mundane area of testing and quality control.

Other attempts to draw a line between basic and applied research on the basis of intentions are equally contentious. Mowery and Rosenberg (1989) describe attempts to use goals – whether the activity is intended to produce an innovation – as artificial and arbitrary. For example, Pasteur's accidental discovery of the pasteurization process, which has become so important to milk, was made while he worked on problems associated with the fermentation of wine.

Since a wide range of expenditures contribute to innovation, businesses tend to think of their research and development capacity in broad terms. They do not always find it relevant to define that capability as one that resides just in upstream activities or to restrict it just to the categories included in the Frascati manual. Moreover, since almost every business has developed a facility for improving the operation of process technology, many will claim a research and development capability, if allowed considerable leeway in their answers.

Since this study focuses on the innovative capacity of Canadian manufacturing firms, we focus here on research capability defined broadly. Research and development, especially in studies that aim at cross-country comparability, is defined rather narrowly, according to the OECD-sponsored Frascati manual. It is argued (Kleinknecht, 1987, 1989; Kleinknecht, Poot, and Reijnen, 1991) that these definitions exclude a significant amount of R&D activity – especially small-firm or informal R&D. Therefore, the 1993 Canadian Survey of Innovation used self-reporting of the existence of an R&D unit without specifying an overly restrictive definition of R&D. Instead, it asked firms whether they innovated and, within that context, whether they had an R&D unit. This allowed respondents to associate R&D in their firm with knowledge creation used for innovation. Since this may, quite appropriately, entail a broader range of expenditure than is used by the Frascati manual, the survey also requested clarification as to the type of R&D unit and whether the type of expenditure qualified for a tax credit. The last comes closest to the narrower R&D definition that is normally used.

The innovation survey elicited responses about a broad range of upstream and downstream knowledge-creating activities that contribute to the innovation process. This broad range had to cover at least those activities included in the standard definition of R&D used by statistical agencies, a definition that has been painstakingly developed to facilitate international comparisons. Therefore, firms were asked about their research and development activities, using familiar terms and providing the standard definitions. However, firms were not restricted to this definition. Contrary to the practice often followed in the collection of the research and development expenditures that strictly follow the OECD Frascati guidelines, firms were not telephoned to ensure that the narrow definition was being strictly applied. Since firms that received the research and development portion of this survey were aware that they were answering an innovation survey, they might well have been expected to include all of the activities that support innovation in their concept of research and development. As a result, the notion of research and development capability that is being addressed here is probably somewhat broader than is normally captured.

Accordingly, the survey focuses on several dimensions of research and development capacity in order to evaluate the nature of the innovation process being reported under the rubric of research and development. The innovation survey begins by investigating the attitude of firms to innovation and technological capabilities with a series of questions on the importance placed on various business strategies that involve spending on research and innovation – such as the extent to which the respondent was competitive with respect to its R&D spending.

These attitudinal questions are complemented with questions that probe the commitment of the respondent to the phenomenon in question. Firms can undertake R&D regularly or only occasionally. On the one hand, research and development can be an ongoing process that tries to reengineer systems or to develop new products from scratch; on the other hand, it can be a reactive solution to problems that arise from production processes or to new product opportunities that arise from customer suggestions. In the former case, R&D is likely to be done continuously. In the latter, it is likely to be done only occasionally. Firms may not set up separate laboratories that constantly search for new products – but they may still devote resources to solving problems or to opportunities when they arise. The innovation survey probed these differences by investigating which type of R&D operation existed – whether the firm conducted R&D on an ongoing basis or just occasionally.[2]

Answers to the questions on the type of R&D operation allow us to better understand the commitment that firms have to research and development. Those doing R&D on an ongoing basis are expected to have a greater commitment to the process. Firms were also asked whether they conducted R&D in a separate research and development group, in other departments, or by contracting it out. This allows us to examine the relationship between a firm's commitment to research and the type of organization used to perform research. The answers also provide measures of the importance of research networks, since sources of research that are external to the firm, whether they be contracts or collaboration, are the manner by which the research capabilities within the firm are bolstered by outsiders.

Firms were also asked whether, within the previous three years, they claimed the income tax credit for research and development that is available in Canada. As indicated, firms responding to the survey may have a much more expansive definition of R&D activity than is adopted by the OECD Frascati manual. As such, the incidence of firms doing *any* research and development reported here provides an upper bound on R&D activity. The percentage of firms utilizing a tax credit provides a lower bound, since the Statistics Canada definition of research and development expenditures is essentially those disbursements that are eligible for the "Scientific Research and Experimental Development" tax credit under the Income Tax Act (Statistics Canada, 1989, p. 52).[3] Firms

---

[2] Palys (1992) stresses the need to elicit information on difficult topics with different types of questions. A parallel to our methodological approach may be found in studies of job satisfaction, which is difficult to measure. In order to develop indices of job quality, questions about satisfaction are accompanied by questions about actions by workers that give more direct measures of satisfaction.

[3] The Industrial Research and Development Statistics reported in Catalogue #88-202 do not come from a comprehensive survey of the industrial population. They are based first on information about which firms have claimed a tax deduction. To this is added information on other firms doing R&D that is derived from lists, such as the Financial Post list of R&D performers. The Innovation and Technology Survey samples all firms with a plant in the manufacturing sector and therefore offers broader coverage. The former may be quite sufficient when it comes to providing expenditure data because the majority of expenditures are performed in large firms. The latter procedure is necessary when a broad overview of the percentage of firms performing R&D is required. Another difference between the two sources of information should be noted. The Industrial Research and

indicating that they perform research and development but not claiming a deduction may find the act too costly, are unaware of the deduction, or have a broader concept of the meaning of research and development than exists under the act.

In the remainder of this chapter, we develop a picture of the type of research and development process at work in the manufacturing sector and relate it to the innovation process.

5.4 RESEARCH AND DEVELOPMENT ACTIVITY

5.4.1 Frequency of Research and Development

Measures of the incidence of R&D require a criterion that can be used to classify a firm as engaging in R&D. A *performer* of R&D is generally defined as a firm that carries out R&D by spending funds directly on the process. A *conductor* of R&D is defined here as a firm that carries out or funds R&D. The latter includes both firms that perform R&D and those that contract it out to others. This study uses the latter, broader concept (a conductor of research) to define firms that are responsible for and pursue R&D, because it is interested in determining how many firms benefit from the research and development process.

Research and development, as is measured in the 1993 Canadian-Survey of Innovation, is pursued by more than half of the manufacturing population (see Table 5.1). Some 65% of manufacturing firms indicate that they conduct R&D, but only 25% indicate that they have an R&D process that was carried out on an ongoing basis.[4] Most of the firms that are responsible for ongoing R&D are also performers in the narrower sense of the word. Less than 1% indicate that they are responsible for ongoing research but list themselves as only contracting out R&D; however, about 5% of firms conducting R&D on an occasional basis indicate that they only engage in contract research.

Development Statistics define performers as those who carry out research. The Innovation and Technology Survey defines a firm as a conductor of research if it either carries out the research or contracts for it. The survey is interested in knowing how many firms have a research program, both those carrying out the research and those funding it.

[4] In the Industrial Research and Development statistics (Catalogue 88-202), performers are defined as those who actually carry out the activity. In the statistics reported here, firms may have included themselves as performers if they contracted out R&D. Thus performers are interpreted more liberally here as those being responsible for R&D or carrying it out.

Table 5.1. *Percentage of Firms Conducting Research and Development*

	All Firms
R&D conducted on an ongoing basis	25 (1)
R&D conducted only on an occasional basis	41 (1)
All firms conducting R&D	65 (1)

Note: Standard errors are in parentheses.

Table 5.2. *Percentage of Firms Conducting Research and Development Using Different R&D Delivery Mechanisms*

	All Firms
Separate R&D department	25 (1)
R&D in other departments	63 (1)
R&D contracted out	22 (1)

Note: Standard errors are in parentheses.

5.4.2 Organization of R&D Facilities

Many firms actively pursue an agenda that includes research and development. This does not imply the widespread existence of separate industrial science laboratories. Research and development can be pursued in a number of different ways – within a dedicated laboratory, throughout the firm in other departments, or by contracting it out.

All three methods are important (see Table 5.2). Of those firms indicating that they conduct R&D, 25% do so in a separate R&D department. More than half (63%) indicate that their R&D is done elsewhere in the firm. Firms that contract out R&D to other organizations make up 22% of those conducting R&D – about the same percentage as have an R&D lab. While there is some overlap among the categories, it is not very significant. About 6% of firms conduct R&D in other departments and contract out research as well. All other combinations are employed by less than 3% of firms.

The type of organization that is adopted is a function of the degree of commitment that the firm makes to the R&D process (see Table 5.3). Some 41% of firms doing ongoing research have separate research departments; less than 15% of firms doing occasional research have a separate R&D department. Despite this difference, it is significant that an ongoing research program does not have to be conducted in a separate

Table 5.3. *Differences in R&D Delivery Mechanisms Between Ongoing and Occasional Performers (% of Firms)*

	Doing Ongoing Research	Doing Occasional Research
Separate R&D department	41 (1)	15 (1)
R&D in other departments	61 (2)	64 (2)
R&D contracted out	15 (1)	27 (1)

Note: Standard errors are in parentheses.

R&D department. Only about 41% of those firms that are conducting an ongoing program establish a separate R&D department.

The other major difference between firms conducting ongoing and occasional research occurs in the use of contract research. Firms that only occasionally conduct research are more likely to contract it out. Contract research then is more likely to be used to handle short-term needs.

There is little difference in the tendency of the two groups (those doing ongoing research in contrast to those occasionally doing research) to conduct their research in departments outside the R&D unit. About 61% firms doing ongoing research use other departments for this purpose; a slightly larger percentage (64%) of firms that only conduct occasional research do the same. It is significant that an ongoing research program does not necessarily equate only with a separate institutional framework for carrying out R&D. It is, nevertheless, the case that a firm making a commitment to ongoing research is more likely to have a separate research division than are those who only do so occasionally.

5.4.3 Collaborative Research and Development

Innovation ideas, as outlined in Chapter 4, come from a variety of external sources – customers, suppliers, and university researchers. Internally, ideas are championed by management and marketing. They are refined by such sources as the production department and research and development laboratories.

Considerable development work is required before ideas can produce new commercial products or workable new production processes. Because innovative ideas emerge from a variety of sources, links between a firm's R&D or production department and external sources will often be used to develop ideas into mature innovations.

The importance of these efforts can be derived from information on the extent of contract or collaborative research. These formal working arrangements provide more concrete evidence on the importance of outside links than just information on the sources of ideas. Contract as well as collaborative research projects are a manifestation of resources being committed and activities being performed that directly link firms one to another in a network of innovators.

The incidence of contract and collaborative research provides a picture of the diversity of the channels that are used to exploit technological opportunities. Contract research allows a firm to incorporate new ideas when it does not have internal expertise. It works particularly well when the incorporation of new ideas and products does not involve tacit or firm-specific knowledge. Since most knowledge of production processes incorporates at least an element of tacit knowledge, having an internal research capacity is often a prerequisite for the ready integration of the outcome of contract research back into the host or sponsoring firm.

Contract research involves third-party transactions and is a market-based transaction. Collaborative research involves a partnership and, therefore, extends the boundaries of the firm. Collaborative research is a substitute for contract research in situations where third-party or market transactions do not work as well as internalization via the creation of a new entity. The knowledge created by research has many properties that make it difficult to transfer in third-party transactions. Collaboration potentially allows the creation and transfer of knowledge more efficaciously. Collaborative research also allows costs to be spread across more parties, thereby permitting the exploitation of economies in scale; it also enhances efficiency when it prevents needless duplication of research effort. By internalizing the externalities associated with new knowledge production in situations where intellectual property rights are weak, it also reduces the appropriability problem – where weak incentives due to inappropriability of the fruits of invention reduce the amount of innovative effort.

Collaborative research has the disadvantage of requiring the costly coordination of efforts with partners who may not share exactly the same objectives. Collaborative research, like contract research, requires that the resulting research output be successfully reincorporated into the firm. These reabsorption costs may be higher than the cost savings that result from the joint research framework. Not all firms will find appropriability to be a problem, or the size of research scale economies to be large enough to offset the disadvantages of collaborative research.

Despite these disadvantages, the intensity of collaborative research is impressive. About 14% of Canadian firms that conduct R&D formed collaborative R&D agreements with other firms at some time in the three

Table 5.4. *Percentage of Firms Conducting R&D with R&D Collaborative Agreements*

	All Firms
All firms conducting R&D	14 (1)
All firms conducting ongoing R&D	20 (1)
Firms conducting R&D with a separate R&D Department	19 (2)

Note: Standard errors are in parentheses.

Table 5.5. *Patterns of Collaboration by Firms (% of Firms Indicating R&D Collaboration)*

Collaboration Partner Type	Region of Collaboration Partners				
	Canada	USA	Europe	Pacific Rim	Any Region
Customers	37 (3)	14 (2)	5 (1)	2 (1)	43 (3)
Suppliers	38 (3)	19 (2)	6 (1)	5 (1)	46 (3)
Affiliated companies	18 (2)	14 (2)	6 (1)	2 (1)	34 (3)
Competitors	8 (2)	2 (1)	3 (1)	.3 (.3)	12 (2)
R&D institution	33 (3)	3 (1)	1 (.5)	.4 (.3)	34 (3)
Universities or colleges	39 (3)	5 (1)	3 (1)	.2 (.3)	41 (3)
Other	6 (1)	2 (1)	1 (1)	0	9 (2)
TOTAL (ANY OF ABOVE)	89 (2)	39 (3)	18 (2)	7 (1)	100

Note: Standard errors are in parentheses.

years prior to the survey (see Table 5.4). Some 19% of those with separate R&D departments did so.

The collaborative agreements extend across a wide range of partners (customers, suppliers, affiliated companies, competitors, R&D institutions, and universities) both in Canada and abroad (see Table 5.5). When all partners are considered together, the top three categories of collaborative agreements are suppliers (46%), customers (43%), and universities (41%). The fewest agreements were with competitors (12%). The notion that the pressures for technological progress are so intense that many competitors are being driven together for research and development purposes is not substantiated by the data.

Collaboration is widespread for those firms that engage in this activity. First, each firm enters into at least two different agreements in total. Second, many firms enter into agreements in more than one region. Some 89% of the firms have agreements in Canada; 39% have agreements with firms in the United States; 18% have agreements in Europe.

The pattern of collaboration differs somewhat for domestic and foreign agreements. Some 39% of firms enter into collaborative agreements in Canada with universities. The second most important group of agreements are with customers and suppliers – with 37% of firms engaging in these exercises. When collaborative partnerships are formed in Canada, both these types of agreements are about twice as frequent as those made with affiliated companies. On the other hand, collaborative agreements with affiliated companies abroad are relatively more important when compared to agreements with both suppliers and customers abroad. This reflects the importance of multinational activity within the Canadian manufacturing sector.

5.5 THE VALUATION OF INNOVATION STRATEGIES BY R&D TYPE

It is straightforward to infer that firms conducting an R&D program value the research process more highly than those not doing so. But additional conclusions about the relative importance of other categories cannot be drawn as easily. Does ongoing research involve a greater commitment to doing research than on an occasional basis? Do those availing themselves of tax credits place a greater emphasis on research?

In order to answer these questions, we make use of the valuation that respondents to the survey placed on how competitive they were with respect to 1) spending on innovation, b) spending on R&D, and c) R&D management. In each case, the scores ranged from 1 'behind'; 2 'somewhat behind'; 3 'about the same'; 4 'somewhat ahead'; to 5 'ahead'. By asking whether a certain class of R&D performers (e.g., those doing R&D on an ongoing or just on an occasional basis) assesses itself as more competitive, we can infer which R&D type is perceived by the industry to be more effective.

The first two columns of Table 5.6 compare the average scores of three different innovation-related activities for firms that do not conduct R&D against those conducting R&D, either on an occasional or a continuing basis. It is not surprising that those having an R&D unit give themselves a much higher score on their competitiveness with respect to both spending on R&D and R&D management. They also do the same for spending on innovation. The second two columns of Table 5.6 contain the average evaluation of each of these strategies for firms conducting occasional and ongoing R&D. The difference between these two groups across all categories confirms that firms doing R&D continuously

Table 5.6. *Average Scores Attached to Innovation Strategies (Mean Score on a Scale of 1–5)*

	R&D Status					
Strategy	None	Any	Occasional	Ongoing	In Separate R&D Department	Claims Tax Credit
Spending on innovation	2.27 (0.05)	2.84 (0.03)	2.58 (0.04)	3.23 (0.05)	2.88 (0.06)	3.15 (0.06)
Spending on R&D	1.96 (0.06)	2.68 (0.03)	2.36 (0.04)	3.16 (0.05)	2.93 (0.06)	3.16 (0.07)
R&D management	1.96 (0.06)	2.67 (0.03)	2.35 (0.04)	3.13 (0.05)	2.95 (0.06)	3.11 (0.06)

Note: Standard errors are in parentheses.

consider themselves more competitive than those only conducting R&D occasionally.

The final two columns contain the mean scores for those with a separate R&D department and those claiming a tax credit. Contrary to expectations, neither of these two groups exhibit a greater commitment to the innovation process than just those reporting an ongoing R&D program. This means that rates of industry R&D incidence that are calculated using the percentage of firms doing ongoing research provide an equally reliable estimate of the competencies of the population that is engaged in meaningful (at least based on self-evaluation) research and development, as do rates of incidence using the percentage of firms with a separate R&D department or those claiming a tax credit.

5.6 INDUSTRY DIFFERENCES IN RESEARCH AND DEVELOPMENT

The research and development effort varies substantially across industries for a number of reasons. First, the scientific climate that is conducive to the discovery of new products differs across industries. Second, the innovation process varies in terms of R&D intensity. The same innovation output requires different amounts of R&D input in industries with different technological opportunities. Third, some industries may be more dynamic in the sense that they more fully exploit the technological advantages that are available.

In order to allow for inherent differences in the scientific climate in each industry, industries are grouped into the three sectors – core, secondary, and 'other' used by Robson et al (1988) – that have been employed in the previous chapters. The differences in the expenditures and activities of each sector are compared.

5.6.1 Frequency of Research and Development

Research and development effort is commonly measured by R&D expenditures.[5] A breakdown of total intramural expenditures taken from published Statistics Canada data for 1993 is provided in column 1 of Table 5.7. The industries with the largest expenditures are electrical and electronic products (which includes telecommunications equipment, business machines, and other electronic equipment), chemicals and chemical products (which includes pharmaceuticals), transportation (which includes both motor vehicles and aircraft), refined petroleum, primary metals (which includes both ferrous and nonferrous metals), and paper.

An alternate measure of research effort taken from the 1993 Innovation Survey is the incidence of firms conducting R&D – the percentage of firms doing any form or a specific form of R&D. Contrary to the expenditure data, this measure of incidence avoids industry differences that are due to the costliness of innovation – industry differences in the R&D expenditure required per innovation produced. While the incidence measure lacks information on the intensity of effort, this may not be a serious deficiency – if the existence of an R&D facility is more likely to be associated with innovation than is the actual amount spent. In a study that examines the factors associated with success in small and medium-sized firms, Baldwin, Chandler et al. (1994) find that it was the existence of an R&D unit, rather than the amount spent on R&D, that was associated with increased profitability and growth at the firm level.

Despite this finding, it is still useful to incorporate into the measure a dimension relating to research intensity. For this purpose, we use the incidence of firms that engage in continuous R&D, since firms with this type of R&D assess themselves as being more competitive or more competent. Therefore, firms are divided here into two groups – those doing R&D on an ongoing basis and those doing so only occasionally (columns 2 and 3, Table 5.7).

The percentage of firms that engage in continuous R&D ranges from a high of 57% in electrical equipment to a low of 13% in printing and in

---

[5] The total of full-time equivalent workers is also used as a measure, especially for cross-country comparisons where it is difficult to compare expenditures because of different currency units.

Table 5.7. *Intensity of Research and Development, by Individual Industry*

Sector	Total Intramural Expenditure 1993 (Cdn. $million) 1	% Firms Conducting R&D		
		Ongoing Basis 2	Occasionally 3	All 4
Core		53 (2)	36 (2)	86 (2)
Machinery	91	50 (4)	38 (4)	85 (3)
Electrical and electronic products	1653	57 (4)	30 (4)	84 (4)
Refined petroleum & coal	120	34 (8)	27 (8)	61 (9)
Chemicals and pharmaceuticals	533	52 (4)	40 (4)	92 (4)
Secondary		27 (2)	42 (2)	68 (2)
Rubber	4	45 (8)	24 (7)	69 (8)
Plastic	17	25 (4)	55 (5)	79 (4)
Primary metal	189	14 (4)	57 (6)	72 (5)
Fabricated metal	39	23 (3)	44 (3)	66 (3)
Transportation equipment	701	46 (4)	31 (4)	76 (3)
Nonmetallic mineral products	12	24 (4)	30 (4)	54 (4)
Other		17 (1)	42 (1)	58 (1)
Food	50	19 (2)	49 (3)	67 (3)
Beverage	10	25 (6)	48 (7)	73 (7)
Leather & allied products	*	25 (7)	62 (7)	87 (5)
Clothing	*	18 (4)	26 (4)	43 (9)
Primary textile	42	35 (8)	29 (8)	63 (5)
Textile products	†	27 (4)	33 (5)	59 (5)
Wood	19	15 (3)	42 (4)	56 (4)
Furniture	4	13 (4)	45 (5)	59 (5)
Paper	101	27 (4)	54 (5)	82 (4)
Printing & publishing	12	13 (3)	40 (4)	53 (4)
Other	101	20 (4)	45 (5)	65 (5)

Note: Standard errors are in parentheses.
* In other industries.
† In primary textiles.

furniture. Industries with the highest incidence of continuous R&D are electrical, chemicals, machinery, transportation, and rubber. Industries where firms are least likely to be doing continuous R&D include clothing, wood, printing, primary metals, and food. There is a rough concordance between expenditure figures and the percentage of firms conducting ongoing R&D. The latter has a correlation coefficient of .7 with R&D expenditures.

The three industry groupings – core, secondary, and other – differ substantially in terms of the percentage of firms that conduct R&D continuously. The core group is highest with 53%, the secondary group is next with 27%, and the tertiary other sector comes last with only 17%. Generally, industries within each group reflect the ranking of the group. Core group industries are most likely to be doing continuous R&D. They occupy four of the seven top places. Secondary industries generally have a higher rank than the other sector.

There are fewer differences across industries in terms of the incidence of occasional R&D. The percentage of firms reporting occasional R&D is 36%, 42%, and 42% in the core, secondary and other groups, respectively. The percentage of firms responsible for continuous R&D and those doing so only on an occasional basis are not correlated at the industry level. Moreover, the incidence of occasional use is negatively correlated ($-.3$) with R&D expenditure. Occasional R&D, then, captures a dimension of the R&D process that is distributed quite differently from those doing ongoing research.

5.6.2 Organization of Research and Development

The source or location of R&D provides a measure of the degree of specialization or cooperation that exists in the production of R&D. Research and development can be done in a specialized R&D facility, by other departments, or in cooperation with others, either by contract or as a collaborative venture. The percentage of firms that conduct R&D in two-digit manufacturing industries via each of these vehicles is presented in Table 5.8.

The first measure – whether a separate R&D facility is used – captures the extent to which the firm has made a commitment to a specialized research facility. While legitimate and productive activity can be done outside of specially dedicated facilities, the existence of the latter is associated with a greater commitment to the R&D process and, as we have shown, with greater R&D competitiveness, because there is a close relationship between the extent to which firms perform R&D on a continuous basis and the creation of separate R&D facilities (a correlation of .7).

The second measure reveals the degree to which R&D is located in other departments alongside other functions, such as production, that are being performed in the firm.

Table 5.8. *Delivery Mechanisms for R&D Activity, by Individual Industry*

Industry	% conducting R&D with			
	R&D Division 1	Other Departments Originating R&D 2	External Contracts 3	Contracting Out or Collaborating 4
Core	38 (2)	57 (2)	25 (2)	39 (2)
Machinery	37 (4)	57 (4)	15 (3)	28 (4)
Electrical & electronic products*	42 (4)	56 (4)	33 (4)	48 (5)
Chemicals & pharmaceuticals	33 (4)	46 (5)	30 (4)	45 (5)
Refined petroleum & coal	31 (9)	61 (10)	31 (9)	51 (9)
Secondary	26 (2)	65 (2)	19 (2)	28 (2)
Rubber	18 (7)	62 (9)	27 (8)	27 (8)
Plastic	19 (4)	61 (5)	27 (5)	36 (5)
Primary metal	20 (5)	73 (6)	19 (5)	29 (6)
Fabricated metal	24 (3)	69 (4)	14 (3)	20 (3)
Transportation equipment	41 (5)	62 (4)	22 (4)	37 (4)
Nonmetallic mineral products	25 (5)	50 (5)	33 (5)	42 (5)
Other	19 (2)	65 (2)	23 (2)	31 (2)
Food	24 (3)	55 (4)	30 (3)	40 (4)
Beverages & tobacco	11 (5)	70 (8)	28 (7)	32 (8)
Leather	8 (4)	84 (6)	8 (25)	9 (5)
Clothing	27 (6)	70 (6)	20 (5)	24 (6)
Primary textile	40 (9)	64 (9)	9 (6)	13 (7)
Textile products	30 (6)	69 (6)	6 (3)	20 (5)
Wood	8 (3)	74 (4)	21 (4)	30 (4)
Furniture & fixtures	13 (5)	79 (6)	9 (24)	13 (5)
Paper	38 (5)	60 (6)	22 (5)	29 (5)
Printing & publishing	7 (3)	62 (5)	36 (5)	39 (5)
Other	36 (6)	55 (6)	16 (4)	32 (6)

Note: Standard errors are in parentheses.
* Includes scientific instruments.

The third measure – the degree to which contracts with outside parties are used – provides an indication of the extent to which innovation relies partly on links outside of the firm. Since firms can also carry out R&D in collaboration with other firms, the extent to which firms enter into either collaborative agreements or contracts is also included. Self-sufficiency occurs when firms neither contract with outside sources nor engage in collaborative arrangements.

The R&D department is most important in the core group of industries, where 38% of firms use a separate R&D unit; this is followed by the secondary (26%) and other groups (19%). Electrical, chemicals, and machinery – all in the core group – are the three industries most likely to have separate R&D units. There is less of a distinction among industries in the secondary and the 'other' sectors with regard to the use of a separate R&D unit. Indeed, textiles, paper, clothing, and food from the other group are more likely to have a separate R&D facility than the majority of industries in the secondary group. Firms in the tertiary other sector may be less likely to engage in continuous R&D, but when they do so, their commitment to separate R&D facilities is as high as for the secondary group of industries.

As was the case with the incidence of occasional R&D, there is less of a difference among industry groups with regards to the use of other departments for the pursuit of R&D. Some 57% of the core group, 65% of the secondary group, and 65% of other industries pursue R&D via this means.

The use of external contracts differs across the industry groups, much as the existence of a separate R&D unit does; however, the differences are not as large. In the core group, some 25% of firms use contracts for R&D, which is slightly above the percentage in the secondary group (19%) and other industries (23%).

When the incidence of either contract or collaborative research is considered together (Table 5.8, column 4), the percentage of firms conducting research that have outside contracts is as large as the percentage of firms with their own R&D department. A little over one-quarter that conduct R&D in the secondary and other groups use outside sources for research – at least as great as the percentage having a separate R&D department. In the core group, outside research is a part of the research program of almost 40% of firms conducting R&D. The high incidence of outside contracts emphasizes the importance of networks for the research process.

Are the various sources complements or substitutes? Are industries that are more likely to devote resources to a separate R&D unit also more likely to network with other firms, or does the reverse hold? While internal R&D departments and outside sources both play an important role, they are not used more intensely in the same industries. The incidence of use of separate R&D facilities and the use of external R&D contracts is not correlated. Some firms need an internal R&D capacity to make the best use of external knowledge; but in just as many cases, firms without an internal facility choose the contract route to satisfy their needs. The two opposing tendencies just offset each other.

In contrast, the use of separate R&D facilities is inversely correlated with the use of other departments for R&D (−.5). These two methods of delivering R&D are substitutes; industries where firms perform R&D in separate units are industries where firms do less R&D in other departments.

5.6.3 Other Research-Related Activities

Measures reported in the previous section provide broad and general indicators that summarize an industry's emphasis on the importance of R&D activity. Industries where firms perform R&D on a continuing basis have a higher level of research incidence than those doing it on an occasional basis. Industries that set up a separate R&D unit are likely to be giving the function more priority than are those that intersperse research among other activities. Contract and collaborative research allows outside expertise to be tapped. It is expected that the most innovative industries will show a greater devotion to R&D in all three of these major dimensions.

These hypotheses were confirmed by the differences found across the core, secondary, and other groups. These industry groups range from the most to the least innovative. The core industries, which have elsewhere been found to be the most innovative, give greater emphasis to ongoing research and are more likely to have a separate R&D facility. The greatest intersectoral differences occur for the incidence of firms conducting R&D on an ongoing basis and the incidence of those setting up a separate R&D unit. There was less of a difference for the incidence of research that was contracted out.

As useful as the statistics on the incidence of ongoing research and the existence of a separate R&D facility are for cross-industry comparisons, they should not be taken to provide exact estimates of the incidence of R&D as defined by the Frascati manual. As was previously argued, they will overestimate the amount of Frascati-defined R&D that is taking place if respondents include more innovation-related activities in the definition than is permitted by the Frascati manual. Thus, they should be treated as placing an upper bound on the incidence of Frascati-defined R&D.

Two other variables can be used to corroborate differences between the core, secondary and the other sector. These are a) the percentage of

Table 5.9. *Alternate Measures of the Importance of Research, by Individual Industry*

Industry Sector	% of Firms Conducting R&D		
	Claiming Tax Credit	Trading Intellectual Property	Collaborating on R&D
Core	36 (2)	29 (2)	22 (2)
Machinery	33 (4)	20 (4)	18 (3)
Electrical and electronic products*	44 (4)	38 (6)	26 (4)
Refined petroleum & coal	34 (9)	29 (9)	28 (8)
Chemical & pharmaceuticals	28 (4)	32 (5)	25 (4)
Secondary	15 (2)	12 (2)	13 (2)
Rubber	5 (4)	7 (6)	6 (4)
Plastic	23 (4)	16 (5)	13 (4)
Primary metal	10 (4)	25 (7)	16 (5)
Fabricated metal	13 (3)	5 (2)	10 (2)
Transportation equipment	18 (4)	17 (4)	19 (4)
Nonmetallic mineral products	17 (4)	9 (4)	13 (4)
Other	9 (2)	11 (2)	11 (2)
Food	10 (2)	11 (2)	15 (3)
Beverage	7 (4)	15 (7)	15 (6)
Leather	4 (3)	11 (7)	3 (3)
Primary textile	22 (9)	7 (6)	5 (5)
Textile products	12 (4)	14 (7)	13 (5)
Clothing	16 (5)	9 (4)	9 (4)
Wood	8 (3)	12 (6)	12 (3)
Furniture & fixtures	5 (3)	9 (4)	4 (3)
Paper	20 (4)	12 (4)	17 (4)
Printing & publishing	2 (1)	24 (6)	7 (3)
Other	20 (4)	19 (7)	17 (5)

Notes: Standard errors are in parentheses. Trading involves the acquisition from or the granting to another firm of the right to use an IPP.
* Includes scientific instruments.

firms taking a tax credit for R&D, and b) the percentage of firms entering into collaborative agreements (see Table 5.9). Both involve activities that leave little question that concrete R&D activity is taking place. These variables characterize the importance and commitment of firms in various industries to the R&D process and place a lower bound on the estimate of the size of the segment of firms doing R&D.

The percentage of firms taking a tax credit for R&D places a lower bound on the incidence of R&D, since some genuinely innovative expenditures considered by firms to be part of product and process development will not qualify as R&D for the tax credit. The percentage of firms that enter into collaborative agreements should bound the research incidence from below because not all types of research will be suited to cooperative exercises.

The percentage of firms trading intellectual property rights (that is, acquiring them from or granting them to other firms) is also included in Table 5.9 to permit a comparison of research incidence with a measure of innovation incidence. Since research and development is defined as innovative activity that might be expected to eventually result in a patent, it might be expected to be related to patent incidence. In order to cast the net somewhat more broadly than patent incidence, intellectual property incidence was defined here to include the use or possession not only of patents, but also of trademarks, industrial designs, and trade secrets. Not all R&D will result in the creation of an intellectual property right, nor will all firms that use intellectual property rights choose to trade them. Nevertheless, if the two processes are connected, a relationship should be evident between the two at the industry level.

The upper and lower bounds of the statistics that measure R&D incidence are closely related. Industries in which firms are more likely to do R&D, especially continuously, are stronger users of R&D tax credits as well (a correlation of .7).

The core industries are most likely to claim tax credits, to trade intellectual property, and to enter into collaborative agreements (Table 5.9). Secondary industries rank behind the core group. Industries in the tertiary 'other' sector come last.

It should be noted that there are two major exceptions to this rule. The paper industry belongs to the other group of industries but resembles industries in the secondary group in terms of its trading of intellectual property and collaborative agreements. It also rates higher in its overall performance regarding ongoing R&D and having a separate R&D department. In Canada, then, paper should probably be included in the second, not the third, tier of innovative industries, based on R&D characteristics.[6]

Each of the variables that measures research incidence is correlated with the others. The correlation between the incidence of tax credit use and collaboration is .8; between trading in intellectual property and R&D collaboration, .8; between tax credits and collaboration, .8. Thus, firms in an industry that use R&D tax credits tend to trade intellectual property more frequently and to enter into more collaborative agreements.

---

[6] The data on innovation rates also suggest this should be done.

There are exceptions that reveal unique strategies at the industry level:

Firms in the rubber industry do not use R&D tax credits to a large extent, nor do they intensively engage in collaborative agreements or trade in intellectual property. Since the industry also has high levels of continuous R&D, this suggests that its research capabilities facilitate its ability to absorb technologies from outside.

Primary metals rank well down the list in terms of its use of tax credits and trading intellectual property. This industry also lags in terms of both continuous R&D and separate facilities – suggesting that its strategy involves using outside sources for its innovative ideas. As Chapter 4 indicated, this industry emphasized the importance of managers as a source for new ideas and competitors as a source for external ideas.

Firms in the beverage industry are less likely to do R&D or set up a separate R&D department. However, they have collaborative agreements, thereby suggesting that external sources of research are important.

5.6.4 Patterns of Collaborative Agreements

The incidence of collaboration for firms conducting R&D varies from a high of 22% in the core sector, to 13% in the secondary sector, to 11% in the other group of industries. Of interest is whether there are differences in the pattern of R&D collaborative agreements across sectors. And is there evidence that these agreements follow the cross-sectoral pattern exhibited by external sources of information that were described in a previous chapter? Or do firms demonstrate a pattern for actual R&D partnerships that differs substantially from the pattern that they describe as having provided them with useful external information? Transfers of information from external sources involve information that is both generic and specific, that involves characteristics of both public and private goods. Actual collaborative partnerships provide evidence on the process that internalizes information. Differences in the patterns of external information and in the use of collaborative partnerships reveal which sources remain only public and which are internalized through collaborative agreements.

The types of collaborative partners vary across sectors (see Table 5.10). These agreements fall into three different groups. First, R&D agreements with universities and R&D institutions help firms keep abreast of recent scientific research and increase the returns to industrial R&D. The

Table 5.10. *Patterns of Collaboration, by Industrial Sector (% of Firms Reporting Collaborative Agreement)*

Collaborative Partner	Core Industries	Secondary Industries	Other Industries
Customers	51 (5)	46 (5)	33 (4)
Suppliers	37 (5)	49 (5)	51 (4)
Affiliated companies	27 (4)	33 (5)	40 (4)
Competitors	7 (2)	10 (3)	18 (3)
R&D institutions	39 (5)	21 (4)	38 (4)
Universities	51 (5)	23 (5)	46 (4)
Other	11 (3)	9 (3)	9 (2)

Note: Standard errors are in parentheses.

increasing financial contribution of the private sector to university research shows that firms recognize the usefulness of collaboration with academic institutions.

Firms in the core sector are closer to scientific research and engage in more of these partnerships than do firms in the secondary sector, and slightly more than firms in the tertiary 'other' sector. But surprisingly, firms in the tertiary other sector make more use of universities than does the secondary sector. The tertiary other sector, as Chapter 4 indicated, is where we found the closest correlation between the use of universities as an external source of ideas and the use of production departments as an internal source.

The second type of collaborative agreement – with suppliers and customers – generally involves applied research relating to ongoing product design and quality control. The core group of industries, in keeping with its role as a supplier of innovations to others, is much more likely to have collaborative agreements with customers than are the other two groups. In accordance with its role as a consumer of capital goods, the other sector is most likely to have collaborative agreements with suppliers.

The third type of R&D collaborative agreement involves other companies that are generally in the same industry. Some agreements of this type are with direct competitors. Here the motivation is often to reduce costs, to set industry standards, or to internalize basic research that otherwise would not be pursued. In this area, industries in the tertiary other sector lead.

Collaborative agreements are also made with affiliated companies. These agreements accomplish essentially the same goals, but are done under one management umbrella. They often involve transfers of technology and innovation across regions. In the case of both foreign-owned and domestically owned multinationals, they permit difficult-to-transmit technical knowledge to be diffused across national boundaries. Collaborative agreements with related firms are generally less common in the core group of industries. Industries in the secondary group rely relatively more on affiliated companies. Industries in the other group have the most agreements with affiliated companies.

How do these patterns differ from the patterns associated with the use of all sources of external ideas presented in Chapter 4 ? In both cases, the core sector relies more on customers than on suppliers, as might be expected because of its key position in the innovation supply chain. In both cases, suppliers are more important relative to customers in the tertiary other sector. The latter is consistent with the other sector's reliance on innovations from the core and secondary sector.

However, there are some noteworthy differences between the cross-sectoral patterns of ideas and the patterns of collaboration. First, while customers provide the same incidence of ideas across all groups, their actual importance in terms of collaborative partnerships is greater for the core than for the other sector. Firms in all sectors obtain ideas from customers, but core firms are more likely to work actively with customer firms in collaborative exercises. This is less likely for firms downstream in the innovation hierarchy, probably because these firms are more likely to serve final consumers. On the other hand, firms downstream in the supply chain are both more likely to use suppliers as a source of ideas and to actively network with suppliers, competitors, and affiliated firms in collaborative exercises.

Second, while Chapter 4 demonstrated that universities are seen to be relatively unimportant source of ideas in general, they are shown here to be a frequent collaborative partner in all sectors. Moreover, they are important both for the core and tertiary other sectors. Ironically, their importance as collaborators, rather than as a general source of ideas, suggests that universities provide information that is more in the nature of a private than a public good. The same is true of R&D institutions.

Third, while competitors were shown to be a relatively important source of ideas in Chapter 4, they are relatively unimportant as R&D collaborators, except in the other sector. Firms engage more often in reverse engineering and imitation than in collaborative agreements with competitors.

Fourth, while Chapter 4 has shown that related firms are more important as a general source of information in the core sector, collaborative agreements with related firms (affiliated companies) become relatively more important outside the core sector. The core sector, then, enjoys the benefits of affiliation without a formal collaborative agreement, whereas this is less likely for the other sectors, where affiliation brings with it active collaboration. Affiliation in the core sector is more likely to be associated with a transfer of ideas without further joint development; but elsewhere, adapting technology to local conditions results in joint development work.

5.7 DIFFERENCES IN THE RESEARCH AND DEVELOPMENT CHARACTERISTICS OF INNOVATIVE AND NON-INNOVATIVE FIRMS

5.7.1 Distinguishing Innovative and Non-innovative Firms

The objective of this study of innovation is to provide measures that can be used to examine differences in the input strategies being followed by firms. This section investigates the relationship between some of these input measures and the output of the process – innovation. It does so by examining the difference in the R&D incidence of the more innovative and the less innovative firms in order to demonstrate the connection between R&D and innovation. This is done to determine how closely innovation relies on R&D activity. More complex multivariate analysis is reserved for a later chapter. Here, we ask what percent of innovators and non-innovators pursue an R&D strategy.

Large firms – those that are profiled by the Business Register – were classified either as innovative or as non-innovative using several questions in the survey.[7] These firms were defined as innovative if they indicated that they had introduced or were in the process of introducing a product or process innovation during the period 1989 to 1991 (question 3.1), if they listed product or process innovations (question 3.2 and question 4.1), or if they reported sales in 1991 resulting from a major product innovation between 1989 and 1991 (question 1.4). Small firms[8] – those not profiled by the Business Register – were deemed to be innovative if they had sales from products resulting from major innovations introduced during that period.[9]

---

[7] It should be reiterated that the firms (IPs) that are described here as large range from 20 to over 500 employees.
[8] These firms generally have fewer than 20 employees.
[9] This difference in the definition of innovation is the result of different questions on innovation being put to large and small firms.

Table 5.11. *Frequency of R&D Activity, by Innovator Versus Non-innovator (% of Firms)*

R&D Is Conducted	Large Firms (IPs)			Micro Firms (NIPs)		
	Innovative	Non-innovative	Difference	Innovative	Non-innovative	Difference
On an ongoing basis	49 (2)	22 (2)	+	45 (3)	17 (1)	+
On an occasional basis	44 (2)	45 (2)	−	43 (3)	39 (2)	−
On either an ongoing or an occasional basis	92 (1)	67 (2)	+	86 (2)	55 (2)	+

Note: Standard errors are in parentheses.

5.7.2 R&D Activity in Innovative and Non-Innovative Firms

This section examines the extent to which R&D incidence is associated with innovation. The most commonly used measures of R&D intensity are ratios of expenditures on R&D to sales. An alternate measure of research effort is the incidence of firms conducting R&D – the percentage of firms doing any form or a specific form of R&D. As a measure of cross-industry differences in the incidence of innovation, this measure of incidence avoids industry differences that are due to the costliness of innovation – industry differences in the R&D expenditure required per innovation produced.[10]

The vast majority of innovative firms engage in R&D activity – 92% of the large firms and 86% of the small firms. A large percentage of non-innovative firms (67% of large and 55% of small) also engage in R&D activity; but the incidence is much lower than for innovative firms (see Table 5.11).

---

[10] While the incidence measure used here lacks information on the intensity of effort, this may not be a serious deficiency – if the existence of an R&D facility is more likely to be associated with innovation than with the actual amount spent. In a study that examines the factors associated with success in small and medium-sized firms, Baldwin, Chandler et al. (1994) find that it was the existence of an R&D unit, rather than the amount spent on R&D, that was associated with increased profitability and growth at the firm level.

Table 5.12. *Delivery Mechanisms for R&D Activity in Innovators Versus Non-innovators (% of Firms)*

R&D Is	Large Firms			Small Firms		
	Innovator	Non-innovator	Difference	Innovator	Non-innovator	Difference
Conducted in a separate R&D unit	36 (2)	25 (2)	+	25 (3)	22 (2)	+
Conducted in other departments	61 (2)	65 (2)	−	65 (4)	63 (2)	+
Contracted out	22 (2)	23 (2)	+	22 (2)	23 (2)	+

Note: Standard errors are in parentheses.

Differences between innovative and non-innovative firms are more evident for continuous than for occasional activity for large firms. Only 22% of large non-innovative firms conduct R&D activity on an ongoing basis, whereas 49% of large innovative firms do so. These differences are equally marked for small firms. On the other hand, there is little difference between innovative and non-innovative firms with respect to the performance of R&D on an occasional basis.

Innovative firms carry out R&D activities in a number of different locations (see Table 5.12). They are more likely to conduct R&D in a separate R&D unit, but there is no real difference between innovative and non-innovative firms in the percentage that conduct R&D in other departments, or via contracts with other parties (such as other companies or institutions). Innovation in both larger and smaller firms, then, is particularly tied to conducting R&D in a separate unit.

Innovation, it is often claimed, is stimulated by firm size because of the economies of scale that exist in the research process. If size confers advantages, collaboration is one way that firms can exploit the scale advantages from large-scale research. The incidence of R&D collaborative agreements differs substantially between innovative and non-innovative firms. Innovative firms outpace non-innovative firms, almost three to one, in terms of the frequency of their R&D collaboration (see Table 5.13). Similarly, even among firms that conduct R&D, innovative firms are more likely to form these partnerships. This difference exists for both larger and smaller firms.

Of firms with collaborative agreements, innovative firms are more likely than non-innovative firms to collaborate with almost all the partner

Table 5.13. *Percentage of Innovators Versus Non-innovators Forming R&D Collaborative Agreements*

R&D Is	Large Firms			Small Firms		
	Innovative	Non-innovative	Difference	Innovative	Non-innovative	Difference
All firms	28 (2)	9 (2)	+	12 (2)	5 (1)	+
R&D conducting	31 (2)	13 (2)	+	13 (3)	10 (1)	+
Continuous R&D conducting	41 (2)	21 (3)	+	16 (4)	13 (3)	+

Note: Standard errors are in parentheses.

Table 5.14. *Source of R&D Collaborative Agreements for Innovators Versus Non-innovators (% of Collaborators)*

R&D Is	Large Firms (IPs)			Small Firms (NIPs)		
	Innovator	Non-innovator	Difference	Innovator	Non-innovator	Difference
Collaboration type						
Customers	34 (3)	24 (5)	+	45 (9)	27 (7)	+
Suppliers	44 (3)	33 (5)	+	45 (9)	23 (7)	+
Affiliates	42 (3)	14 (4)	+	38 (9)	8 (4)	+
Competitors	8 (2)	5 (3)	+	15 (7)	7 (4)	+
R&D institutions	32 (3)	22 (5)	+	31 (9)	18 (6)	+
Universities	38 (3)	27 (5)	+	34 (9)	26 (7)	+
Region of partner						
Canada	73 (3)	58 (6)	+	84 (7)	57 (8)	+
USA	49 (3)	26 (5)	+	41 (9)	9 (4)	+
Europe	26 (3)	11 (4)	+	17 (8)	3 (3)	+
Pacific Rim	8 (2)	0	+	9 (6)	3 (3)	+

Note: Standard errors are in parentheses.

types investigated – customers, affiliated companies, competitors, R&D institutions, universities or colleges, or other partners (see Table 5.14). The incidence of these collaborative agreements is generally higher for partners both in Canada and abroad. Consequently, not only are innovative firms more likely to forge collaborative agreements, but they also form a greater number of them in total, across most partner types and across almost all regions (For more detail, see Tables 5.15–5.18).

Table 5.15. *Regional Patterns of R&D Collaborative Agreements in Large Innovative Firms (% of Collaborators)*

Collaboration Partner Type	Region of Collaboration Partner					
	Any Region	Canada	USA	Europe	Pacific Rim	Other
Customers	37 (3)	28 (3)	22 (3)	5 (1)	4 (1)	.0
Suppliers	48 (3)	33 (3)	19 (3)	5 (1)	1 (1)	.7 (1)
Affiliated companies	46 (3)	20 (3)	23 (1)	14 (2)	4 (1)	.2 (1)
Competitors	9 (2)	5 (1)	2 (1)	3 (1)	1 (1)	.0
R&D institutions	35 (3)	33 (3)	8 (2)	2 (1)	1 (1)	.0
Universities/colleges	41 (3)	40 (3)	6 (2)	3 (1)	1 (1)	.0
Other	7 (2)	4 (1)	1 (2)	2 (1)	0	.7 (1)
Any type of collaboration	100	80 (3)	53 (3)	29 (3)	9 (2)	2 (2)

Note: Standard errors are in parentheses.

Table 5.16. *Regional Patterns of R&D Collaborative Agreements in Large Non-innovative Firms (% of Collaborators)*

Collaboration Partner Type	Region of Collaboration Partner					
	Any Region	Canada	USA	Europe	Pacific Rim	Other
Customers	35 (7)	32 (7)	17 (5)	9 (7)	0	3
Suppliers	48 (7)	29 (6)	30 (6)	5 (7)	0	0
Affiliated companies	20 (6)	9 (4)	12 (4)	3 (7)	0	2
Competitors	8 (4)	5 (3)	6 (3)	0 (7)	0	0
R&D institutions	32 (7)	32 (7)	0	4 (7)	0	0
Universities/colleges	39 (7)	30 (6)	6 (3)	0 (7)	0	0
Other	7 (3)	7 (3)	0 (7)	0 (7)	0	0
Any type of collaboration	100	84 (5)	38 (7)	16 (5)	0	5

Note: Standard errors are in parentheses.

5.7.3 R&D Differences Across Novelty Types

Firms that are more innovative are more likely to have conducted R&D. Of issue, however, is the extent to which the incidence of R&D differs within the group of firms that are innovative – whether the relationship between R&D and innovation is sufficiently strong to show up even within the innovative subset.

To investigate this issue, firms[11] were divided on the basis of the novelty of their innovation. Not all innovations are equally important. Some

---

[11] These data are available only for the larger firms (IPs).

Table 5.17. *Regional Patterns of R&D Collaborative Agreements in Small Innovative Firms (% of Collaborators)*

Collaboration Partner Type	Region of Collaboration Partner					
	Any Region	Canada	USA	Europe	Pacific Rim	Other
Customers	49 (11)	40 (10)	13 (7)	5 (5)	0	0
Suppliers	49 (11)	46 (11)	25 (9)	11 (7)	10 (6)	0
Affiliated companies	41 (11)	30 (10)	11 (7)	0	0	0
Competitors	16 (8)	10 (6)	3 (3)	8 (6)	0	0
R&D institutions	34 (10)	34 (10)	0	0	0	0
Universities/colleges	37 (10)	32 (10)	5 (5)	5 (5)	0	0
Other	11 (7)	6 (5)	5 (5)	0	0	0
Any type of collaboration	100	92 (6)	45 (11)	19 (8)	10 (6)	0

Note: Standard errors are in parentheses.

Table 5.18. *Regional Patterns of R&D Collaborative Agreements in Small Non-innovative Firms (% of Collaborators)*

Collaboration Partner Type	Region of Collaboration Partner					
	Any Region	Canada	USA	Europe	Pacific Rim	Other
Customers	47 (10)	47 (10)	3 (3)	3 (3)	3 (3)	0
Suppliers	40 (10)	40 (10)	9 (6)	3 (3)	6 (5)	0
Affiliated companies	14 (7)	8 (5)	6 (5)	3 (3)	0	0
Competitors	13 (7)	13 (7)	0	0	0	0
R&D institutions	32 (9)	32 (9)	0	0	0	0
Universities/colleges	46 (10)	46 (10)	3 (3)	0	0	0
Other	12 (6)	9 (6)	3 (3)	0	0	0
Any type of collaboration	100	100	15 (7)	6 (5)	6 (5)	0

Note: Standard errors are in parentheses.

innovations are at the frontier of technological leadership; others are adaptations of the innovation of others. To draw a distinction between firms based on the importance of their innovations, the sample was divided into three categories – firms that indicated they had recently introduced a major innovation and that described the innovation as a world-first, firms that had recently introduced a major innovation but did not describe is as a world-first, and all other firms in the survey. Various measures of research incidence for each group are presented in Table 5.19.

Both innovative categories are more likely to conduct ongoing R&D, to have a separate R&D department, to claim tax credits, and to engage in

Table 5.19. *Percentage of Firms Performing R&D, by Type of Innovation*

	Firm Types		
R&D category	No Innovations	Non-World-First Innovation	World-First Innovation
All Firms			
Ongoing R&D	22 (2)	50 (2)	70 (5)
Occasional R&D	45 (2)	47 (2)	30 (4)
Conductors of R&D			
Separate R&D Department	16 (1)	33 (1)	49 (5)
Other R&D location	43 (2)	59 (2)	58 (5)
Contract Research	15 (1)	20 (1)	33 (5)
Claim Tax Credit	10 (1)	33 (1)	67 (5)
Collaborative Research	9 (1)	29 (1)	43 (5)

Notes: Larger firms (IPs) only. These generally have more then 20 employees. Standard errors are in parentheses.

collaborative research than are the non-innovative categories. Equally important, there are significant differences between those firms that report a world-first innovation and those with some other type of innovation. These differences indicate a close connection between innovation success and the probability of engaging in R&D.

It is noteworthy that the incidence of R&D in the group with a world-first is quite high. Some 70% of this group perform ongoing research, and over 99% do either ongoing or occasional research. Over 49% of the world-first group have a separate R&D department, and over 67% claim a tax credit. Finally, almost 43% engage in collaborative research. Firms that are responsible for world-first innovations are invariably more R&D intensive.

5.8 CONCLUSION

Innovation data by itself is not sufficient to delineate the intensity of the innovation process since it can only indicate the degree of success. The rate of innovation will depend upon how many firms are actively seeking to commercialize new products and processes and their rate of success in this activity. An economy where some 30% of firms innovate in any three-year period may be one where all firms attempt to innovate and only 30% succeed in any three-year period, or it may be one where only

30% try and most succeed. These two are quite different. The intensity of R&D helps to distinguish these two cases.

Previously, it was reported that about one in three firms was found to be introducing innovations. The R&D statistics at first glance suggest that innovative activity is more widespread. Over two-thirds of firms were conducting some form of R&D. However, most of this activity was done only on an occasional basis. Only about one-quarter of firms conducted R&D on an ongoing basis. It is this form of R&D that is most closely associated with innovation and that is associated with firms scoring themselves as being more competitive with regards to their R&D. Of those firms conducting ongoing R&D, only about 40% perform their R&D in a separate R&D department; less than 10% of firms regularly perform R&D and do so in a dedicated facility. It, therefore, is the case that the Canadian economy is characterized by a relatively small group of innovators and of specialized R&D performers.

The R&D process, then, consists of a small core of firms conducting R&D regularly in a dedicated facility and a large group who do so only occasionally and most often do so in a facility that performs activities other than just R&D. In addition, the process has a considerable external component. Some 22% contract out R&D, and another 14% have collaborative agreements with other firms. Indeed, the percentage of firms that either have an outside contract or perform R&D on a collaborative basis is just as high as those with a separate R&D department. It is, therefore, the case that research networks are extremely important. They take place with both customers and suppliers. Universities are particularly important in this regard.

Differences in R&D intensity across industries correspond to our expectations based on the interindustry hierarchy that data on innovation intensity revealed. A core set of industries is more R&D intensive along almost all dimensions. These industries also collaborate with customers downstream. Other industries are less R&D intensive and tend to collaborate with their suppliers upstream in the core industries.

Finally, there are significant differences between innovators and non-innovators in the extent to which they conduct R&D, especially continuing R&D. Innovators are much more likely to be conducting R&D, on a continuing basis in a dedicated R&D facility.

In conclusion, the Canadian innovation system is one where only a subset of all firms innovates and where only a segment pursues a long-run R&D strategy. This is not an economy where many firms are investing in R&D and only a few succeeding. This is a system where only a small group invest in ongoing R&D in a separate facility, and their participation in this activity is strongly linked to innovation. Not all innovators conduct R&D. But more do so than non-innovators. And the more novel the innovation, the more likely the firm is to conduct R&D. Finally, the R&D intensity varies dramatically across industries, being much higher in the innovation-diffusing industries than in the innovation-receiving industries.

SIX

Effects of Innovation

6.1 INTRODUCTION

This chapter examines the effects of innovation on the organization, activity, and performance of innovating firms and their employees.

Understanding the process of innovation requires answers to two main questions. What is it that is being done? How is it being accomplished? A study of innovation needs to understand what the knowledge-creation process produces and what inputs are used. Previous chapters have examined each of these in turn – outlining the different types of innovative outputs and then examining which inputs (associated with R&D or engineering and production departments) are employed to produce innovations. In this chapter, we examine select aspects of how innovation affects the firm.

A number of issues are investigated. We ask whether changes in the production process allow the firm to improve its ability to exploit scale economies or to improve its specialization. We investigate whether and how the innovation affects the ultimate objectives of the firm – its size and profitability.

Firms innovate in order to increase their profitability, which can occur via reductions in costs, improvements in sales, or a combination of both. The two issues are related. The general economic objectives relating to market share and profitability of innovative activity are accomplished by decreasing production costs, by increasing product line diversity, or by improving the quality of the product.

Since innovation is not universal, it is important to understand the specific effects that Canadian entrepreneurs associate with their innovation, because they delineate both the advantages and impediments to the innovative process. The magnitude of both benefits and impediments determine whether innovation is undertaken and the extent of innovation.

Some innovations, such as the automobile, have far-reaching effects on industry and society. These social effects of innovation may in the long term surpass the immediate effects on the innovating firm that concern us here. However, the objective of this chapter is only to investigate the impact of innovation on the innovating firm.

Innovation may have effects that are common across different types of innovations. The common effects that are investigated in the first section of the chapter include such items as the organizational structure of the firm, its ultimate profit and market-share goals and objectives, and the means that each firm utilizes to attain its goals.

Innovation can have several impacts on the organization of production in the firm. It can affect the organization of production, either in terms of exploitation of scale economies or in terms of specialization of function and organizational flexibility. Process innovations can reduce costs through the exploitation of volume economies. Or they may permit firms to better customize their products to suit niche markets and offset the disadvantage of small-scale production. The first section of the chapter investigates which of these two effects has dominated in Canada.

While some effects of innovation are generalizable across all types of innovation in a firm, other effects are innovation-specific. In particular, the means through which innovation affects profits and market share will vary across specific innovations. An innovation can improve profitability by improving a firm's product quality, its interactions with customers and suppliers, its market share, and its need for skilled workers. Each innovation is likely to accomplish these goals differently. Some will improve flexibility and response times to changing consumer demands. Others will allow for cost reductions and changes in profit margins.

Many of these specific effects are likely to vary across types of innovations. Firms find it difficult to generalize these specific impacts of innovation to the level of the firm because these effects often differ by innovation, and most innovators produce more than one innovation.[1] Therefore, the 1993 Canadian Survey of Innovation asked respondents to focus in this area on the effects of their most profitable innovation. These effects are examined in the second section of the chapter. The focus

---

[1] This issue became evident during preproduction testing of the survey.

of the second section is on how these effects vary across innovation types and sectors.

Previous chapters have demonstrated that the core, secondary, and tertiary other sectors differ in terms of their rates of innovation – that is, the extent to which they create original new products and processes. The core sector tends to produce new products and material that are then incorporated downstream in the secondary and tertiary other sectors. But we have been careful not to label the downstream sectors as less innovative. Incorporating new products into the production process is difficult and potentially yields benefits for consumers of new products produced by these sectors that are as large as those it yields for consumers of innovations produced by the core sector.

By examining the extent to which the specific benefits associated with innovations differ for innovative firms across sectors, we ask whether firms in the secondary and tertiary sectors are any less likely to have found that innovation positively affects profitability, or market share, or customer satisfaction.

Three other issues that fall within the general topic of the impact of innovation are addressed in this chapter. The first is the impact of outside forces associated with regulation on innovation itself. The second is the impact of innovation on the demand for labour. Finally, the extent to which innovation is connected to export activity is investigated.

Some innovations are introduced in response to government regulations, standards, and certification requirements. The impact of government regulation on innovation, especially as regards environmental regulation, has been a source of controversy. On the one hand, some view government regulation as imposing significant costs and, therefore, hindering productivity growth. On the other hand, this is sometimes counterbalanced with a more optimistic view that environmental standards can trigger profitable innovation. The third section examines whether those innovations that permit firms to better meet regulatory requirements are any less profitable than others.

Technical change is sometimes blamed for having a negative effect on employment. It is also sometimes said to reduce the need for skilled labour. Introduction of new or improved products and processes can have a direct impact on employment in the innovating firm and an indirect impact on employment in the rest of economy. Apart from their overall effect on employment, innovation and technical changes may also affect the composition of an innovating firm's workforce. Other studies (Berman, Bound, and Griliches, 1993; Baldwin and Rafiquzzaman, 1999) have focused on whether innovation has more recently increased the demand for skilled, relative to the demand for unskilled, labour in the manufacturing sector. Evidence about the impact of innovation on employment and the skills of workers in the innovating firm is provided in the fourth section of this chapter.

Finally, we briefly examine the connection between trade and innovation. The patterns of international trade are in part determined by technological competition. Innovation can reduce unit costs, introduce new or differentiated products, and change the nature of the comparative advantage that some industries have. If so, we would expect to find the most successful innovators becoming the more successful exporters. An overview of the connection between innovation and export intensity at the firm level completes our investigation of firm-specific effects of innovation.

6.2 CHANGES IN ORGANIZATION OF PRODUCTION BROUGHT ABOUT BY INNOVATION IN GENERAL

Innovation may permit or force firms to exploit volume economies and lead to increased plant specialization or to an increase in the scale of plant. In both cases, innovation reduces unit costs by permitting a firm to take advantage of scale effects, either at the level of the product line or at the plant level. Technological innovation may also serve to overcome the disadvantage of low-volume operations by allowing rapid changes or increased production flexibility. The former, scale-enhancing effect of innovation may be more prevalent in economies with large markets, where scale economies can be fully exploited. The latter may be more important in countries like Canada, which have smaller markets.

Five categories of effects that the firm experiences from innovation in general are examined here. These are 1) increased plant specialization, 2) increased scale of plant, 3) reorganization of work flows, 4) increased production flexibility, and 5) increased speed of response.

The categories can be divided into those affecting the main objectives of innovation in quite different ways:

(1) those reducing unit production costs via increased specialization, increased scale of production, and the reorganization of work flows, and
(2) those increasing sales because they improve the innovator's ability to react efficiently to changes in demand and to serve market niches

in a cost-effective manner – either via increased speed of response to customers or increased production flexibility.

Firms with products that rely primarily on price competition will use innovations to reduce production costs through increased specialization and economies of large scale. Alternately, or at the same time, firms may innovate in order to increase sales by responding rapidly and with competitive prices to specific demands of customers. The latter require improved flexibility of the production process. Reorganization of work flow may be a concomitant side effect of either form of innovation.

The most frequently reported general effect of innovation is increased production flexibility and increased speed of response to customers (see Table 6.1). Increased scale of production and specialization are second in importance, but still highly significant. Innovation has both the major effects that were outlined in section 6.1. Nevertheless, increased production flexibility is cited most frequently. Canadian firms, which ingest technology that is often developed for larger markets, most frequently innovate in order to provide for the increased flexibility required of smaller markets.

It is also important to note that the reorganization of plant work flows and the functions of different groups occur quite frequently. Innovations do not simply involve the introduction of new equipment. Reorganizations of internal processes are frequently required to facilitate the process. In some cases, this simply amounts to a physical realignment of equipment; but, in many cases, it involves the adoption of new practices, such as concurrent engineering, to fully exploit the opportunities that the new equipment provides (Baldwin, Sabourin, and West, 1999).

The general effect of innovation varies by size of firm. Since larger firms are by their nature less flexible than smaller ones, larger firms used innovations to increase their production flexibility more often than did smaller ones. The employment-weighted estimates in Table 6.1 for production flexibility are significantly higher than the company-weighted estimates.

Large firms are also more likely to note that innovation served to enhance plant specialization. It has often been argued that the small Canadian economy operating behind tariff protection was characterized by production runs that were too short and product lines that were too diverse (Eastman and Stykolt, 1967; Caves, 1975). Baldwin, Beckstead, and Caves (2001) find evidence that trade liberalization increased plant specialization in the early 1990s. The evidence from the innovation survey confirms that this was a major goal of innovation in larger firms at that time.

Table 6.1. *Changes in Production Organization Associated with Innovation, by Industrial Sector (% of Innovators)*

	All Sectors, Employment Weighted	All Sectors, Company Weighted	Core Sector, Company Weighted	Secondary Sector, Company Weighted	Other Sector, Company Weighted
Increased plant specialization	47.1 (2.1)	36.2 (2.0)	35.9 (3.6)	33.4 (3.8)	38.3 (3.3)
Increased scale of plant production	44.9 (2.1)	44.8 (2.1)	43.0 (3.7)	42.7 (4.0)	47.7 (3.3)
Reorganization of work flows and/or functions	41.2 (2.1)	39.6 (2.1)	36.1 (3.6)	35.5 (3.8)	45.1 (3.3)
Increased production flexibility	63.2 (2.0)	54.6 (2.1)	54.5 (3.7)	46.2 (4.0)	61.0 (3.3)
Increased speed of response	56.1 (2.1)	53.6 (2.1)	47.2 (3.7)	52.3 (4.0)	59.1 (3.3)
None of the above	12.0 (1.4)	14.5 (1.5)	19.7 (3.0)	17.3 (3.0)	8.8 (1.9)

Notes: Columns sum to greater than 100 because multiple responses were allowed. Standard errors are in parentheses.

Larger firms, then, are more likely to report multiple effects. They use innovation more often to increase plant specialization in order to improve flexibility and also to increase the speed of response to customer needs. On the other hand, both large and small plants report equally that innovation generally allows them to better exploit scale economies. Large firms are therefore more likely to innovate in order to develop the ability to improve their flexibility.

The relative ordering of flexibility versus scale effects is similar across all sectors – core, secondary, and tertiary other. However, the tertiary other sector places the greatest emphasis on production flexibility, thereby suggesting that excessive diversification has been highest in this sector.

The uniform ranking of the effects at the sector level hide some significant differences at the industry level. Increased plant specialization was reported by more than half of firms in food, beverage and tobacco, rubber, textile, primary metals, and, perhaps surprisingly, in the pharmaceutical industry. Increased specialization was associated with increased scale of production in pharmaceuticals and in food, beverage, and tobacco industries (reported by 90% and 62% firms, respectively). Increased scale of production was also reported by more than half of firms in rubber, fabricated metals, and petroleum refining.

In industries producing mainly consumer products belonging to the tertiary other sector, innovation was most often seen to have increased the firm's flexibility and its speed of response to customers. Between 80% and 90% of firms in traditional industries, such as leather and clothing, printing and publishing, and rubber products, reported that the main effect of innovation activity was an increased speed of response to customers. This effect of innovation was fairly widespread, reported by about three-quarters of firms in the industries mentioned. Innovation also increased flexibility in a large percentage (72.5%) of electronic firms.

Although these general effects were captured only at the level of the firm, by cross-tabulating these effects against the firms' reporting that they had a product in contrast to a process innovation, we can infer whether the organizational impacts differed between these two groups. This is done in Table 6.2 for those firms *only* reporting a process innovation that did not involve new products, for those reporting *only* a product innovation that did not involved a process innovation, and for the remainder – those who were reporting some combination of product and process innovations.

Almost all of the production-related effects are associated more often with pure process innovations than with pure product changes (Table 6.2). Firms that only introduced process innovations at the same time as they

Table 6.2. *Changes in Production Organization Associated with Innovation, by Type of Innovation (% of Innovators)*

	Only Product Innovation Without Change in Man. Technology	Combination of Product/process Innovation	*Only* Process Innovation Without Product Change
Increased plant specialization	13.5 (3.7)	47.8 (2.7)	27.6 (4.6)
Increased scale of plant production	27.7 (4.8)	53.4 (2.7)	35.6 (4.9)
Reorganization of work flows and/or functions	14.1 (3.7)	46.2 (2.7)	47.5 (5.1)
Increased production flexibility	46.3 (5.4)	60.9 (2.7)	48.7 (5.1)
Increased speed of response	27.1 (4.8)	59.2 (2.7)	55.6 (5.1)
None of the above	33.3 (5.1)	10.1 (1.6)	11.0 (3.2)

Note: Standard errors are in parentheses.

produced unchanged products increased plant specialization more often than did firms that only introduced a product innovation. However, increased plant specialization is most often associated with a product innovation that accompanies a change in manufacturing technology. The new production technology introduced by these firms also led to the reorganization of work flow and enabled them to increase their flexibility and shorten their response time to changing consumer demand.

These results corroborate findings of the 1989, 1993, and 1998 surveys of advanced technology (Baldwin, Diverty, and Sabourin, 1995; Baldwin and Sabourin, 1995; and Baldwin, Rama, and Sabourin, 1999). These surveys show that the highest adoption rate and also the fastest growth of all advanced technology groups were found in inspection and communication, design and engineering, and manufacturing information systems. These technologies increase the flexibility of production processes from the design and engineering stage, through planning and automated control and inspection of materials, to production process control and final product inspection. The adoption of advanced manufacturing technologies, such as inspection and telecommunications technologies, has had widespread effects on firms, including changes in organizational structure. Communication technologies facilitate the collection of information, and control technologies allow that information to be used to manage production processes. Together, these technologies are labour-enhancing; they enable firms to augment the capabilities of skilled workers by quickly responding to changing consumer requirements and by better tailoring their products to specific consumer needs.

6.3 EFFECT OF THE MOST PROFITABLE INNOVATION ON A FIRM'S DEMAND, SHARE OF THE MARKET, FACTOR COSTS, AND PROFITABILITY

The effects of innovation activity reported in the previous section cover the general impact of all innovation activity in the firm. Inevitably, large firms introduce many innovations. When this occurs, broad generalizations about the effect of innovation will be inaccurate if the effects vary across innovations. Not only are averages potentially misleading when there are large variances in the underlying observations; important information that shows how various types of innovation have different impacts is lost when innovation is treated as a single generic event.[2]

In order to overcome this problem, respondents were asked to describe their firm's most profitable innovation and its effects. The remainder of this chapter analyzes this information. Other chapters examine other differences in the characteristics associated with the most important innovation.

The effects that are investigated range, on the one hand, from outcomes associated with ultimate objectives (the what) – such as increases in market share and improved profit margins – to causes of these outcomes (the how) – such as improved quality of products, improved interaction with customers and suppliers, reduced lead times, and extended product range (see Table 6.3).

6.3.1 Innovation and Outcomes

Innovation improved profit margins in some 63% of the cases, and it increased domestic market share for 66%. For some 40% of major innovators, it also increased share in foreign markets.

By far, the more prevalent cause of these improvements (73%) was an improvement in a firm's interaction with its customers. This accords with our earlier finding that, in general, innovative activity had the greatest effect on the firm's speed of response to customers and increased

---

[2] This is a problem that many of the European innovation surveys have.

Table 6.3. *Effects of Innovation on Profit, Factor Costs, and Demand, by Industrial Sector (% of Innovators)*

Effects Reported	All	Sector		
		Core	Secondary	Other
Outcomes				
Improved profit margin	62.7	66.2	57.4	64.1
Increased share in domestic market	65.5	63.6	62.7	69.2
Increased share in foreign market	39.3	48.5	37.8	33.4
Instruments				
Improved interaction with customers	72.9	74.9	72.2	72.0
Improved quality of products	60.5	54.8	56.2	68.2
Extended product range	56.1	60.9	47.3	59.0
Reduced lead times	31.6	27.4	27.5	38.1
Improved interaction with suppliers	24.9	27.3	21.0	26.1
Mean Standard Error	2.1	3.7	4.1	3.4

production flexibility and that customers provide one of the main sources of information for innovation (see Chapter 4).

The second most important associated effect of innovation was an improvement in quality of products (61%). Improvements in product diversity (extending the product range) came third at 56%. The least important effects involved reduced lead times (32%) and improved interactions with suppliers (25%).

In previous chapters, we have shown that innovative activity is more intense in the core than the tertiary 'other' sector. Therefore, it is worth noting that firms in the core sector are no more likely to report any of the specific benefits than are firms in the secondary and tertiary other sectors. The largest proportion of firms that experienced increased domestic market share are in the tertiary other sector – industries with a low intensity of R&D.[3] This was accomplished through a greater incidence of quality improvement and reductions in lead time than in other sectors. Improved interaction with customers or suppliers was slightly higher in the core sector. But the important conclusion to be drawn is that perceived benefits associated with innovation are generally widespread across sectors.

---

[3] It should be noted that survey respondents did not provide detailed data on the profitability of innovation, only its incidence. It is, therefore, not possible to assert unequivocally that innovations introduced by firms in the other sector were more profitable than those introduced in the core sectors.

Table 6.4. *The Percentage of Innovative Firms That Improved Their Market Share, by Size Class*

	Size Class (Employees)			
Effect of Innovation	0–20	21–100	101–500	500+
Increased domestic market share	63.9	70.0	62.6	56.4
	(9.1)	(3.5)	(4.2)	(4.0)
Increased foreign market share	9.6	36.6	45.4	49.7
	(5.6)	(3.7)	(4.3)	(4.1)

Note: Standard errors are in parentheses.

The one exception to this conclusion is that innovations in the core sector have a much greater impact on foreign market share (Table 6.3). Innovation in this sector is most likely to lead to greater exports.

The impact of innovation on the ability of the firm to increase its market share and profitability does not differ dramatically by size of the firm. Improvements in domestic market share are reported just as often by smaller firms as by larger firms. On the other hand, since the propensity to export increases with firm size, an innovation's contribution to improved foreign market share increases with the size of the firm (see Table 6.4).

6.3.2 Innovation and Sources of Reduction in Production Costs

Process innovation increases profitability through the reduction of production costs. These reductions arise from decreases in capital, materials, labour, or energy requirements. They also occur as the result of improvements in working conditions or in technical capabilities that allow for general cost economies.

For 51% of innovations, general technical improvements are reported as the most frequent effect of innovation (see Table 6.5). The most important specific saving occurs in the area of labour requirements. Almost one-third of firms reported a reduction of labour costs in production (32%). Material, design, and energy cost savings are less frequent (19%, 13%, and 9%, respectively). Savings on capital cost are reported least frequently (6%). Thus, judging from the distribution of the frequency of various categories of cost savings, innovators have been introducing labour-saving technical change. The incidence of innovations with savings in unit labour costs increases with the labour intensity of sector; it is lowest in the core sector and higher in the secondary and tertiary 'other' sectors.

Table 6.5. *Effects of Innovation on Factor Costs, Working Conditions, and Technical Capabilities, by Industrial Sector (% of Innovators)*

	\multicolumn{8}{c}{Sector}							
Effects	All		Core		Secondary		Other	
Improved technical capabilities	50.8	(2.3)	56.0	(3.9)	49.6	(4.4)	47.6	(3.6)
Reduced labour requirements	32.3	(2.1)	23.9	(3.4)	36.7	(4.2)	35.4	(3.4)
Reduced material requirements	19.0	(1.8)	19.3	(3.1)	16.7	(3.3)	20.5	(2.9)
Reduced design requirements	12.9	(1.5)	11.4	(2.5)	15.6	(3.2)	12.0	(2.8)
Reduced energy requirements	9.2	(1.3)	6.8	(2.0)	9.2	(2.5)	11.0	(2.3)
Reduced capital requirements	5.9	(1.1)	5.1	(1.7)	6.5	(2.2)	6.0	(1.7)
Improved working conditions	30.3	(2.1)	26.3	(3.5)	29.8	(4.0)	33.9	(3.4)

Note: Standard errors are in parentheses.

Innovation, then, has a greater effect on the cost structure of the tertiary other and secondary sectors through its effect on labour unit costs. The novelty of innovations may be greater in the core sector, but it is the cost structure of the downstream sectors that is primarily influenced by the innovative machinery, equipment, and materials that are absorbed in the downstream sector from the upstream sectors.

Differences in the originality of the innovation are not significantly associated with differences in the incidence with which a reduction of labour unit costs is reported. But world-first innovations are associated with a greater frequency of decreases in their nonlabour costs of production – reductions in material, energy, and capital requirements. Firms that introduced Canada-first innovations most frequently report design-cost economies.

In contrast to labour-cost savings, which are spread about evenly across size categories, the incidence of material-cost saving is higher in larger firms. Smaller firms benefited more often than the larger ones from innovation-induced economies in design costs. That innovation in smaller firms reduced design costs more often than in the larger ones corroborates findings that smaller firms have been adopting advanced technology for design and engineering more rapidly than have larger firms (Baldwin, Sabourin, and Rafiquzzaman, 1996, Table 11).

6.4 INNOVATION AND GOVERNMENT REGULATION

In spite of a trend toward deregulation of economic activities, environmental and health regulation has increased in recent years. The conventional wisdom among economists is that environmental regulations impose significant costs and slow productivity growth, thereby hindering the ability of firms to compete in international markets.[4] An alternate view (Porter and van-der-Linde, 1995) is that properly crafted environmental standards can trigger innovation offsets, allowing companies to improve their productivity.

The empirical evidence from the United States supports the conventional view. Jaffe and Palmer (1994) established that environmental compliance increases R&D expenditures. However, they found little evidence that industries' inventive output, and by implication innovation, is related to compliance costs. A study of 445 manufacturing industries by Robinson (1995) suggests that regulation diverts economic resources and managerial attention away from innovations that are productivity enhancing.

The 1993 Canadian Survey of Innovation can be used to shed light on the Canadian experience. Respondents indicated not only whether innovation affected their profitability, market share, and product quality, but also whether their most significant innovation improved their ability to respond to government regulatory requirements with respect to environmental or health and safety regulations. About 20% of respondents indicated that innovation helped them satisfy environmental and health and safety regulations. Larger firms were more likely than the smallest firms to indicate that innovation helped them respond to environmental regulations.

In order to test whether these types of innovations had differential effects on firms, we divided all innovations into those mentioning that they served to satisfy regulation and those not making any such mention. By comparing the associated effects in the two groups, we can infer whether innovations associated with regulatory compliance are inherently unproductive (see Table 6.6). We asked whether those innovations that enhance regulatory compliance are more or less likely to be associated with such benefits as gains in market share or increased profitability.

Innovations that affected health and safety regulations do not reduce the likelihood of improvements in profitability and market share that accompany most innovations. These regulatory-compliant innovations

---

[4] Rosenberg (1982) quotes Denison as claiming that pollution and safety programs reduced U.S. TFP (total factor productivity) by as much as 1.4% in 1975.

Table 6.6. *Effects of Introducing Innovation in Response to Government Regulations, by Industrial Sector (% of Innovators)*

Did the Innovation Facilitate Regulatory Compliance?	All		Core		Secondary		Other	
	No	Yes	No	Yes	No	Yes	No	Yes
Outcomes								
Improved profit margin	61.3	66.9	63.6	71.4	57.4	57.4	62.6	69.4
	(2.6)	(4.0)	(4.7)	(6.0)	(5.0)	(8.5)	(4.0)	(6.8)
Increased share in	65.2	66.5	59.8	71.5	62.9	62.0	70.6	64.4
domestic market	(2.6)	(4.0)	(4.8)	(6.0)	(4.9)	(8.3)	(3.8)	(7.1)
Increased share in	37.9	43.4	46.6	52.3	38.7	35.2	31.5	39.8
foreign market	(2.6)	(4.2)	(4.9)	(6.6)	(4.9)	(8.2)	(3.8)	(7.2)
Instruments								
Improved quality	55.7	74.3	46.7	71.2	55.6	58.0	61.7	90.9
of products	(2.7)	(3.7)	(4.9)	(6.0)	(5.0)	(8.5)	(4.0)	(4.3)
Improved working	22.6	52.4	19.5	40.0	18.7	66.2	27.6	55.7
conditions	(2.2)	(4.2)	(3.9)	(6.5)	(4.0)	(8.1)	(3.7)	(7.3)
Reduced lead times	29.9	36.6	26.2	29.7	25.2	35.3	35.9	45.5
	(2.5)	(4.1)	(4.3)	(6.1)	(4.4)	(8.2)	(4.0)	(7.3)
Improved interaction	70.6	79.8	71.1	82.7	69.0	82.6	71.4	74.1
with customers	(2.4)	(3.4)	(4.4)	(5.0)	(4.7)	(6.5)	(3.7)	(6.5)

Note: Standard errors are in parentheses.

were more likely to have resulted in improved working conditions, better product quality, improved interaction with customers, and reduced lead times in all sectors than innovations that did not improve compliance with regulations (Table 6.6). On this basis, it cannot be argued that innovations that improved compliance with regulations were necessarily unproductive.

In addition, regulatory-compliant innovations are more likely to be associated with both improved profit margins and market share for all sectors taken together. Even here, there is no strong evidence of a regulatory burden. In the core sector, the profit margins are higher and the effects on both domestic and foreign market share are positive. There is no significant difference between the profit margin and market share effects of the two types of innovations in the secondary sector. In the tertiary 'other' sector, there is an improvement in profit margins and foreign market share accompanied by a decrease in domestic market share for regulatory compliant innovations.

We conclude that environment and health regulation do not universally have a deleterious impact. Indeed, innovation that improves regulatory compliance is uniformly associated with improvements in the quality of product, working conditions, interaction with customers, and reduced lead

Table 6.7. *Effects of Innovation on the Number and Skill Requirements of Workers in the Firm (% of Innovators)*

Effects	Decrease	Increase	No Change
Number of production workers	12.5 (1.5)	35.5 (2.2)	49.0 (2.3)
Number of nonproduction workers	3.9 (0.9)	23.4 (1.9)	58.5 (2.2)
Skill requirements of workers	0.8 (0.4)	60.3 (2.2)	38.4 (2.2)

Note: Larger firms (IPs) only. These firms are generally larger than 20 employees.

times. Moreover, it usually improves both the margins and the market share of the firm.

6.5 EFFECT OF INNOVATION ON EMPLOYMENT AND SKILLS OF WORKERS

Technical change affects the demand for workers in several ways. In the first instance, it may do so by reducing the unit labour requirements. However, a reduction in unit labour requirements does not necessarily lead to a reduction in the demand for labour. If a reduction in unit costs is reflected in lower prices and higher demand for the firm's product, the latter may offset the effect of a decline in unit labour demand, and the overall demand for labour in a particular firm may remain unchanged or even increase.

Respondents to the innovation survey outlined the effect of their most significant innovation on unit labour costs and whether it increased employment (both production and nonproduction workers), decreased it, or left the total demand for labour in their firm unchanged.

Almost half of the firms reported unchanged employment (see Table 6.7). More importantly, the proportion of firms reporting that innovation increased the number of employees in their firm was about three times higher than the proportion of those that reported a decrease. Even though innovation reduced blue-collar (production worker) employment in 13% of firms, it increased this type of employment in 36% of firms. White-collar workers (nonproduction workers) saw decreased employment in less than 4% of firms and an increase in about 23% of firms (figures not tabulated here). On this evidence, there is no strong presumption that innovation leads to a decline in employment – at least in the innovating population.[5]

---

[5] Ideally, we would also like to have data on the magnitude of changes; but a survey is not a good instrument for the collection of this information.

This finding should be set in the context of other work on the effect of technological innovation. Other studies of technological innovation have investigated the extent to which innovation on the process side has been accompanied by increases in employment. Baldwin, Diverty, and Sabourin (1995) and Baldwin and Sabourin (2001) have both reported that firms introducing new advanced technologies increased their market share, productivity, and profitability. Moreover, these firms (especially smaller ones) also tended to increase their share of employment at the industry level.

The difference in the impact of innovation on white- and blue-collar workers implies that technical change has been skill enhancing. This is confirmed more directly by firms' responses to the effect of innovation on skill requirements (Table 6.7). For some 60% of firms reporting major innovations, there was an increase in the skill requirements of workers; in 38%, there was no change in skill requirements. Less than 1% of firms reported a decrease in skill requirements as a result of innovation.

6.5.1 Employment Effect Depends on the Type of Innovation

The effect of innovation on employment varies according to the type of innovation. The chief motivation for process innovations is a reduction in the costs of production. Firms introducing only process innovations are more likely to reduce employment than are only product innovators. Some 16% of firms that only introduced a pure process innovation reported that their employment of production workers decreased (see Table 6.8), compared to only 1% for those who only reported a product innovation. It would, however, be wrong to conclude that process innovations always destroy jobs. In fact, for every process innovator that sheds blue-collar labour, there is more than one that increases it.

The largest proportion of job-creating innovations is reported by firms that introduced innovations involving both product and process features. These more complex innovations led 40% of firms that introduced them to create new production jobs, while 27% increased their white-collar employment. In both cases, a much smaller percentage decreased employment than increased it. The net effect is positive both for production and nonproduction workers, regardless of whether company- or employment-weighted estimates are used.

Both the complex product/process and the pure process innovations are more likely to require an increase in skill levels.

Table 6.8. *Effects of Innovation on the Number and Skill Requirements of Workers, by the Type of Innovation (% of Innovators)*

Type		Decrease	Increase	No Change
Number of production workers				
Type of innovation				
Only process with no product change	Company weighted	15.5 (4.0)	21.5 (4.5)	59.9 (5.4)
	Employment weighted	26.3 (4.9)	16.9 (4.1)	49.9 (5.5)
Only product with no change in man. technology	Company weighted	0.9 (1.1)	35.0 (5.5)	57.5 (5.7)
	Employment weighted	0.2 (0.5)	28.2 (5.2)	70.7 (5.2)
Combination of product/process	Company weighted	15.5 (2.0)	40.3 (2.8)	42.1 (2.8)
	Employment weighted	16.9 (2.1)	35.1 (2.7)	46.2 (2.8)
Effect of innovation on skills				
Only process with no product change	Company weighted	0.0	62.6 (5.3)	36.7 (5.3)
Only product with no change in man. technology	Company weighted	2.7 (1.9)	40.8 (5.6)	56.6 (5.7)
Combination of product/process	Company weighted	0.6 (0.5)	66.1 (2.7)	32.6 (2.7)

Notes: Larger firms (IPs) only. Standard errors are in parentheses.

6.5.2 Discriminant Analysis of Employment Effects

We have seen that an innovation has many effects, a change in employment being only one of them. Since an innovation is but one element of competitive strategy, its impact on employment may depend on a host of other variables, some firm-specific, others related to the industry sector and to market conditions.

To situate the effect of innovation, its type and, originality on employment in the context of a firm's overall characteristics, strategy, and environment, we explore the relationship between innovation and its effect on employment using discriminant analysis. The use of discriminant analysis serves to isolate the primary relationships between firm characteristics and the effect of innovation on employment. The analysis determines the characteristics that serve to best discriminate between firms that increased their employment after innovation and those that decreased it.

Our information on an innovation's effect on employment is limited to three possible outcomes: no change, an increase, or a decrease. The last two outcomes being more interesting than the first, we focus on finding a linear combination of explanatory variables that best discriminates between firms where innovation led to increased employment and those where it has decreased employment. The analysis serves to assign each firm to one of the two groups. This 'predicted membership' is then compared with the actual situation, providing a measure of the accuracy of the classification that results from the estimation. The composition, the size, and the sign of weights (coefficients) of the discriminant function show which variables determine whether an innovation increased or decreased employment.

6.5.2.1 Firm Characteristics

SIZE. A measure of firm employment is included to test whether the employment effect is associated with size. Larger firms are more likely to be in the latter stage of the life cycle and to introduce process innovations, which typically aim at reducing the cost of production, hence labour, costs.

Size is measured here by the total number of employees in a firm, including both production and nonproduction workers. Firms are classified as belonging to one of three size categories – fewer than 100 employees, 100 to 499 employees, and 500 employees or more. Based on this classification, three binary variables have been constructed to capture size effects.

NATIONALITY OF OWNERSHIP. A measure of nationality of ownership is used to test whether foreign or domestic firms are more likely to respond to innovation by increasing employment. Canada, because of its size and proximity to the United States, has a mixture of both Canadian-owned and foreign-owned firms. Elsewhere in this study, we find that the nationality of ownership matters, as regards R&D activity, innovation, and self-reported employment effects of innovation.

A binary variable – taking a value of 1 if the firm is foreign-owned, and 0 otherwise – is included to investigate whether innovations introduced in foreign-controlled firms tend to increase or decrease employment more or less than in locally owned firms.

UNIONIZATION. A measure of unionization is used to test whether the labour movement serves to protect its members from the deleterious effects of job losses associated with innovation. Unions are sometimes seen to oppose innovation suspected of reducing employment. The percentage of employees covered by a collective agreement is used as an indicator of unionization.

DEVELOPMENT STRATEGIES OF THE INNOVATING FIRM. Whether a firm is able to exploit the potential associated with innovation will depend on the extent to which it develops a number of associated capabilities (Baldwin and Johnson, 1996b, 1998b). There are a series of questions in the 1993 Innovation Survey that examine the extent to which firms have developed a number of complementary strategies and firm-specific competencies. Each of these may influence how effective a firm is in exploiting an innovation and, therefore, the extent to which the output-expanding effect of innovation will lead to increases in employment.

Respondents rated (on a scale of 1 – not important – to 5 – crucial) the importance of various factors regarding their contribution to the overall development strategy of the firm. Five areas are of particular interest to this section – human resources, improved management, more efficient use of production inputs (including 'cutting labour cost'), marketing, and technology.

Several strategy variables for each of these areas are constructed from the managers' responses to a set of questions about the importance they give to strategies in these areas. For *management strategy*, four questions were used – the importance given to management-incentive compensation schemes, innovative organizational structure, improved inventory control, and improved process control. For *market strategy*, three questions were used – the extent to which a firm introduced new products in present markets, current products in new markets, or new products in new markets. Under *technology strategy*, three questions were used – the importance of developing new technology, improving technology developed by others, and improving on their own existing technology. For *production strategy*, four factors were used – the importance of using new materials, using existing materials more efficiently, improved inventory control, and improved process control. Under *human resource strategy*, two questions were used – the importance that a firm gives to continuous staff training, and innovative remuneration schemes.

An aggregate score for each of the strategies was constructed by summing the scores of their constituent factors. For example, the sum of the scores of three factors – the importance of developing new technology, improving technology developed by others, and improving on their own existing technology – was used as the aggregate score for technology strategy. Since the number of factors varies across strategies, the results were standardized to correct for this by averaging the scores in each category.

INDUSTRY EFFECTS. Operating conditions in various industries are affected by different technological opportunities, different stages in the life cycle, and different rates of growth of demand. These industry characteristics are interrelated and not only influence innovation but may also have an effect on employment. The industry sector in which a firm operates – core, secondary and other – is included to account for the industry-specific effects.

COMPETITIVE CONDITIONS. Since the degree of competition that a firm faces is hypothesized to influence its profitability and the pressures to be competitive with regard to costs, competitiveness is captured by a separate variable. Firms facing strong competitive pressures may innovate in order to reduce operating costs and, in turn, may be more likely to use innovation to keep labour costs down. Since the intrinsic concept that we want to measure is the degree of competition faced, and concentration is a poor proxy for this (Baldwin and Gorecki, 1994), we choose to measure potential competition by the number of competitors that a firm tells us it faced. Firms are grouped according to whether they faced 5 or fewer competitors, 6 to 20 competitors, or 20 or more competitors. Three binary variables are used to capture each of these effects.

INNOVATION-RELATED VARIABLES. By their nature, process innovations are more likely to lead to reduced employment than are product innovations. We classify innovations into the three categories used throughout the study: product innovation without a change in manufacturing technology, product innovation with a change in manufacturing technology, and process innovation without a change in product. We also use the actual features of their innovation – whether the innovation involved new functional parts, new production techniques, new intermediate products, new materials, and an increase in the scale of plant or the new organization of work effort.

INNOVATION IMPEDIMENTS AND EFFECT OF INNOVATION ON SKILL REQUIREMENTS. The last set of discriminant variables involves impediments to innovation programs as reported by innovating firms and the change in skill requirements induced by innovation. The most frequently reported difficulty is the lack of skilled personnel. Innovation often increases skill requirements (Baldwin, 1999). Impediments to innovation in terms of worker shortages may result in the types of innovations that are labour-saving.

6.5.2.2 Results of the Discriminant Analysis

The discriminant function and related statistics are presented in Table 6.9. The analysis selected explanatory variables by a stepwise procedure, by

Table 6.9. *Parameter Estimates of the Function Discriminating Between Firms in Which Innovation Increased or Decreased Employment*

Variables	Coefficient*	Standardized Class Mean† – Decreased Employment	Standardized Class Mean† – Increased Employment
SIZE (100–500)	−0.466	−0.213	+0.703
SIZE (500–2,000)	+0.152	+0.580	−0.019
OWNER	−0.284	−0.197	+0.065
SECTOR – OTHER	+0.218	+0.053	−0.017
COMPETITORS > 20	−0.460	+0.013	−0.004
COMPETITORS 6–20	−0.154	−0.0776	+0.025
PROCESS INNOVATION	−0.463	−0.207	+0.068
WORLD-FIRST INNOVATION	−0.062	−0.048	+0.016
NEW MANAGEMENT STRATEGY	+0.412	+0.169	−0.056
Features of innovation			
NEW FUNCTIONAL PARTS	−0.005	+.040	−0.013
NEW PRODUCTION TECHNIQUES	−0.235	−0.148	+0.049
USE OF NEW INTERMEDIATE PRODUCTS	−0.237	−0.126	+0.042
USE OF NEW MATERIALS	+0.508	+0.096	−0.032
INCREASE IN SCALE OF PLANT	+0.048	−0.071	+0.023
NEW ORGANIZATION OF WORK	−0.039	−0.004	+0.001
Impediments to innovation			
LACK OF INFORMATION ON TECHNOLOGY	+0.090	+0.035	−0.011
BARRIERS TO INTERFIRM COOPERATION	+0.231	+0.107	−0.035
BARRIERS TO COOPERATION WITH UNIVERSITIES	+0.142	+0.087	−0.029

Note: Larger firms (IPs) only.
Statistics:
  Number of cases and class means on canonical variables:
  Increased employment $n = 168$ mean +0.174
  Decreased employment $n = 61$ mean −0.527
  Wills' lambda = 0.688 $F = 5.2$ Num. degrees of freedom 18; $Pr > F = 0.0001$
  The canonical correlation is 0.558; the test that the canonical correlation is zero is rejected
    $F = 5.3$.
  $Pr > F = 0.0001$
* Canonical Discriminant Analysis (SAS): Pooled within-class standardized canonical coefficients.
† Canonical Discriminant Analysis (SAS): Pooled within-class standardized class means.

Table 6.10. *Percentage of Observations Classified into Employment-Change Groups*

	Decreased	Increased	Total
Decreased (%)	77	23	100
Increased (%)	29	71	100
TOTAL (%)	42	58	100
ERROR COUNT RATE (%)	23	29	26

including only those variables whose partial F value was significant at an α level of less than 15%.

The actual and predicted classification appears in Table 6.10. The table shows the 'actual' group membership and the group membership predicted on the basis of the discriminant score. Some 47 out of 61 firms (77.05%), were correctly assigned to the 'decreased' group and 119 firms out of 168 (70.83%) to the 'increased' employment group.

To interpret the results in Table 6.9, note first that the mean on canonical variables is +0.174 for the group of firms that increased employment and −0.527 for those that decreased employment. The difference is statistically significant at the 0.0001 level. Firms that belong to a category that has a negative sign on the discriminant coefficient are likely to experience a decrease in employment. The opposite is true for firms that belong to a category appearing with a positive coefficient. Innovation in these firms is likely to be associated with an increased employment. The absolute value of the size of each discriminant coefficient (column 1 in Table 6.9) indicates the importance of the given predictor on the discriminant score.

All predictor variables included in the discriminant function except "management strategy" are binary. The results are therefore expressed relative to the default value, that is, relative to the category left out. The reference case is a Canadian-owned firm, employing fewer than 100 persons, operating in the core sector, which introduced a product innovation without a change in manufacturing technology, which was either a world- or a Canada-first.

Innovations introduced by medium-size firms are more likely to have led to a decrease in employment, in contrast to larger firms (employing 500 to 2000 persons), where employment was more likely to increase. The effect of the size of firm on the increase or decrease of employment is stronger in the medium-size firm category (employing between 100 and 500 persons) than in the larger category.

Innovations introduced by foreign-owned firms tend to be associated with a decrease in employment compared to domestic firms.[6]

When a firm operates in the tertiary 'other' sector, innovation is more likely to be associated with increased employment.

Characteristics of the innovation matter. Process innovation is associated with decreased employment. This may be due to introduction of new production techniques, new intermediate products, and less importantly by a new organization of work. Innovations that introduce new materials are associated with increased employment. Increased scale of plant has little effect. Originality of innovation does not have a strong discriminating influence, though other things being equal, world-first innovations tend to decrease employment.

Experiencing various difficulties in the innovation process does not have a negative effect on employment; for instance, the positive sign on the variable indicating that a firm experienced difficulties in cooperating with other firms indicates that in these cases, employment actually increased. This reinforces our finding that impediments are generally greater in firms that are expanding and having to overcome problems (Baldwin and Lin, 2002).

Among firm strategy variables, only improved management strategy has a statistically significant discriminating effect. Innovation in firms that reported improved management practices (improved management incentives via compensation schemes, innovative organizational structure, improved inventory control, improved process control) tends to be associated with increased employment.

More competition serves to increase the likelihood that jobs will be lost. Unionization does not serve to discriminate between firms that increased or decreased employment in response to innovation.

The multivariate approach reveals that the effects of innovation on employment growth are extremely heterogeneous. Compared to the firm producing product innovations, firms with process innovations are more likely to have decreased employment. More competition is also associated with less employment. Foreign-owned firms that innovate are also associated with less employment. On the other hand, firms that face more impediments are in the process of increasing employment as a result of innovation. Sweeping generalizations about the effect of innovation on employment do not emerge from the microdata.

---

[6] It must be remembered that the survey only indicates whether jobs were gained or lost, not the number of jobs gained or lost.

Table 6.11. *Export Incidence and Export Intensity of Innovative Firms (Employment Weighted)*

Type	All	Core	Secondary	Other
% of innovative firms that exported	46	61	38	39
	(2)	(4)	(5)	(4)
Exports/sales ratio	29	38	27	22
	(2)	(4)	(5)	(4)

Note: Standard errors are in parentheses.

6.6 INNOVATION AND EXPORT SALES

We have shown that innovators, particularly those with world-firsts, felt that innovations enabled them to increase their share in foreign markets. Corroboration comes from information on an innovation's contribution to export sales. Firms reported the value of domestic and foreign sales resulting from their most important innovation.[7]

Less than one-third of innovating firms provided information on sales of their most important innovation, and less than half of these reported export sales for the innovation. The export incidence, that is, the percentage of innovating firms providing information on export sales, is highest in the core sector (see Table 6.11). This is what would be expected from the product-cycle hypothesis (Vernon, 1966) and earlier studies in the United States (Mansfield et al., 1982) and in Canada (Hanel and Palda, 1982). Firms in the high-tech core sector export more frequently than do those in the other two sectors.

The survey data also provide information as to whether the industry origin of an innovation has an influence on its export/sales ratio. Most of the smallest firms did not export and are not included in the sample of larger innovating firms that reported both sales of the innovation and export sales thereof. The difference in the export/sales ratio across industries corroborates earlier industry studies. Innovations created by firms in the core sector have a higher exports/sales ratio than innovations from the other two sectors. Firms included in the core sector are more R&D intensive than those of the two other industrial sectors. This provides additional evidence that there is a positive association between R&D intensity

---

[7] This question had a lower response rate than most of the other questions, but the results are reported here because they substantiate other results on differences in the sectoral importance of exports.

and export propensity of Canadian manufacturing industries (Wilkinson, 1968; Hanel 1976).

6.7 CONCLUSION

The innovation process in Canada has its greatest impact on a firm's ability to respond flexibly to customer needs. Adoption of advanced manufacturing information systems has increased the flexibility of production processes from design and engineering, through planning and automated control and inspection of materials, to fabrication and final product inspection. Innovations increased production flexibility and sped up the response to changing customer requirements in more than half of all firms. These firms accounted for two-thirds of total industry employment.

Innovation also served to reduce unit costs through the exploitation of product line and plant scale economies. But these impacts were listed less frequently than improvements in flexibility. It is thus flexibility, rather than volume costs, that innovation in Canada primarily affects. This difference in relative emphasis was even greater in larger than smaller firms.

The specific effects of the most profitable innovations that are covered by the 1993 Canadian Survey of Innovation include improved quality and an extended range of products, improved interaction with customers and suppliers, reduced lead times, and an increased share of the domestic market. These innovation-specific effects are widespread. They can be found across all sectors and all firm-size groups. Differences in intensity of innovation do not correspond closely to differences in perceived effects arising from innovation. Firms in the tertiary other sector reported many of these effects of innovation just as frequently as did firms in other sectors.

Innovation has also enhanced profitability by improving production capabilities in more than half of innovating firms. The most frequent outcome was a reduction of labour costs, but other variable costs (energy, materials, and inputs) were reduced as well. Firms in the tertiary other sector most frequently report that innovation reduced unit labour costs.

It is therefore not surprising that innovating firms operating in the tertiary other sector reported increased profitability just as often as did firms in the secondary and core sectors. This pattern underlines the important economic contribution of diffusion of innovation from high- to low-tech sectors and of the diffusion of technological change through imitation. Original innovation may not occur as frequently in the downstream sectors, but innovation is generally listed as being profitable just as frequently in these industries.

Conclusion

Some innovations are introduced as a response to government regulations. The analysis of the impact of government regulations on innovation and on firm profitability suggests that broad negative generalizations about the effect of regulation on innovation are risky. Innovations associated with regulatory compliance increased product quality on average just as much as those innovations that are not associated with regulatory compliance. They were also more frequently associated with an increased share of domestic and foreign markets and with increased profit margins. This was particularly the case for innovations in the core sector.

While innovations reduced unit labour costs in many firms, they were more frequently associated with increases than decreases in employment. Firms that reported increases in the employment of production workers substantially outnumbered those firms where innovation led to a decline in employment. The employment creation of nonproduction jobs was even more one-sided. Innovation also improved working conditions in almost one-third of innovating firms and increased the need for more skilled workers in almost two-thirds of all innovators.

The multivariate discriminant analysis confirmed that process innovations with unchanged product are more likely to have a negative effect on the employment of production workers than are product innovations. Other negative impacts were to be found in more competitive industries and in foreign-owned firms.

Last but not least, innovations lead to increases in foreign market share. A significant proportion of the sales from innovations was exported. The more original, world-first, and Canada-first innovations had a higher proportion of export sales than did the imitative innovations. Larger and foreign-owned firms recorded a higher export/sales ratio than did smaller and domestically owned firms.

SEVEN

Innovation and Research and Development in Small and Large Firms

7.1 INTRODUCTION

Considerable economic research has been devoted to establishing whether small and large firms differ with regard to the rate of innovation or their R&D activity. On the one hand, this research was seen to have implications for aggressive American antitrust policies that focused on large firms that performed what was perceived to be a disproportionate amount of scientific research (Scherer, 1992). But more recently, the literature has focused more on the need to develop special support for R&D in small firms (Rothwell and Zegveld, 1982; Acs and Audretsch, 1990).

Since the share of total employment in Canada accounted for by small firms has been increasing (Baldwin and Picot, 1995), attention in Canada has been focused on the need for policies to facilitate more innovation in this sector. The growth of the importance of small firms has led to a reexamination of the adequacy of science and technology policies, in general, and research and development R&D subsidies, in particular, that are available to this group.

If an informed decision is to be made on whether aid for small firms' R&D efforts requires policies that are distinct from those designed for large firms, it is essential to assess the differences in the R&D capacity and innovation capabilities of small and large firms. For this reason, this chapter examines whether variations exist in the R&D profile and in the tendency of small and large firms to innovate.

The chapter uses data on both the R&D and innovation profile of small and large firms to compare differences between the two groups.

Introduction

Previous work in many countries has relied either on data on R&D (Soete, 1979: Kleinknecht, 1987), which is an input to the innovation process, or on patents (Chakrabati and Halperin, 1990), which is one output of the innovation process. Cohen and Levin (1989) have noted the need to move beyond the use of input measures to a more general measure of innovative output than is provided by patent data. Differences in the propensity to patent across industries (Scherer, 1983) make patents an imperfect measure of innovative output. This has led, more recently, to studies using broader measures of innovation – either specific counts of new products derived from technical journals (Acs and Audretsch, 1990) or from innovation surveys (Kleinknecht et al., 1991). This chapter makes use of data drawn from the 1993 Canadian Survey of Innovation. The advantage of this particular source is that it focuses both on innovation outcomes and the types of processes used to generate the innovations. It also allows us to measure the intensity of different types of innovation.

Most previous studies have focused on whether there are economies of scale in the R&D function or whether R&D expenditures increase more than proportionately with firm size (e.g., Soete, 1979).[1] This literature implicitly treats firms as entities that are almost homogeneous – differing only in terms of size and R&D propensity. In reality, firms are heterogeneous with regards to strategies pursued. Baldwin, Chandler, Le, and Papailiadis (1994) and Johnson, Baldwin, and Hinchley (1997) demonstrate that small and medium-sized firms differ substantially in terms of their innovative stance and how they carry out innovations. These differences also extend across size classes. Acs and Audretsch (1990), Link and Bozeman (1991), and Cohen and Klepper (1992) recognize that large and small firms bring different skills to the innovation process.

The existence of differences in R&D intensities between small and large firms, then, does not provide evidence that there is a need for special policies to aid small firms in this area. Ultimately, it is important to understand how a company's emphasis on R&D affects its innovativeness. Small firms may be just as innovative as large firms, but they may innovate in unique ways. In particular, they may not require R&D facilities to the same degree as do large firms. In order to illustrate the distinctions in the role that R&D plays in small versus large firms, this chapter examines differences in the sources of ideas that are used for innovation in the two groups.

---

[1] See Scherer (1992) for a summary of the literature.

Any examination of the causes of innovation must recognize that R&D is only one of the routes that can be used to generate innovations. The innovation system is complex: Some firms rely on traditional R&D laboratories, whereas others develop alliances and joint ventures that allow them to tap into scientific work being done elsewhere. R&D labs are frequently large and costly, and economies of scale associated therewith may prevent small firms from constructing their own facilities very frequently.

When firms are confronted with scale economies in any crucial input (including R&D), numerous solutions are utilized to overcome the problem. Smaller firms can contract with third parties. Or they form collaborative ventures with competitors. Both of these options permit costs to be shared and, therefore, offer potential solutions to the scale problem – though, in the case of R&D, both are second-best solutions for two reasons. First, it is costly to integrate the results of collaborative research back into the firm. Second, aligning the objectives of partners who are competitors is often difficult because of an inherent disparity in goals.

A firm may also solve the problem by forming partnerships with other firms that are either upstream (suppliers) or downstream (customers) of itself. These arrangements not only offer advantages with regards to cost sharing but also permit the alignment of the goals of each firm.

In comparing small and large firms, it is therefore important not to presume that firms in different size classes must duplicate one another in all respects. Small firms possess advantages in some areas and disadvantages in others. Because of their size, small firms may suffer unit cost disadvantages, although even here these may be offset by the arrangements just described; but, more importantly, small firms may have offsetting advantages in terms of flexibility and response time to customer needs. It is important to recognize the differences in firms that are inherent in a heterogeneous environment. Policy intervention in the area of small firms should not be directed at creating miniature replicas of large firms. Rather, it should be focused on areas that offer solutions to problems that small firms have with the innovation process. Therefore, this chapter not only examines the connection between innovation and R&D, but also looks at impediments that small firms have with innovation in general.

The chapter asks whether there are different patterns of innovation in small firms and whether fewer rely on R&D for their innovation ideas. It also recognizes that the issue of R&D effectiveness is important. Therefore, it focuses on the efficacy of the R&D process in those firms that are pursuing this strategy. It asks whether those small firms that perform

R&D are more or less likely to report innovations than are larger firms, whether the innovations that are reported tend to be product or process innovations, and whether the innovations vary in importance. It asks not only whether R&D is more likely to lead to innovation but also whether innovation is more likely to be tied to R&D in small and large firms. Finally, it explores differences in the problems that impede innovation in large and small firms.

Since this chapter compares large and small firms, it is important to note the definitions that will be used. Four groups are chosen. The first class consists of microfirms – those fewer than 20 employees; the second class consists of small firms – those with 20 to 99 employees; the third class consists of medium-sized firms – those with 100 to 499 employees; the fourth consists of large firms – those with over 500 employees. While many studies group all firms below 100 employees together, the microfirms were separated out in this analysis because their profile often differs from that of the other small firms. However, not all questions in the survey were sent to all small firms, in particular, the microfirms. The number of questions on the survey was reduced for firms that are not profiled by Statistics Canada's Business Register. These are mainly but not exclusively microfirms. Therefore, subsequent tables reported herein vary in terms of their coverage – with most of the firms in the micro-class sometimes being excluded. Generally, when a question covers only the larger firms, the smallest size class – containing the microfirms – is excluded.

7.2 DO SMALL FIRMS SUFFER FROM AN INNOVATION GAP?

Innovation consists of the commercialization of a product or process that is new to the firm. Innovation has different dimensions and can be measured in different ways. The Canadian Survey of Innovation and Advanced Technology does this in two ways in order to test whether answers to specific issues (like size-class differences) are sensitive to the way in which innovation intensity is measured.

There are two ways that innovation can be measured from the survey. First, there is the percentage of firms that had introduced an innovation. The survey asked firms whether they had introduced *a product or a process* innovation in the three years prior to the survey. A product innovation was defined as the commercial adoption of a new product – minor product differentiation was to be excluded. A process innovation was defined as the adoption of new or significantly improved production processes. In both cases, the definition stressed that minor innovations

Table 7.1. *Percentage of Firms with Innovations, by Size Class*

	Measure	Size Class (Employees)				
		All Firms	0–20	21–100	101–500	500+
1)	Firms that introduced product or process innovations	33 (1)	31 (1)	39 (2)	41 (3)	61 (3)
2)	Firms reporting sales from major product innovations	22 (1)	19 (1)	29 (2)	36 (3)	37 (3)
3)	Firms reporting sales from major or minor product innovations	41 (1)	36 (1)	50 (2)	62 (63)	61 (3)

Note: Standard errors are in parentheses.

were to be excluded. Second, a measure of intensity is available since the survey asked firms for the percentage of sales in 1993 that came from *a major* product innovation introduced between 1989 and 1991 and the percentage of sales that came either from *a minor or a major* product innovation.

The measures of innovation by size class are presented in Table 7.1. The percentage of firms that indicated they either introduced or were in the process of introducing an innovation between 1989 and 1993 – the year of the survey – is given in row 1. This measure of the probability of being recently innovative (Table 7.1, row 1) shows substantial differences across size classes. Only 31% of microfirms were innovative using this criterion, while 61% of firms with more than 500 employees were innovative. The percentage of firms that report sales from a major innovation is given in row 2. It, too, shows differences between the microfirms and the largest firms. However, with this measure, there is little difference between the medium-sized firms and the largest firms. The percentage of firms that reported sales from either a major or minor innovation is given in row 3. Once more, there is a major difference between the microfirms and the largest firms; once again, there is very little difference between the medium-sized and the largest firms.

These numbers show that the microfirms and the small firms are less likely to innovate than are the medium-sized and largest firms. But they generally do not show the largest class to be significantly more innovative than the medium-sized classes – at least with regards to product innovations (Table 7.1, rows 2 and 3). The largest firms become significantly more innovative than medium-sized firms only when process innovations are added to the picture (Table 7.1, row 1).

Table 7.2. *The Percentage of Small and Large Firms That Introduced, or Were in the Process of Introducing, an Innovation in 1989–91, by Individual Industry*

SIC	Industry	0–500 Employees	500 + Employees
10–12	Food, beverage, and tobacco	30 (3)	61 (7)
15&16	Rubber and plastic	49 (5)	81 (14)
18&19	Textiles	33 (5)	91 (12)
17&24	Leather & clothing	10 (3)	48 (22)
25&26	Wood and furniture	27 (3)	44 (18)
27	Paper	39 (6)	64 (10)
28	Printing and publishing	30 (3)	53 (13)
29	Primary metal	24 (6)	52 (13)
30	Fabricated metal	33 (3)	40 (13)
31	Machinery	43 (4)	24 (15)
32	Transportation equipment	33 (4)	70 (9)
33	Electrical and electronic products & instruments	50 (5)	81 (8)
35	Nonmetallic mineral products	30 (4)	54 (19)
36	Petroleum refining and coal	51 (9)	100 (0)
37&3,741	Chemicals & pharmaceuticals	44 (5)	76 (7)
39	Other	43 (5)	

Note: Standard errors are in parentheses.

These size class differences extend across almost all industries, as Table 7.2 indicates. The largest firms – those with more than 500 employees – have a larger percentage of innovators in every industry except machinery. Other comparisons taking into account the employment of innovating and non-innovating firms show that largest firms in all industry

Table 7.3. *Percentage of Innovators with Product Versus Process Innovations, by Size Class*

Type	All Innovators	Size Class (Employees)			
		0–20	21–100	101–500	500+
New product with no change in manufacturing technology	35 (2)	28 (9)	35 (4)	32 (41)	42 (4)
Both new product & new process	44 (2)	26 (8)	44 (4)	46 (4)	53 (4)
New process without new product	46 (5)	41 (11)	45 (4)	44 (5)	57 (4)
Product in progress	20 (2)	14 (7)	18 (3)	22 (4)	28 (4)
Product/process in progress	32 (2)	20 (8)	33 (4)	30 (4)	35 (4)
Process in progress	23 (2)	5 (4)	18 (3)	28 (4)	36 (4)

Notes: Large firms (IPs) only. These generally have more than 20 employees. Standard errors are in parentheses. A firm may fall into more than one category.

groups innovate more frequently than small and medium-sized enterprises (SMEs). These results cast doubt at the conjecture of Patel and Pavitt (1991) that the relatively high proportion of Canadian patents awarded to individual inventors reflects the innovative strength of small firms in Canada.

A breakdown showing the percentage of firms that are product innovators or process innovators and those that combine product and process innovation demonstrates that the innovative activity of small firms differs from large firms in all three dimensions (see Table 7.3). The three smallest groups are quite similar to one another with respect to the incidence of product or process innovations, but largest firms have a higher probability of innovating in each of these areas. However, the incidence of joint product/process innovation increases monotonically from the smallest to the largest size class.

The intensities reported in Table 7.3 refer to the probability of reporting any of the categories and therefore allow for multiple answers. If, instead, we examine the incidence of reporting *only* a product or *only* a process innovation, then the small firms report both of these at much higher rates than do large firms. Small firms, then, are more likely to be specializing in one aspect of the innovation process; large firms are combining different types of innovations.

Table 7.4. *Number of Product and Process Innovations Introduced, by Size Class*

Type	Size Class (Employees)			
	All Innovators	21–100	101–500	500+
Product innovation without change in man. technology	3.4 (0.6)	3.8 (1.0)	2.9 (0.5)	4.2 (0.8)
Product innovation with change in man. technology	2.4 (0.3)	3.1 (0.6)	1.7 (0.2)	2.9 (0.7)
Process innovation without product change	1.9 (0.1)	1.7 (0.2)	2.1 (0.2)	2.4 (0.3)
Products in progress	2.6 (0.3)	1.8 (0.3)	3.5 (0.6)	4.0 (1.0)

Notes: Larger firms (IPs) only. Standard errors are in parentheses.

The percentage differences across size classes are generally much larger for innovations in progress than for innovations actually introduced. This points to differences in the continuity of the innovation process. Large firms are constantly working on innovations and have an inventory of projects at any one point in time. Small firms survive because of quickness and flexibility in their general operations. This also extends to their innovative capabilities. They have fewer innovations in the pipeline because they introduce them more quickly.

The importance of innovation in any particular size class depends not just on whether a firm is innovative but also on how innovative the firms in that size class are. Measures of incidence capture the former. Measures of intensity capture the latter. One measure of intensity is the number of innovations produced per innovator (see Table 7.4). Small innovators do not differ significantly from medium-sized innovators with respect to the number of either new products or processes produced. Indeed, the smaller innovators produce more product innovations per firm than do medium-sized innovators, although they generate fewer process innovations per firm. There is also no significant difference between smaller innovators and innovators in the largest group with respect to product or combined product/process innovations. Smaller innovators do produce a significantly lower number of process innovations per firm than do large innovators.

The importance of an innovation also depends on its significance, which is measured here in two ways: first, by the degree of novelty of the innovation, and second, by its importance to the firm in terms of the percentage of sales that the innovation generates.

Table 7.5. *Novelty of Innovation, by Size Class (% of Innovations)*

		Size Class (Employees)		
Type	All Innovators	21–100	101–500	500+
World-first	16	11	18	30
	(2)	(2)	(4)	(4)
Canada-first	33	36	33	31
	(2)	(4)	(4)	(4)
Other	51	54	50	39
	(2)	(4)	(4)	(4)
TOTAL	100	100	100	100

Notes: Larger firms (IPs) only. Standard errors are in parentheses.

Table 7.6. *Distribution of Sales, by Innovation Category and Size Class (% of Total Sales)*

		Size Class (Employees)			
	All Firms	0–20	21–100	101–500	500+
Unchanged product sales	80	82	78	78	77
	(1)	(1)	(1)	(2)	(2)
Minor product improvement sales	12	11	13	19	15
	(.5)	(.5)	(1)	(2)	(2)
Major innovation product sales	8	8	9	10	8
	(.5)	(.5)	(1)	(1)	(1)
TOTAL	100	100	100	100	100

Note: Standard errors are in parentheses.

Novelty is investigated here using the distinction between innovations that are world-first, Canada-first or 'other' (see Table 7.5). The percentage of small firms that reported world-firsts (11%) is less than for medium-sized firms (18%), which in turn is less than for large firms (30%). Small firms, on the other hand, are more likely to implement Canada-firsts and other types of innovations.

Although the innovations of small firms are more likely to be of the imitative type, this should not be interpreted to mean that they have less of an impact on the firm – as measured by the percentage of sales generated by sales of the innovative product. The percentage of sales generated by major innovations is about the same in microfirms and small firms as it is in large firms (see Table 7.6, row 3). While small firms may have a tendency to innovate less frequently than large firms, innovation, when

Table 7.7. *Exports as a Percentage of Sales, by Innovation Type and Size Class*

Type	Size Class (Employees)			
	All	1–100	101–500	500+
All	16 (2)	15 (2)	25 (4)	37 (5)
World-first	33 (6)	29 (9)	33 (10)	41 (10)
Canada-first	34 (4)	29 (6)	45 (9)	44 (08)
Other	16 (3)	14 (4)	14 (4)	30 (10)

Notes: Larger firms (IPs) only. Standard errors are in parentheses.

it occurs, has just as large an effect on the small firm. A large firm has a greater breadth of product lines and is continuously seeking innovations; but each innovation in a large firm has less of an effect at the margin since the total effect across size classes is much the same, even though large firms have slightly more innovations per firm. It is also the case that there are no systematic differences across size classes in the importance of minor innovations.

Finally, we examine the export intensity associated with innovation. Exporting is a costly activity requiring firms to engage specialized human resources. Owing to fixed overhead costs related to exporting activity, it is to be expected that larger firms, which can spread the export-related costs over a larger volume of production, will export a higher proportion of their sales than will smaller ones.

The propensity to export increases with the size of the firm (see Table 7.7). The smallest firms (employing fewer than 100 persons) export 15% of their sales, the largest ones 37%. More original innovations enhance exports. World-first and Canada-first innovations are exported more than the less original 'other' innovations (Table 7.7). In fact, the Canada-first group has a slight edge over the world-first group in the medium-size firm category, but given the limited number of observations, the difference is not statistically significant. The association between originality of the innovation and its export sales is best illustrated in the largest firm category. The export share for world-first innovations is 41% but only 30% for the more imitative other ones.

In summary, small and medium-sized firms are generally less likely to have introduced a major innovation than is the largest group. The differences between the smallest and the largest size classes are more pronounced when the frequency of joint product/process innovations is considered, whether these be completed or in-progress innovations. The difference between the largest and the middle-size classes is largest for process innovations. Despite these differences for the population as a whole when just innovators are examined, there are a number of similarities across size classes. Small innovators are not significantly less likely to produce product or product/process innovations. Small and medium-sized innovators produce the same number of product innovations per firm as do large firms, although there are differences for process innovations. The impact of the innovation, in terms of sales, is about the same for small, medium, and large innovators. But small firms are less likely to have produced world-first innovations; they are more likely to be introducing changes that have already been put in place by others in Canada. They are less likely to be exporting their innovation.

None of this means that small firms are ineffective innovators. A determination of effectiveness requires a comparison of innovative outputs to inputs. This topic is addressed in the next two sections.

7.3 SOURCES OF INNOVATIONS

The innovation process differs across size classes with respect to both the importance of outputs and the type of inputs that are used. Small and large firms follow unique innovation paths. One of the differences is the source of ideas for innovation.

Innovations are most commonly thought of as resulting from the activities of research and development divisions; but they also originate from the engineering groups who are responsible for the production process. In addition, they may result from vertical linkages with either suppliers or customers. Customers often facilitate innovation by specifying new qualities that are required of the inputs that they purchase, as well as by working closely with suppliers to develop the new products. Suppliers can also provide innovations when they develop new uses for their products and actively work with their customers to demonstrate these new uses. While this is especially true of suppliers of machinery and equipment, it is also true of intermediate materials inputs. Innovative ideas also come from firms that are neither customers nor suppliers. Related firms pass on knowledge. This is one of the reasons for diversification – especially by multinationals. Knowledge is also passed between unrelated firms in the form of licenses for new technology or patents.

While all of these sources provide knowledge that is used for innovation, the knowledge provided by each takes quite different forms. Some is easily transmitted from one party to another because the concepts are easily described. This is codifiable knowledge. But other information is more tacit and less codifiable. Knowledge can also differ in its specificity. It can be generic, in that it is applicable to a wide range of situations, or it can be highly specific to the particular circumstances of a firm (Nelson, 1987, p. 75).

Tacit or firm-specific information is associated with higher transfer and transaction costs. In the case of high transaction costs, arm's-length market transactions are often replaced with other arrangements. Von Hippel (1988, Chapter 6) outlines how some firms trade in informal know-how. But much know-how has characteristics that make it hard to trade. When market transactions are difficult, alternate institutions evolve to solve the problem (Williamson, 1985). One alternative is to extend the boundaries of the firm via acquisitions and mergers so as to internalize the difficulties of transactions in tacit knowledge. One theory of the multinational firm (Caves, 1982) is based on the argument that it is this type of firm that is used to transfer firm-specific knowledge. While the transaction is still costly (Teece, 1977), doing so via intrafirm transfer is sometimes less costly than the arm's-length market alternative.

Information sources differ in terms of the extent to which they provide codifiable nonspecific knowledge. Customers and suppliers provide information that is either codifiable or nonspecific. R&D labs generate information that is less easy to transfer and is often firm-specific (Rosenberg, 1990). While pure research often has certain characteristics associated with a public good in that it provides codifiable knowledge to others, the development component of R&D is much more firm-specific – often being engaged in making a product work.[2] Intrafirm networks for the transmission of R&D knowledge evolve when market transfers are less efficient than internal transfers – where information is difficult to evaluate because of its tacit, specific nature.

Differences in the networks that support innovation in small and larger firms are revealed by the frequency of use that is made of the various sources of ideas for innovation. While R&D is the main source of new ideas for 44% of firms, customers are the source for 46% of ideas (see Table 7.8). However, small and large firms place very different emphases on these two sources. Small firms use R&D much less frequently than do large firms (34% and 64%, respectively). On the other hand, small firms rely more than large firms on customers for their innovations (49% and

---

[2] For a discussion, see Cohen and Levinthal (1989)

Table 7.8. *Main Sources of Ideas for Innovations, by Size Class (% of Innovators)*

Source	All Firms	Size Class (Employees)		
		21–100	101–500	500+
Management	53	55	54	35
	(2)	(4)	(4)	(4)
R&D	44	34	54	64
	(2)	(3)	(4)	(4)
Sales/marketing	43	43	48	37
	(2)	(4)	(4)	(4)
Production	35	38	44	26
	(2)	(4)	(4)	(3)
Suppliers	28	24	36	27
	(2)	(3)	(4)	(4)
Customers	46	49	47	38
	(2)	(4)	(4)	(4)
Related firms	15	12	19	25
	(3)	(2)	(3)	(3)
Trade fairs	17	18	18	13
	(2)	(3)	(3)	(3)

Notes: Larger firms (IPs) only. Standard errors are in parentheses.

38%, respectively). The greater emphasis on customers in small firms is also accompanied by more emphasis on the sales and marketing department (43% and 37% in small and large firms, respectively). Since the sales department is closely tied to the customer, this difference also stresses the importance of a linkage between small firms and their customers.

Large firms are more likely to be tied to an external network that is provided by related firms. About 25% of large firms receive ideas for innovations from a related firm, while only 12% of small firms do likewise. These intrafirm transfers often involve the transfer of the fruits of R&D labs of sister organizations (Teece, 1977). Thus, large firms use internal R&D sources more frequently and tie into an external network that is regulated by intraorganizational ties. Both of these findings suggest that large firms are more likely to depend upon specialized research facilities – either within their own firm or in associated firms. Large firms experience the advantage of specialization of function. Size allows firms to develop specialized R&D facilities. Large firms are also less likely to depend upon managers per se for ideas than are small firms.

In contrast, small and medium-sized firms rely less on R&D, although they place relatively more stress on the technical capabilities of their production department than do large firms. Their innovations come not so much from specialized, separate laboratories as from generalized facilities associated directly with the production process.

This accords with the view of Mowery and Rosenberg (1989) that attributed considerable importance to production personnel rather than R&D personnel in the innovation process. They argue that many advances are made first on the assembly line or in the fabrication process. It is only later that these innovations are more fully explored in R&D labs where, for instance, attempts are made to understand the composition of new materials so as to be able to mass-produce them.

There are several reasons why it is these technological breakthroughs where small firms concentrate their innovation efforts. First, these breakthroughs may be particularly common for the types of processes in which small firms specialize. Second, they may occur because the comparative advantage of large firms lies in the production of the type of knowledge that originates in R&D facilities, since the costs of conducting R&D for large firms are lower because specialization of function means that large firms will enjoy cost advantages in the pure R&D function. This cost differential leads to the development of a network that links small and large firms. As shown in Chapter 4, small firms place more importance on external contacts with their customers – firms that tend to be larger. In addition, smaller and medium-sized firms rely more on management than does the largest group – again probably because smaller size militates against specialization of function and requires management to substitute for a specialized R&D department.

Innovation involves the creation of both new products and new processes. New products often require new technologies. Indeed, some 60% of establishments in the survey that introduced new computer-based fabrication technologies did so in order to facilitate major innovations in the firm. Data on the source of new technologies that are used in process innovations (Table 7.9) confirm the size-class differences found for innovations as a whole. For the population, research is the least important source of technology, it is surpassed by experimental development and production engineering. What is more important, while small firms use R&D facilities less than do large firms (25% and 45%, respectively), this is not the case for production and engineering departments.

The frequency with which experimental development or production engineering is listed as a source of innovation does not vary in a systematic way across size classes. Once again, unrelated firms are a more important source of innovation for small firms (23%) than for large firms (14%),

Table 7.9. *Main Sources of Technologies Associated with Innovations, by Size Class (% of Innovators)*

Source	All Firms	Size Class 21–100	101–500	500+
Research	40	25	27	45
	(2)	(3)	(4)	(4)
Experimental development	52	51	50	57
	(2)	(4)	(4)	(4)
Production engineering	48	50	66	52
	(2)	(4)	(4)	(4)
Related firms	13	14	12	23
	(1)	(3)	(3)	(4)
Unrelated firms	18	23	23	14
	(2)	(3)	(4)	(3)
Customers	11	14	20	9
	(1)	(3)	(4)	(2)
Suppliers	28	22	38	26
	(2)	(3)	(4)	(4)
Trade fairs	24	17	11	12
	(2)	(3)	(3)	(3)

Note: Standard errors are in parentheses.

while large firms are more likely to rely on related firms than are small firms (23% and 14%, respectively).

In the case of ideas for both the sources of innovation and the sources of new technology, suppliers are quoted with about the same frequency by both small and large firms. In both cases, smaller firms are more likely to find customers an important source of ideas. Since smaller firms are generally supplying larger firms, this implies that ideas for innovation spread from the larger to the smaller firms via a type of partnership that is based on the mutual dependence that exists between customers and suppliers.

7.4 RESEARCH AND DEVELOPMENT ACTIVITY

7.4.1 Frequency of Research and Development

Measures of the incidence of R&D require a criterion that can be used to classify a firm as engaging in R&D. A *performer* of R&D is defined as a firm that directly carries out R&D. A *conductor* of R&D is defined here as a firm that carries out *or* funds R&D. The latter would include both firms that perform R&D and those that contract it out to others. This study uses the broader concept (a conductor of research) to define firms that are responsible for and pursue R&D, because it is interested in measuring how many firms benefit from the research and development process.

As described in Chapter 5, the concept that is used here is somewhat broader than that found in some surveys. Research and development, especially in studies that aim at cross-country comparability, is defined rather narrowly, according to the OECD-sponsored Frascati manual. It is argued (Kleinknecht, 1987, 1989; and Kleinknecht et al. 1991) that these definitions exclude a significant amount of R&D activity – especially small-firm or informal R&D. Because the 1993 Canadian Survey of Innovation was designed to provide a comparison of firms across a wide size spectrum, it used self-reporting of the existence of an R&D unit without specifying an overly restrictive definition of R&D.

Research and development, as is measured in the survey, is pursued by about two-thirds of the manufacturing population (see Table 7.10). However, a large percentage (41%) carry out R&D only on an occasional basis; only 25% indicate that they have an R&D process that was carried out on an ongoing basis.

Small and large firms differ in their tendency to conduct any form of R&D. Large and medium-sized firms are quite similar – some 87% to 90% of both conduct some form of R&D. Small firms are not far behind at 76%. Microfirms lag far behind at 59%.

There are much greater differences in the extent to which firms of different sizes conduct R&D on a regular basis. Some 58% of firms over 500 employees conduct R&D on an ongoing basis; but this is true of only 43% of medium-sized firms, 36% of small firms, and only 19% of firms with

Table 7.10. *Percentage of Firms Conducting Research and Development, by Size Class*

	All Firms	Size Class (Employees)			
		0–20	21–100	101–500	500+
R&D conducted on an ongoing basis	25 (1)	19 (1)	36 (2)	43 (3)	58 (3)
R&D conducted only occasionally	41 (1)	40 (2)	42 (2)	45 (3)	35 (3)
All firms conducting R&D	65 (1)	59 (2)	77 (2)	87 (2)	90 (2)

Note: Standard errors are in parentheses.

Table 7.11. *Delivery Mechanism for Research and Development, by Size Class (as % of Firms Conducting R&D)*

Type of R&D Facility	Size Class (Employees)				
	All Firms	0–20	21–100	101–500	500+
Separate R&D department	25	23	23	34	54
	(1)	(2)	(2)	(3)	(3)
R&D in other departments	63	60	72	59	47
	(1)	(2)	(2)	(3)	(3)
R&D contracted out	22	24	18	23	26
	(1)	(2)	(2)	(2)	(3)

Note: Standard errors are in parentheses.

fewer than 20 employees. In contrast, the smaller and middle-sized groups are more likely to do some R&D on an occasional basis than are the firms in the largest size class. In keeping with their superior flexibility in general, smaller firms respond to opportunities as they arise by conducting R&D on an occasional basis.

7.4.2 Organization of R&D Facilities

Many firms actively pursue an agenda including research and development. Research and development can be pursued in a number of different ways – within a dedicated laboratory, throughout the firm in other departments, or by contracting it out.

Large firms enjoy the specialization of labour that is associated with size; they are much more likely to conduct R&D in a dedicated department than are small firms (see Table 7.11). Some 54% of firms with more than 500 employees have a separate unit; the percentage of the middle and smaller classes that do so diminishes steadily until less than a quarter of the smallest two groups have a separate R&D department. This means that small and medium-sized firms that conduct R&D are more likely to do their R&D as part of the work of other departments. There is less of a difference for small and large firms to rely on outside companies for contract R&D than there is for them to build a separate R&D department. Small and medium-sized firms resemble one another closely in that from 18% to 23% contract out R&D, while 26% of the largest firms do so.

The type of organization that is adopted to conduct R&D is a function of the degree of commitment that the firm makes to the R&D process (see Table 7.12). Large and small firms differ in terms of their organization of

Table 7.12. *Delivery Mechanisms for Research and Development, by Type of R&D Performer and by Size Class*

	Size Class (Employees)				
	All Firms	0–20	21–100	101–500	500+
Of firms conducting ongoing research (% with)					
Separate R&D department	41	38	39	51	79
	(2)	(3)	(3)	(4)	(3)
R&D in other departments	61	62	65	55	38
	(2)	(3)	(3)	(4)	(4)
R&D contracted out	15	13	17	13	26
	(1)	(2)	(2)	(3)	(4)
	All Firms	0–19	20–99	100–499	500+
Of firms conducting occasional research (% with)					
Separate R&D department	15	16	10	18	15
	(1)	(2)	(2)	(3)	(5)
R&D in other departments	64	59	79	64	63
	(2)	(2)	(2)	(4)	(5)
R&D contracted out	27	29	20	34	26
	(1)	(2)	(2)	(4)	(4)

Note: Standard errors are in parentheses.

the R&D function only if they are doing ongoing research. In this group, 79% of large firms use a separate R&D department, while only 38% of microfirms do the same. The two smallest size classes are more likely to conduct their R&D in other departments than they are to have a separate R&D department. There is no significant difference in the percentage of large and small firms that contract out research. When occasional conductors of research are examined, few significant differences are found. Here, small and large firms are equally likely to use a separate R&D department or to contract out research to others.

In conclusion, small firms differ primarily from large firms in that they are more likely to be performing R&D occasionally and, therefore, are less likely to use a separate, dedicated R&D facility. This suggests that the innovation process in small firms differs from that in large firms in that it is less continuous.[3] This may be the result of an agglomeration effect that

---

[3] The evidence that the difference between small and large firms for innovations in progress is much larger than for innovations introduced (Table 7.3) supports this interpretation.

results from the network between large firms and their smaller suppliers. Large firms often assemble inputs from myriads of smaller suppliers. If each of the suppliers and the larger assembler has an equal probability of innovating at a point of time, the large firm will experience more change because it is ingesting modifications by the purchase of products from many suppliers. This, in turn, means that it is more likely to develop specialized R&D facilities to coordinate and control this change.

7.4.3 Collaborative Research and Development

Considerable development work is required before ideas can produce new commercial products or workable new production processes. Because innovative ideas emerge from a variety of sources, outside links between a firm's R&D or production department and external sources will often be used to turn ideas into successful innovations. These links can be made through either contract or collaborative research.

Contract research involves third parties and is a market-based transaction. Collaborative research involves a partnership and, therefore, extends the boundaries of the firm. Collaborative research is a substitute for contract research where third-party or market transactions do not work as well as internalization via the creation of a new entity. But collaborative research has the disadvantage of requiring the costly coordination of efforts with partners who may not share the same objectives.

Despite these problems, a substantial number of firms find that the advantages of collaborative research outweigh the disadvantages. While about 22% of firms contracted out R&D, some 14% of Canadian firms that conduct R&D formed collaborative R&D agreements with other firms at some time in the previous three years (see Table 7.13). This form of R&D is more likely to be found in large firms. About 46% of large

Table 7.13. *Percentage of Firms Conducting R&D with Collaborative Agreements, by Size Class*

Population	Firm Size (Employees)				
	All Firms	0–20	21–100	101–500	500+
All firms conducting R&D	14	10	16	27	46
	(1)	(1)	(2)	(2)	(3)
R&D conducted on ongoing basis	20	14	20	36	51
	(1)	(2)	(3)	(4)	(4)

Note: Standard errors are in parentheses.

firms that conduct R&D enter into collaborative agreements; less than 10% of microfirms initiating R&D do the same. The percentage engaging in R&D collaborative agreements increases to 51% for large firms that conduct ongoing R&D.

7.5 THE LINK BETWEEN R&D AND INNOVATION

Large firms are more likely to utilize R&D facilities and are more innovative than small firms by a number of standards. Nevertheless, small firms may not be ineffective performers of R&D. While small firms produce major innovations less frequently than do large firms, they also make less use of R&D facilities. The effectiveness of the research and development process in smaller firms requires a comparison of their differences with large firms both with respect to innovativeness and R&D intensity. Ideally, this requires information that would allow a comparison of the value of innovative output to the value of the resources used in the innovative process.

Some researchers have chosen to do this by using employee size to deflate the intensity of innovations (Acs and Audretsch, 1990). When they compare the share of employment in small firms to the share of their innovations, they find that the latter generally exceeds the former and argue that this makes small firms more efficient innovators.

If this were to be done here, small firms would also be found to be more effective innovators than large firms. The smallest group of firms are about one-half as likely to innovate as are the large firms; but they account for only about one-fiftieth the employment of the large group.

Unfortunately, comparisons such as these presume that the amount of resources devoted to R&D is proportional to employment in the firm.[4] Yet we know small firms are less likely to create a separate R&D lab or to perform R&D on an ongoing basis. They are also more likely to obtain their innovation ideas from other, larger firms. Therefore, a comparison of the share of innovations made by small firms to their share of the number of employees is likely to be biased against large firms.[5]

---

[4] Quite different results were obtained for the U.K. by Rothwell and Zegveld (1982) when the small-firm share of innovations was compared to the small-firm share of R&D and the small-firm share of employment.

[5] Perhaps even more problematic with this type of comparison is that it can be interpreted quite differently to suggest that large firms are more important innovators. A large firm, after all, has both a larger share of employment and of sales. If large firms account for more than their numerical share of employment, so do they of sales. Thus, their innovation, in that it serves a larger market, could be said to have a 'greater' impact on the consumer.

In this section, we will take an alternate approach to examining the efficacy of innovation. Efficacy is measured here by examining differences across size classes in the probability that a firm conducting R&D will innovate – whether R&D is more likely to produce an innovation in smaller firms than it is in larger firms (see Freeman, 1971). Alternately, this section also examines whether an innovation is more closely tied to R&D in larger or smaller firms by investigating whether the percentage of innovative firms that conduct R&D differs across size classes. The first is a type of productivity measure that is derived by asking whether R&D is more likely to result in innovation in smaller or larger firms. The second investigates the extent to which R&D is a necessity in the innovation process – the extent to which innovation is closely tied or linked to R&D.

It is recognized that this is just one out of many approaches that could be taken to measure efficacy. Efficacy might be investigated by asking whether a dollar of R&D in small firms produced a greater impact on a dollar of sales of innovative products than in large firms. This is a more ambitious task that awaits further data development. Instead, this section focuses on the more immediate task of interpreting the meaning that should be attached to differences in both the intensity of innovation and the intensity of R&D across size classes.

In order to investigate the effectiveness of R&D, firms were separated into those that conducted R&D occasionally and those that did so on an ongoing basis. Then, the percentage of those that were innovative in each group was calculated for each size class (see Table 7.14). Innovative firms are defined as those reporting either a product or process innovation.

Of those doing R&D continuously, 57% of firms are innovators based on their responses that they produced a product or process innovation in

Table 7.14. *Innovative Intensity for Conductors of R&D, by Size Class (% of Firms)*

		Size Class (Employees)		
	All Firms	21–100	101–500	500+
A) Conductors of ongoing R&D				
(i) product/process innovators	57	53	54	79
	(2)	(4)	(4)	(3)
B) Conductors of occasional R&D				
(i) Product/process innovators	39	42	39	49
	(2)	(3)	(4)	(5)

Notes: Larger firms (IPs) only. Standard errors are in parentheses.

the three years prior to the survey. Generally, the percentage of small and medium firms reporting innovations is lower than that for large firms.

When just those firms that conduct R&D occasionally are examined (panel B, Table 7.14), innovation rates are all below the comparable rates for those firms that conduct R&D on an ongoing basis (panel A, Table 7.14). But these comparisons should be avoided. Occasional R&D should not be expected to result in innovations as frequently as continuous R&D.

A fair comparison for the conductors of occasional R&D requires a longer time period over which innovation is measured so as to equate the R&D inputs across the two types of R&D performers. That is not available to us here. Despite this, it should be noted that there are few significant differences in the efficacy of occasional R&D among small, medium, and large firms.

Instead of examining whether R&D leads to innovation, one can investigate whether innovation is more closely tied to R&D in small or in large firms. Since we know smaller firms are more likely to obtain their ideas from sources other than R&D, it is possible that focusing just on firms that engage in R&D biases comparisons of efficiency in favour of large firms.

To investigate this issue, innovative firms were separated from non-innovative firms and the percentage within each group that conducted R&D was calculated (see Table 7.15). This ratio measures whether innovation only or primarily occurs where there are R&D facilities. Some 47% of innovators conduct R&D on an ongoing basis; only 17% of non-innovators do so. Some 88% of innovators conduct some form (ongoing or occasional) of R&D, but only 57% of non-innovators do so. R&D performance by itself does not guarantee successful innovations – especially over three-year time horizons. But success is almost invariably associated with some form of R&D performance.

For innovators, the percentage of firms that conduct R&D is about the same for small, medium, and large firms – although the extent to which the R&D was done regularly or only occasionally differs substantially across different size classes. Small innovators have about the same dependence on R&D; they differ from large firms in that they conduct their R&D only occasionally, rather than continuously. For non-innovators, smaller firms are slightly less likely to conduct R&D than middle- and large-size firms.

A comparison of the relative usage of R&D by size class between innovators and non-innovators reveals that R&D is slightly more likely to be connected to innovation in small firms than in large firms. For example,

Table 7.15. *R&D Intensity for Innovators and Non-innovators, by Size Class (% of Firms)*

	All Firms	Size Class (Employees)		
		21–100	101–500	500+
Innovators (reporting product or process innovation)				
Ongoing R&D	47	52	50	70
	(2)	(3)	(3)	(3)
Occasional R&D	44	43	42	29
	(2)	(3)	(3)	(3)
TOTAL	88	92	91	94
	(1)	(2)	(2)	(2)
Non-innovators				
Ongoing R&D	17	25	33	30
	(1)	(2)	(4)	(6)
Occasional R&D	40	41	48	49
	(1)	(3)	(4)	(7)
TOTAL	57	65	82	79
	(1)	(2)	(3)	(5)

Notes: Larger firms (IPs) only. Standard errors are in parentheses. Note that the population of large firms used in this table has a higher probability of doing R&D than the population as a whole that is used for Table 7.10.

the proportion of large innovative firms that conduct R&D at 94% is larger than for large non-innovative firms (79%) – a ratio of the former divided by the latter is 1.19 (Table 7.15). For small firms, this ratio increases to 1.41. This suggests that the appropriate model to describe the nexus between R&D and innovation differs for the two groups. For the group of large firms, a random chance model is more appropriate – a model where most large firms conduct R&D but where only some are successful in the three-year period being examined. On the other hand, the fact that innovation is more closely tied to R&D activities in smaller than in larger firms would suggest a sorting or heterogeneity model for R&D in smaller firms.[6] Many small firms do not conduct R&D, since they obtain their innovations from other sources. Those that do so are relatively successful in terms of their ability to produce innovations.

In summary, small and medium-sized firms may conduct less R&D than the largest firms, but when they do so, there is little evidence to suggest

---

[6] This difference is even more marked if we define innovation using the sales, rather than the incidence of innovation question.

that they are less efficient R&D performers. Moreover, it appears that innovation is more closely tied to R&D performance in small firms than in large. In larger firms, both innovators and non-innovators alike are conducting R&D. In smaller firms, innovators are more likely to be conducting R&D than are non-innovators.

There is, therefore, considerable heterogeneity in the small-firm segment. Some small firms develop an R&D capability to produce new products that will allow them to grow and to supplant existing firms. Other small firms draw on their customers and suppliers for innovative ideas – ideas that are the result of R&D done by these other firms. It is here that R&D spillovers occur.

7.6 IMPEDIMENTS ASSOCIATED WITH INNOVATION

Other research (Baldwin, Chandler et al., 1994; Baldwin and Johnson, 1999a) that uses special surveys of small and medium-sized firms confirms that R&D is associated with success in small firms. An R&D innovation strategy has a significant impact on a firm's success. It is associated with greater increases in market share and greater profitability. It is closely associated with producing successful innovations. These studies also provide information relevant for policy formation. The small and medium-sized firms in the surveys that are most successful make greater use of government programs in those areas that support broad framework policies, such as R&D and export programs. These are policies that complement private-sector success strategies. Firms that give greater stress to innovation are more likely to make use of these programs.

While R&D strategy is important, it is not the only route taken by small firms to achieve success. An R&D strategy is more important in manufacturing than in the service sectors where innovations are less connected with traditional investments in machinery and equipment and process technology. In the service sector, technology is important; but so too is the development of human capital (Baldwin and Johnson, 1996b). Policy that ignores the development of human capital ignores the importance of the contribution that investments in training make to those industries where most of the innovation capital possessed by a firm consists of human capital, rather than machinery and equipment.

Equally important, it should be recognized that small firms suffer major problems in areas where the existence of externalities has led governments to develop support policies. In a population that is heterogeneous, small firms generally suffer disadvantages in information collection, processing, and analysis. It is proficiency in these areas, as well as economies of scale associated with the production process, that have allowed large firms to grow and prosper. Large firms have developed methods to create and transmit tacit knowledge – through the establishment of R&D labs and the forging of ownership links across units. By way of contrast, small firms are more likely to rely on codifiable knowledge that is transmitted from customers and suppliers. As a result, they face quite different problems. Relatively greater deficiencies are commonly seen to exist for small firms with regards to information on market development (especially export markets) and on new technologies. Large firms have developed extensive information networks via intercompany relationships and via extensive links to both customers and suppliers. In contrast, small firms focus relatively more on links with unrelated parties, which tend to be capable of transmitting easily codifiable information. They have a less developed system of acquiring the type of tacit knowledge that large firms transmit from sister firm to sister firm within the same organization. As a result, small firms are also seen to suffer relatively greater problems when it comes to developing interfirm cooperative agreements.

These deficiencies are confirmed by the emphasis given to different types of impediments investigated by the 1993 Canadian Survey of Innovation. Training problems are cited most frequently as impediments by firms of all sizes (see Table 7.16); but there is very little difference in the severity of this problem across size classes. Human resources provide the most critical source of imbedded information for innovation, and firms of all sizes believe that lack of skilled personnel is the most important problem that they face. For most of the other problems, the smaller firms more frequently indicate that a particular impediment is a problem than do large firms. Both a lack of information on technologies and a lack of technical services are seen to be greater problems by micro-, small-, and medium-sized firms than by large firms. Firms in the three smaller groups also perceive that interfirm cooperation is a problem more frequently than do large firms, who benefit from intrafirm cooperative agreements. This should be set in the context of the lower frequency of interfirm R&D collaboration of smaller firms (Table 7.14). The differences in problems associated with information acquisition are generally greater for technology and technical services than for market information – except in the case of the very smallest microfirms.

Innovation involves changes in both products and processes. One of the greatest differences between the innovative behaviour of small and large

Table 7.16. *Impediments to Innovation, by Size Class (% of Firms)*

	Size Class (Employees)				
Category	All	0–20	21–100	101–500	500+
Lack of skilled personnel	46	44	51	47	43
	(2)	(3)	(4)	(4)	(4)
Lack of information on technologies	30	31	31	33	21
	(2)	(3)	(3)	(4)	(3)
Lack of information on markets	39	43	30	31	30
	(2)	(3)	(3)	(4)	(4)
Lack of external technical services	21	21	23	14	13
	(2)	(3)	(3)	(3)	(3)
Barriers to interfirm cooperation	20	21	19	15	6
	(2)	(3)	(3)	(3)	(2)
Barriers to university cooperation	8	9	6	5	7
	(1)	(2)	(2)	(2)	(2)
Government standards	32	34	31	21	31
	(2)	(3)	(3)	(4)	(4)

Note: Standard errors are in parentheses.

firms lies in the degree to which small firms lag behind larger firms in advanced manufacturing-process technology like CAD engineering systems, robots, flexible manufacturing systems, LAN systems, and automated material handlings (Baldwin and Sabourin, 1995). This process technology is an important key to success (Baldwin, Diverty, and Sabourin, 1995). The fact that small firms perceive that they suffer information gaps in the technology area and that it is an important factor in conditioning success in small firms suggests that one of the highest payoffs for those government policies that focus on externalities and additionality lies in this area.

7.7 CONCLUSION

The debate over the appropriate function of government policy for R&D subsidies brings into focus the different roles that are played by large and small firms in the innovation process. Small firms, it is often claimed, have different tendencies to use R&D facilities than do large firms and, therefore, are less innovative.

The chapter demonstrates that, in Canada, small firms are less likely than large firms to introduce new products and processes. Similar differences are observed between large and small firms with regards to the frequency of R&D activity. Although there are fewer innovators in the smaller size classes, those smaller firms that do innovate resemble larger innovators in several dimensions. There are few differences in the number of product innovations per innovative firm across size classes. In addition, major product innovations in small innovative firms account for just as large a percentage of their sales as in larger innovative firms. There is, nevertheless, a tendency for smaller firms to produce fewer innovations that are extremely novel.

Large and small innovators do, however, rely on different sources of ideas for innovation. Ideas for innovation come from a number of sources – R&D, customers, the marketing or sales department, the production division, or suppliers. Large firms use R&D much more frequently that do small firms. Large firms are also more likely to rely on an external research network, through a relationship with a related firm, or to engage in collaborative research with other firms. On the other hand, small firms rely relatively less on an R&D department and relatively more on the technical capabilities of their production departments than do large firms. Small firms also depend on networks – but these networks rely more heavily on customers and their marketing departments for innovations.

While small firms rely less on the R&D department and more on linkages to other unrelated firms, the differences for the percentage of firms in a size class that conducts any form of R&D – on a continuous or an occasional basis – are not large. In contrast, small firms are much less likely to conduct R&D on a continuous basis. Small firms are significantly less likely to set up a separate R&D unit and much less likely to take advantage of R&D tax subsidies. There is, therefore, more of a difference in the way that R&D is conducted in small firms than whether it is conducted at all. Large firms have regularized the R&D process in order to shape their environment, whereas small firms use it to exploit opportunities in their environment when the need arises. Many of these opportunities are brought to their attention by their customers or suppliers. Small firms exhibit the same flexibility in their R&D that they show in many of their other operations. These differences suggest that large firms, as a whole, have mastered the need to create and acquire noncodifiable information that is difficult to incorporate into the firm. Small firms, as a whole, utilize information sources that are more easily transmitted through supplier relationships.

These differences in the sources of ideas, when taken together, suggest that the information and transmission processes vary considerably across size classes. Large firms rely more on tacit, noncodifiable information that is developed within the firm, either in their own R&D laboratories or in the R&D laboratories of related firms. Some smaller firms do the same, but this group as a whole does not do so with the same frequency. The rest of the small-firm group relies on information that is more easily codifiable and transmitted from customers and suppliers. The information networks of small firms differs from those of larger firms.

Differences in the intensity of innovation across size classes should not be taken to imply that small firms are relatively less important innovators. For one thing, intensity should not be confused with volume of innovation. While small firms may be less intensive innovators, their numerical importance is so much greater than large firms that small firms account for a larger proportion of innovations.

In the same vein, differences in intensity of innovation and R&D activity by themselves do not imply that small firms are less efficient innovators than are large firms. When the efficacy of the R&D process is compared across small and large firms, few differences emerge. In those firms that conduct R&D, the percentage that report an innovation is about the same for small, medium, and large firms. There is a higher success rate for those with an ongoing R&D operation than for those conducting occasional R&D, irrespective of the size class; but this is partly a statistical phenomenon that is likely related to differences in the type of R&D performed. Smaller firms tend to be more likely to conduct occasional R&D than continuous R&D, which causes small firms to look less productive if all those who conduct any form of R&D are aggregated together.

An examination of the extent to which innovation comes only from R&D shows that innovation is more closely associated with R&D in small firms than in large firms. In large firms, both innovators and non-innovators alike are conducting R&D. In small firms, product innovators are more likely to be conducting R&D, but non-innovators are less likely to be conducting R&D. More large firms are doing R&D, and these firms sort into those that are successful and not successful conductors of R&D – where success is assessed as the production of an innovation. On the other hand, smaller firms sort more cleanly into those doing R&D who are successful innovators and those who can survive by finding their innovations elsewhere.

Together, these comparisons suggest that attempts to assess the relative efficacy of innovation activity in small and large firms are misplaced because of the heterogeneity of the small-firm population. Small firms can be divided into two groups. The first group consists of firms that resemble large firms in that they perform R&D and generate new products and processes, primarily through their own efforts. The second are those who rely upon customers and suppliers for their sources of ideas for innovation. Large firms, by way of contrast, tend to rely more heavily on R&D. While they too rely on networks for ideas, their networks focus more heavily on relationships with other sister firms.

Evidence shows that despite differences in R&D intensity, the success of small firms depends critically on their innovation capabilities – especially in the areas of R&D. Small firms also benefit from the R&D done in large firms because a larger proportion of their innovations are the result of liaisons with their customers. All of this means that subsidies for R&D directly aid the most dynamic small firms that are conducting R&D and also aid this group indirectly because of the spillovers from large firms to small firms.

Despite the importance of R&D for innovation, there are other areas where public policy makers can develop special policies for small firms. These are areas where small firms have indicated that they face problems. Small firms perceive that externalities are relatively important in the area of information on technologies, on markets, and on technical services. They also perceive that there are significant barriers to interfirm cooperation. This is a particularly serious problem since this is the method that they use most frequently for developing new ideas for innovation.

EIGHT

Innovation Regimes and Type of Innovation

8.1 INTRODUCTION

Measuring innovative output is difficult since innovations can differ in many different dimensions. More importantly, economic growth results from innovations of different types. Economic growth clearly results from pathbreaking innovations, such as the computer, lasers, or new chemical entities. But innovations that are not at the frontier can also substantially contribute to growth. Economic growth can result when a firm introduces new products or processes that have already been adopted in other countries but that need to be adapted to special national circumstances. It can occur when firms adopt processes from other industries. Both of these imitative types of innovations involve substantial novelty. Finally, economic growth occurs from more imitative innovations – when firms succeed in improving the products of those who pioneered a new product or process but who could not develop it quickly enough. Economic progress is also enhanced by competitive pressures that ensure that good ideas become good commercial products or processes. Sometimes these changes may simply modify the product to better satisfy consumer demands. At other times, they may involve superior production processes that lower production costs and reduce consumer prices.

Innovation is, therefore, best thought of as a continuous process, whose characteristics often change over the length of the product life cycle. Gort and Klepper (1982) delineated four phases of the product life cycle. In the first, new products emerge and a small number of firms work to develop a commercial product. In the second stage, there is rapid entry and exit. At this stage, a large number of firms experiment with new product and process designs. Entry is relatively easy but so too is failure. In the third stage, entry slows, exit continues at high levels, and firms grow larger. It is at this stage that firms begin to move down their cost curve as process innovations allow the successful to exploit scale and scope economies. The final stage of the product life cycle is characterized by a fairly stable market structure, with both entry and exit now reduced to lower levels and where oligopolistic competition prevails.

Innovation is important in all phases of the product life cycle. Yet the development of the entirely new product probably only dominates the first stage. Throughout the other stages, improvements occur as firms build on a body of common knowledge or on the work that others have already commenced. In this world, much progress is incremental and few innovations are completely original. Nevertheless, when taken together, the incremental changes have a significant effect.[1]

Our research must not fall into the trap that studies of R&D once did. Much is now made of the fact that innovation was once regarded as a linear process that stretched from R&D labs to final commercialization of products. Mowery and Rosenberg (1989) have argued that the R&D division often was not the discoverer of new products; rather, innovations were often discovered by technologists or by engineering staff, and the research and development staff were given the primary responsibility for developing an understanding of the underlying science, and then the engineering production staff had to solve the problem of producing the new material in large quantities. In this system of innovation, knowledge creation and technological progress are interwoven in different divisions of the firm.

The same is true at the industry level. Some innovations come from new firms. But the failure rate of new firms is large. The new firms are supplanted by others. Those that survive and grow often combine with others who have skills which the newcomers lack. This may be done either via a takeover that allows related firms to share knowledge or through collaborative ventures with other parties that require the two firms to coordinate their objectives. As one firm supplants another in this process, innovations build on other innovations. There are few innovations that are truly 'original' or 'first' or 'unique'. Or more importantly, even when there is such an entity, its study would tell us little about nine-tenths of the innovation process.

---

[1] Hollander (1965) details that much of Dupont's productivity growth came from incremental improvements in production technology.

8.2 THE DEFINITION OF INNOVATION

Innovation studies, then, need to come to grips with the variety of innovations that take place. We can do so with the 1993 Canadian Survey of Innovation, which focused both on a firm's innovative capacities and the characteristics of the firm's major innovation.

This allows us to classify the characteristics of the innovators by the importance of their most important innovation. Some innovations are pathbreaking – like lasers and transistors. Others involve only incremental improvements, but may have a significant impact when cumulated over a long period. It is important not to restrict our interest to just the former, since much of economic well-being stems from the smaller incremental types of innovations. Therefore, firms who reported that they had at least one major innovation are divided into three groups – depending upon whether their most important innovation was a world-first, the first of its kind in Canada, or neither of these two.

Then, the characteristics of the innovation regime are documented separately for these groups. This allows us to investigate whether the innovation regime differs by the nature of the innovation. The innovation regime is defined by the effects and benefits that arise from the innovation, the source of ideas for both the innovation and new technologies, the intensity and type of the R&D process associated with the innovation process, the impediments to innovation that firms faced, the importance of intellectual property rights to firms as a means of protecting their innovation, and the effect of innovation on the demand for skilled workers. Each of these characteristics of the innovation process is investigated for the most novel of the innovation types – world-firsts – and for other types. When warranted, the latter category is broken into two subgroups – Canada-firsts and other innovations.

8.3 THE DATA

The data presented in this chapter are based on the population of the larger firms (IPs)[2] which indicated that they either had or had not introduced an innovation. While we refer to these as large firms, they vary in size from about 20 to over 500 employees. They exclude the set of quite small firms that are generally fewer than 20 employees and for whom Statistics Canada does not maintain a profile on their central list of companies that is used for developing survey frames.

---

[2] These are generally greater than 20 employees.

Much of the detail reported here is based on the subset of 573 firms that indicated they were innovative and that provided details about their most important innovation. Firms were asked for information on their most important innovation that was introduced during the period from 1989 to 1993. Their most important innovation was defined as the one having the largest impact on their profits. The degree of novelty was established by having firms categorize their most important innovation as a world-first, a Canada-first, or 'other'. In the following sections, reported data are employment weighted.[3] Thus, the proportion of firms that are shown to have a given characteristic (for example, those who have process innovations) represents the proportion of total employment accounted for by large manufacturing firms with that characteristic (i.e., a process innovation).

In this chapter, we report employment-weighted percentages rather than the company-weighted percentages that we generally use in other chapters. The employment-weighted percentages tell us the percentage of total employment that falls in a particular category. For example, the percentage of firms reporting a world-first innovation is only 5%, but 13% of total manufacturing employment was in firms doing so. We report the employment-weighted percentages here because innovation is a large-firm phenomenon and we wish to examine the relative importance of different types of innovations; therefore, it is particularly appropriate to take into account differences across size classes here.

8.4 HOW PREVALENT IS INNOVATION?

Innovation involves the successful commercialization of an invention that enables firms to produce new goods or services, to improve on existing ones, or to improve the way existing or new products are produced or distributed.

Canadian manufacturing firms are intensively involved in the innovation process. Some 39% (company weighted)[4] of all the large (IPs) firms either introduced an innovation over the period of 1989 to 1991, or were in the process of introducing an innovation in 1992–1993.[5] These innovative firms accounted for 57% of employment (employment weighted).[6]

---

[3] The employment used for this purpose is the consolidated manufacturing employment of the business entity that controls the firm answering the survey.
[4] A company-weighted result is the percentage of firms in the population that possess a given characteristic.
[5] If we had used the sample of all firms – both IPs and NIPs – the innovation rate is 33%.
[6] An employment-weighted result reveals the percentage of total employees accounted for by firms in the population that possesses a given characteristic.

How Prevalent Is Innovation?

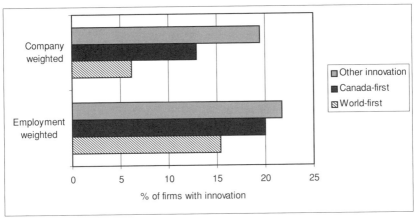

Figure 8.1. Innovation Intensity, by Novelty Type

Innovations cover a range of new products and processes that differ in terms of novelty. Of all larger firms, 15% (weighted by employment) introduced a world-first innovation (see Figure 8.1). These firms were able to introduce innovations that involved the use of new advanced technologies, or the development of distinct products, or some combination of the two that was highly unique. Some 20% of all large firms indicated they introduced an innovation that was a first within Canada. The remainder (22%) produced innovations that were imitative.

If we consider just the innovating population, some 27% of *innovators* (weighted by employment) produced world-firsts, some 35% produced Canada-firsts, and the remainder (38%) introduced innovations that were just new to themselves but that had already been introduced elsewhere in Canada.

Although the data presented are for the larger, more complex firms accounting for the majority of economic production in the Canadian manufacturing sector, these firms vary considerably in terms of size. In order to examine whether or not the intensity of innovation differs across size classes, the large firms were divided into three employment-size categories: firms with fewer than 100 employees, firms with between 100 and 500 employees, and firms with more than 500 employees.

The overall rate of innovation of the largest size classes is above that of firms between 100 and 500 employees; moreover, the larger of the two size classes has a higher rate of world-first innovations and Canada-first innovations (see Table 8.1). Firms in the smallest size class are least likely to introduce innovations of this nature. Imitative innovations are

Table 8.1. *Innovation Intensity, by Novelty of Innovation and Size Class (% of Firms, Employment Weighted)*

Innovation Type	Size Class (Employees)		
	21–100	101–500	500+
World-first	5 (1)	8 (1)	20 (2)
Canada-first	14 (1)	14 (2)	24 (2)
Other	22 (1)	27 (2)	21 (2)

Notes: Larger firms (IPs) only. These are mainly firms with more then 20 employees. Standard errors are in parentheses.

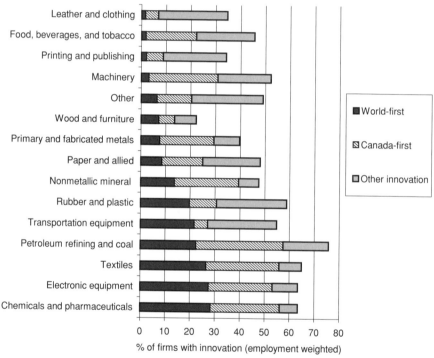

Figure 8.2. Innovation Intensity, by Individual Industry

introduced at approximately the same rate by firms in the two smaller groups and are less likely to be introduced by firms in the largest group.

Substantial differences in the intensity of innovation exist across industries. In Figure 8.2, industries are ranked in descending order according to the percentage of firms that introduced a world-first innovation.

The leading industries are electronic products and chemicals, with 28% of total industry employment belonging to firms introducing a world-first innovation. Also at the forefront is petroleum refining. Industries from the secondary group that lead overall are transportation equipment and rubber, as well as nonmetallic mineral products. Textiles, an industry in the tertiary group that is generally less innovative, comes second here with 27% of employment in firms that introduced world-first innovations. Food and beverages, printing and publishing, and leather and clothing have the lowest intensities, with 8% to 12% of employment belonging to firms introducing a world-first.

If industries had instead been ranked according to the amount of innovations introduced, regardless of the significance of the innovation, machinery moves up substantially. This industry may not produce many world-first innovations, but it is actively involved in introducing innovations. This is an industry where process innovation is important and where incremental change that is rarely a world-first is continuously taking place.

In subsequent sections of this chapter, firms will generally be divided into one of two categories – those that produce world-first innovations and all other innovators – because the main differences of interest occur between these two groups. Differences between these categories will be used to show how characteristics of innovative firms vary by the significance of the innovation produced. Although the innovative behaviour of firms will vary depending on the size of their workforces, this study will focus on the differences between world-first innovators and all other innovators, regardless of firm employment size. However, by using employment weights, recognition is given to the fact that large firms tend to be more innovative.

8.5 TYPES OF INNOVATION

The innovation systems of countries differ in their tendency to concentrate on product innovations in contrast to process innovations. Of interest here is the extent to which the novelty of a firm's most important innovation affects the likelihood that it concentrates on products rather than processes. Theories of international trade (Vernon, 1966) have posited that some countries may be superior to others in developing new products. Some countries, then, will be seen as specializing in product innovations because they are comparatively well endowed with those inputs (related to R&D) that are used at the leading edge of the product cycle. Imitators are seen to specialize in process innovations that result from their being comparatively well endowed with those factors that allow them to produce mature products at low cost. The same dichotomy may also apply to firms. This section, then, asks whether non-world-first innovations rely on process innovations relatively more than world-firsts.

It must be recognized that the division of innovations into products versus processes involves a distinction that is difficult to make in practice. A product innovation is the commercial adoption of a new product.[7] Product innovations may be accompanied by technological change when the underlying manufacturing processes are modified in order to produce the innovative product. These modifications often, though not exclusively, involve quite radical change and therefore qualify as process innovation. We know, for instance, that over 50% of advanced fabrication technologies – flexible manufacturing systems, robots – that are investigated in the SIAT were introduced in order to produce new products. Thus, in the case where a product innovation involves a change in technology, the innovation is referred to here as a combined product-process innovation.

A process innovation is the adoption of new or significantly improved production methods. These methods may involve changes in equipment or production organization or both. They may be intended to produce new or improved products, which cannot be produced using conventional plants or production methods, or to increase the production efficiency of existing methods.

Firms are classified according to whether they described their innovations as either process, product, or combined product-process innovations (see Figure 8.3). Some firms introduced several different innovations and, therefore, chose two or in some cases all three of the innovation types as being descriptive of their innovative behaviour.

Overall, Canadian innovative firms favour pure process over pure product innovations. Some 65% of innovators introduce purely process innovations, compared to only 46% with purely product innovations. However, a large percentage (57%) introduce combined product-process innovations. The innovation process does not split neatly into two groups because innovation on both the product and process side go hand in hand in such a large percentage of the population. The importance of

---

[7] Changes in products that are purely aesthetic (such as changes in color or decorations), or which simply involve minor design or presentation alterations to a product while leaving it technically unchanged in construction or performance, are not considered to be product innovations.

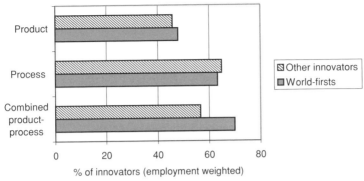

Figure 8.3. Product-Process Innovations, by Novelty Type

technological competence to the innovation process is emphasized by the fact that pure process innovations are more important than pure product innovations, and process innovations are at work in at least 80% of the innovating population (i.e., 80% of firms introduced process or combined product-process innovations).

World-first innovators are most likely to introduce combined product-process innovations, with process-only innovations following closely behind. On the other hand, 'other' innovators are more likely to concentrate on pure process innovations. The concentration on some form of process innovation, either by itself or as part of a product innovation, is evident in both innovative-firm segments.

A higher percentage of world-first than non-world-first innovators pursue the two innovation types that involve product innovations (the percentages can sum to more than 100% since a firm can report more than one type). The largest and only significant difference between the two segments occurs for the combined product-process category. Some 70% of world-first innovators introduce combined innovations, while only 56% of other innovators do so. World-first innovators are different in that they emphasize changes both to manufacturing technologies and to end products produced. It is not so much the emphasis on products or processes that distinguishes the world leaders from the followers as it is the ability of the leaders to master both product conception and new production processes and to produce more than one type of innovation.

This result emphasizes the importance of both world-first and non-world-first innovations. As was stressed earlier, follower-type innovations are extremely important in contributing to economic progress. While they are not the very first innovations on the market, they are not simply duplicates of the innovations of others. These innovations involve new product improvements *or* new process changes almost as frequently as world-firsts. They tend to involve process innovation slightly more frequently than do world-firsts.

8.6 FEATURES OF INNOVATION

While the world-first innovators differ from the other group primarily in the stress that they place on innovations that involve both products and processes, there still may be a difference of emphasis that each places on the type of new products and processes produced. In order to clarify this issue, we examined specific features of both product and process innovations in more depth.

Product innovations differ in a number of dimensions. On the one hand, they may be completely new products in that they satisfy fundamentally new functions. On the other hand, they may satisfy the same basic functions as existing products, but they may allow improved performance at an equivalent or lower cost. Quality improvements may come about through the use of higher performance components or materials, or the development of a more complex product that consists of a number of integrated technical subsystems.

Process innovations also take different forms. On the one hand, they may involve completely new production processes that are based on radically new production technologies or on changes in the organization of the production process. On the other hand, they may simply involve increases in the extent to which the production line is automated. The latter occurs when existing capital equipment is used in greater quantities, but in traditional ways.

Both world-first and other innovators note that developing new production techniques is one of the most important facets of their innovative behaviour (see Figure 8.4). It is cited by 64% of world-first innovators and 55% of other innovators. Most of the innovation effort by both groups of firms is geared towards developing new manufacturing (production) techniques.

Some 37% of non-world-first innovators indicate that increased automation is important, while significantly less (25%) of world-first innovators list this feature. Automation involves increased mechanization and is less radical than completely new production techniques. Non-world-first innovators, therefore, are more likely to focus on the type of incremental improvements in process innovations that are associated with increasing

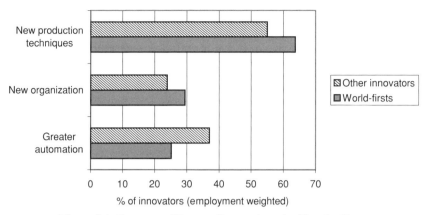

Figure 8.4. Features of Process Innovations, by Novelty Type

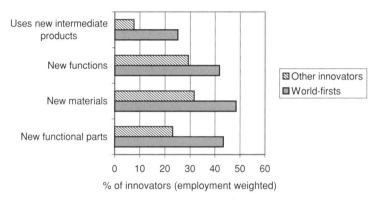

Figure 8.5. Features of Product Innovations, by Novelty Type

the amount of automation, and world-firsts are significantly more likely to stress the development of new production technologies.

World-first innovators incorporate a number of new product features into their innovations more often than do other innovators (see Figure 8.5). The use of new parts, the inclusion of new materials in end products, and the development of products with new functions are all named by between 42% and 48% of world-first innovators. By way of contrast, significantly fewer non-world-first innovators list these as characteristics associated with their product innovations. The most important category for this group is the inclusion of new materials in innovative products.

In summary, firms that produce world-first innovations differ most from firms producing other types of innovations with regard to the emphasis that they place on developing new product features. A change in process technology often accompanies product change and, therefore, world-firsts also place greater stress on production techniques – but the difference is less than for product features. This confirms the major difference between the two classes of innovations described previously. World-first innovations occur in firms that take a two-pronged approach to innovation in that they stress both process and product change.

8.7 THE BENEFITS OF INNOVATION

Innovations are often seen to be the key to a firm's success. However, not all innovations are the same. World-first innovators are more likely to focus on new product characteristics, and their innovations are more likely to involve new production techniques that coincide with the introduction of new products. Thus, innovations of different types might be expected to have quite different effects on each group of firms.[8] This does not occur as often as might be expected.

Product innovations may change the nature of existing products either by improving their quality or by extending the product line that a firm offers. Process innovations may reduce lead times when they allow design, development, and production to be compacted into a shorter time period. They may increase the technological prowess of a firm when the innovation involves changes in the production process. Finally, they may influence the quality of work by improving working conditions.

Both world-first and other innovators experience these changes frequently (see Table 8.2). Indeed, other innovators are slightly more likely to have experienced many of these effects than are world-first innovators. Some 57% of world-first innovators report an improvement in product quality. Even more Canada-firsts (66%) do the same. The other group of innovators (60%) indicating an improvement in product quality fall slightly above world-firsts but behind Canada-firsts in this area.

Some 9% of world-firsts report reductions in lead time, while 32% of Canada-firsts and 28% of all other innovators do the same.

Other innovators are slightly more likely than either world-firsts or Canada-firsts to report that they have improved working conditions and that they have extended their product line.

---

[8] Contrary to the divisions in other sections of this chapter, non-world-firsts are divided into Canada-firsts and other innovations because of the intrinsic interest in demonstrating whether there are major differences between them and, therefore, whether their aggregation disguises important differences.

Table 8.2. *Effects of Innovation, by Novelty of Innovation (% of Innovators, Employment Weighted)*

Effect	Innovator Types		
	World-First	Canada-First	Other
Improved product quality	57 (5)	66 (4)	56 (3)
Improved technological capabilities	56 (5)	62 (4)	52 (3)
Improved working conditions	13 (4)	24 (3)	26 (3)
Reduced lead times	9 (3)	32 (4)	28 (3)
Extended product range	54 (5)	61 (4)	68 (3)

Notes: Larger firms (IPs) Only. Standard errors are in parentheses.

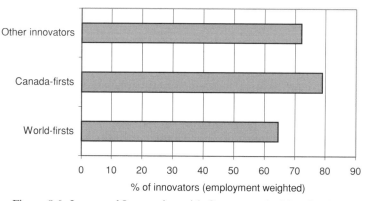

Figure 8.6. Improved Interaction with Customers, by Novelty Type

Changes in both the quality and diversity of products will affect relations with customers. All three classes of innovators indicate that improved interactions with their customers were attributable to the innovations they introduced (see Figure 8.6). Some 65% of world-firsts report this benefit, while 79% of Canada-firsts and 72% of other innovators experience the same benefits.

Finally, the benefits of innovation extend to both gains in output and in profitability. In the results of a separate study that linked administrative data on firms' performance (market share and profitability) to their response to questions designed to gauge their innovativeness, firms that adopted a more innovative strategy by developing new products and processes were found to be more successful than the less innovative firms (Baldwin, Chandler, et al., 1994). The more innovative firms increased

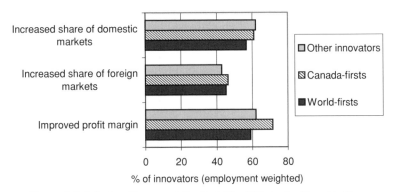

Figure 8.7. Market-Share and Profitability Effects, by Novelty Type

their market share and their profitability relative to the less innovative firms.

The results of this survey confirm the earlier findings. Innovation is associated with increases both in market share and profitability (see Figure 8.7). Some 57% of world-first innovators indicate that their innovation serves to increase their share of the domestic market, 45% increase their share of foreign markets, and 59% increase their profit margins as a result of their major innovation. All innovators, whether they be Canada-firsts, world-firsts, or others have similar experiences. The other group reports slightly more frequently than world-firsts that it increases domestic market share and improves profit margins, but these differences are not statistically significant. They report just as frequently that they increase foreign market share. Canada-firsts do slightly better than both other groups when it comes to increasing their profit margins – differences that are statistically significant.

While firms that produce all three types of innovation report increasing their share of foreign markets as a result of innovation, the effect may differ across the various types of innovations. More original innovations have a better chance of penetrating export markets. One measure of success is the percent of the sales due to an innovation that is exported. Firms with world-first innovations report the highest percentage of export sales at 37% (see Table 8.3).[9] Canada-firsts come close behind at 34%, thereby indicating that this group is really indistinguishable from the most novel as far as export tendency is concerned. Finally, the export-to-sales ratio for the most imitative category is lowest at only 27%. Nevertheless, it is not zero. Even here, innovation has an effect on exports.

---

[9] These are the weighted averages of export/sales ratios for those firms reporting exports.

Table 8.3. *Exports as a Percent of Sales, by Novelty of Innovation (Employment Weighted)*

Type	All
All	29 (2)
World-first	37 (6)
Canada-first	34 (4)
Other	27 (3)

Notes: Larger firms (IPs) only. Standard errors are in parentheses.

In summary, while innovations may differ in terms of their originality, each group of innovators reports benefits with about the same frequency. These results are significant since they confirm the picture drawn in the introduction regarding the importance of the different types of innovation. At each stage in the life cycle of the product, innovation serves to enhance product performance. The results presented in this section also show that innovation at each stage also serves to enhance firm performance. In one sense, this is not surprising. Firms, on average, will engage in activities that are beneficial to themselves. This section shows that each type of innovation increases the profitability and the market share of the innovators. Since the results confirm that each type of innovation has similar benefits (though each is accomplished slightly differently), they emphasize the need to avoid focusing on one type of innovation exclusively and to better understand the entire innovation system.

8.8 INTERNAL SOURCES OF INNOVATION IDEAS

Innovation ideas originate from sources both within and outside the firm. The main sources of ideas within firms include research and development units, sales and marketing staffs, management, and workers in production areas. Research and development units have traditionally received the greatest emphasis as a source of innovation ideas, though increasingly the importance of other sources, when used in conjunction with R&D, has been recognized.

The most advanced type of innovations are often regarded as requiring a support system involving laboratories that perform fundamental research. There are, however, other important sources of innovation within a firm. Technological capabilities in modern corporations owe as much to engineering capabilities as to pure science. Indeed, it has been argued that the prowess of North American firms owes more to the development of engineering capacities than to the adoption of formal science laboratories (Rosenberg and Nelson, 1994).

Because of the importance of R&D to basic knowledge and the importance of the latter in creating brand new products, R&D is expected to be a critical input for world-firsts. On the other hand, it might be expected that non-world-first innovations make greater use of the engineering capabilities within the firm. Such firms are adapting, reengineering, and improving existing ideas. These skills are more likely to come from the production or engineering departments or from other areas of the firm than from the research division.

Research and development efforts are, by far, the most important source of information used by world-first innovators, with 88% of firms relying on this source to facilitate innovation (see Figure 8.8). In fact, no other internal source of ideas is used by more than 38% of world-first innovators. Research and development efforts, thus, are paramount to the successful introduction of world-first innovations.

Research is also important to non-world-first innovators. These firms also rely heavily on research and development efforts (55%); but they rely significantly more heavily on ideas from management (48%) than do world-firsts. Ideas from production areas and from sales and marketing staffs are also used by approximately 37% of these firms. Overall, non world-first innovators differ from their world-first counterparts in that they are more apt to rely on multiple internal idea sources. They do some research, but they tend to focus relatively more on the type of

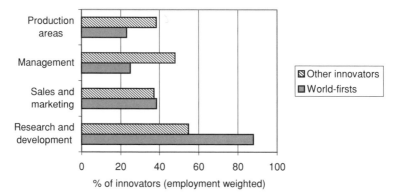

Figure 8.8. Internal Ideas for Innovation, by Novelty Type

engineering skills that lead to product improvements, rather than product breakthroughs.

In summary, world-first innovators were, as expected, relatively more likely to rely on research and development. Other innovators do not ignore this source, but they bring into play more alternative sources – in particular, the applied technology sources that are so important for process development work.

8.9 EXTERNAL SOURCES OF INNOVATION IDEAS

While considerable emphasis has traditionally been given to such internal sources of innovation ideas as the research and development department, it is recognized that external networks are also essential to the innovation process. These sources help to diffuse knowledge about new products and processes from one firm to another. Some of this is done between parties that are linked as part of the product chain – through affiliates, for example. Some firms provide information to others within their vertical product chain in an attempt to improve their own competitive position. Firms can benefit from suggestions from their customers since the latter are often looking for improved inputs that will allow them to better compete with other firms.

Many external sources for ideas are available to innovators. Some of these are complements to firms' internal research and development units – consultants, private R&D institutions, and government development agencies. Each provides outside sources of R&D. Other external sources include suppliers, customers, and related firms. Customers and suppliers aid one another because they have a symbiotic relationship. Related firms, especially within multinational families, provide an efficient conduit for the transfer of proprietary knowledge from one firm to another. Trade fairs and conferences allow suppliers of technologies to provide information about these technologies to potential customers. Finally, competitors are monitored for new ideas and new products, which are then copied by using such techniques as reverse engineering.

Many of the external sources provide information and ideas for innovation free of charge, or at a cost lower than the value of information to the innovating firm. The extent to which innovative firms use these 'technological spillovers' (Griliches, 1979) varies by the type of innovation. Ideas for the most original, world-first innovations are harder to come by outside of the firm. Therefore, the world-first innovations draw on ideas from external sources less frequently than do less original innovations.

Table 8.4. *External Sources of Ideas for Innovation, by Novelty of Innovation (% of Innovators, Employment Weighted)*

Source	Innovator Types	
	World-First	Canada-First and Other
Suppliers	33 (5)	30 (2)
Customers	32 (5)	44 (2)
Related firms	26 (4)	22 (2)
Consultants	22 (4)	14 (2)
Private R&D institutions	14 (3)	6 (1)
Government development agencies	14 (3)	1 (1)
Competitors	4 (2)	32 (2)
Trade fairs/conferences/meetings	4 (2)	16 (2)

Notes: Larger firms (IPs) only. Standard errors are in parentheses.

Nevertheless, the external sources that are used frequently by both world-first and other innovators are the same – suppliers, customers, and related firms (see Table 8.4). These three sources are used by between 22% and 44% of innovative firms. That the non-world-first innovators would rely heavily on outside parties might be expected since this group of innovators does not possess internal R&D facilities. But the fact that even the world-first innovators frequently use these sources indicates the importance of the relationship that develops among all types of innovative firms and their customers and suppliers.

Non-world-first innovators do use certain sources more frequently than do world-firsts. The largest such difference occurs for ideas obtained from competitors. About 32% of non-world-first innovators use some form of reverse engineering for their innovation ideas while only 4% of world-firsts do so. Trade fairs and conferences are also used more frequently by non-world-firsts.

There are several areas in which world-firsts rely more on outside sources of information for innovation. The world-first innovators more frequently use information from public and private R&D institutions, the patent office, universities, and colleges than do their less original counterparts. They also rely more heavily on related firms. External idea sources that are tapped into more frequently by world-first innovators include those which complement their own internal R&D sources. Public institutions that provide technological infrastructure (Tassey, 1991; Lundvall, 1992) play a significant, even though quantitatively limited, role in the creation of the most original world-first innovations. World-first innovators rely somewhat more heavily on internal R&D and on R&D provided by external sources; they rely less on customers. However, like non-world-first innovators, they combine ideas from a variety of sources to generate innovations.

8.10 THE IMPORTANCE OF R&D ACTIVITY

Research and development activity is an important source of ideas for innovation in the case of both world-first and other innovators. Both sets of firms use experimental development as a critical internal source of information for new technologies. World-firsts place a heavy emphasis on the research component in addition to experimental development for this purpose.

This difference in emphasis is accompanied by a difference in the frequency with which the two types of innovators conduct R&D. Some 98% of the world-first innovators perform R&D on either an ongoing or an occasional basis. Some 95% of the other group of innovators perform some R&D.

R&D may be performed continuously or only occasionally (see Table 8.5). Firms that perform R&D continuously have a stronger commitment to innovation. Of world-first innovators, 84% perform R&D on a continuous basis, while less (57%) of the other group of innovators do so. By way of contrast, only 16% of world-first innovators perform R&D on an occasional basis, while 42% of other innovators do so. Thus, world-firsts have a greater commitment to the R&D process.

Research and development activity may be pursued in a variety of ways. It can be done through separate R&D facilities, or in other departments

Table 8.5. *R&D Use, by Novelty of Innovation (% of Innovators, Employment Weighted)*

	Innovator Types	
Source	World-First	Canada-First and Other
R&D performed on ongoing basis	84 (4)	57 (2)
R&D performed occasionally	16 (4)	42 (2)
R&D contracted out	36 (5)	37 (2)
R&D done in other departments	53 (5)	53 (2)
Firm has a separate R&D unit	72 (5)	53 (2)

Notes: Larger firms (IPs) only. Standard errors are in parentheses.

of the firm, or via contracts with other firms. Some firms will use a combination of these methods.

All three methods are important for those world-first innovators that perform R&D (Table 8.5). Some 72% have a separate R&D department, 53% perform R&D in other departments, and 36% contract R&D out to other firms (firms may use more than one source at any given time). World-first innovators use separate R&D units significantly more frequently than do non-world-first innovators. While 72% of world-first innovators conduct at least some of their R&D through such dedicated units, only 53% of non-world-first innovators do so.

In summary, research and development is important to both groups of innovators, although both the commitment to continuous R&D and its organization in a separate R&D department suggest a very different type of R&D agenda in world-first innovators. This confirms the differences in emphasis that were previously described with respect to the sources of ideas for innovations.

8.11 INTERNAL SOURCES FOR NEW TECHNOLOGIES

An ingredient that is essential to most process innovations is the development of new technologies. New technologies involve the use of new production equipment, new production techniques, and new organizational structures. As was the case for sources of innovation ideas, firms make use of both internal and external sources for the development or application of these technologies.

There are three main internal technology sources available: the research portion of R&D, the development portion of R&D (experimental

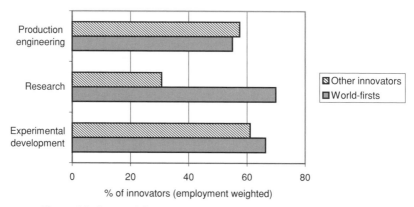

Figure 8.9. Internal Sources of Technology Ideas, by Novelty Type

development), and the production engineering process. With the emphasis of world-firsts on extremely novel products and processes, it might be expected that their technology sources would come more frequently from the research side of R&D. This is the case (see Figure 8.9).

Nevertheless, it is not the only area on which world-first innovators rely heavily. World-first innovators tend to use all three sources frequently, with anywhere from 55% to 70% of firms obtaining ideas from each (Figure 8.9).

While both research *and* development are important for world-firsts, production engineering – the division of the firm that is devoted to operations – receives about the same emphasis as research. World-firsts tend to take a multifaceted approach to the production of internal ideas for technologies that are used in their innovation process.

Of the three sources of innovation ideas, non-world-first innovators tend to use production engineering and experimental development frequently (57% and 60% of firms, respectively). They do not, however, make much use of research. This other group once more tends to rely on its engineering to solve more practical problems. Non-world-first innovators, then, tend to use research much less frequently than do world-firsts and to use experimental development and production engineering at about the same rate.

The fact that world-first innovators rely on research much more than their counterparts lends credence to earlier findings that showed research and development to be the most important source of innovation ideas for world-first innovators.

Non-world-first innovators concentrate on finding more efficient means of implementing previously developed innovations. Towards that end, their focus is not so much on the research required to conceive of new technologies but, rather, the development of better ways to use existing techniques. Much of this work can be accomplished through experimental development and production engineering. This group of firms is less oriented to research and more to solving assembly line problems.

8.12 EXTERNAL SOURCES OF NEW TECHNOLOGIES

The pattern of differences that is found for the use of external sources of new technologies by world-firsts is very similar to that found for external sources of innovation ideas for innovation. World-firsts are more likely to turn to a network of outside sources for technology that complements

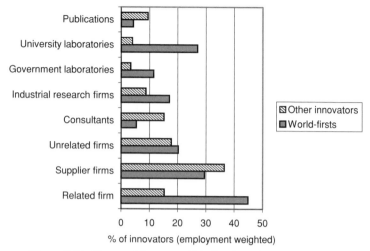

Figure 8.10. External Sources of Technology, by Novelty Type

their own internal R&D sources – much as they do for innovation ideas in general. They are more likely to use related firms, industrial research firms, and university laboratories than are non-world-firsts (see Figure 8.10).

In contrast, non-world-firsts rely less on almost all of the external technology sources. This is in marked contrast to their search patterns for ideas in general. Much of these differences may be related to the difference in emphasis that world-firsts place on internal sources of R&D. It may simply be that firms that use R&D departments make use of very different external networks and that this is what conditions the differences between the two group of innovators.

In order to examine this possibility, the two groups of innovative firms were divided into those that conducted research and development and those that did not do so, and the sources for technologies were compared. The external sources of technology are found to vary, depending on whether the innovator makes use of research as a source of internal technology and whether the firm is a world leader or essentially an imitator. Table 8.6 depicts the external sources of technology for those innovators that use research and for innovators that do not use research.

World-first innovators that rely on research are slightly more likely to make use of outside technology sources than are non-world-firsts that use research labs. But these two groups rely on many of the same outside sources. World-firsts with a research lab focus most frequently on related firms (43%), universities (38%), suppliers (27%), and unrelated firms (21%). Non-world-firsts that make use of research also rely on both

Table 8.6. *External Sources for Technology, by R&D and Novelty of Innovation (% of Innovators, Employment Weighted)*

Sources	Firms Conducting Research		Firms Not Conducting Research	
	World-First	Other	World-First	Other
Related firms	43 (7)	10 (2)	49 (7)	18 (2)
Supplier firms	27 (7)	40 (4)	34 (7)	35 (3)
Unrelated firms	21 (6)	26 (4)	18 (6)	14 (2)
Consultants	2 (2)	21 (3)	14 (5)	13 (2)
Industrial research firms	17 (6)	5 (2)	17 (5)	10 (2)
Government laboratories	14 (5)	7 (2)	6 (4)	2 (1)
University laboratories	38 (7)	5 (2)	1 (2)	4 (1)
Publications	1 (1)	10 (2)	13 (5)	9 (2)

Notes: Larger firms (IPs) only. Standard errors are in parentheses.

suppliers and unrelated firms for technologies. They differ in that they place less emphasis on the services offered by universities and more emphasis on the findings of outside consultants than do world-firsts with research labs.

8.13 PROTECTION FOR INTELLECTUAL PROPERTY

Innovation requires substantial investments for the commercialization of new ideas and inventions. These investments create the knowledge-based assets that are at the heart of the knowledge economy and are less likely to be made if innovations are easily copied by others. Intellectual property rights offer protection for those knowledge-based assets that are associated with innovation. Intellectual property rights are legally enforceable rights over an innovative product or process. They can take the form of patents, trademarks, trade secrets, industrial designs, copyrights, or integrated circuit designs.

A patent gives the inventor the exclusive right to produce an original invention for a limited period in return for the public disclosure of information about the innovation. Trademarks are devices or words legally registered as distinguishing a manufacturer's goods. The Industrial Designs Act protects the ornamental aspects of goods. Copyrights give the author of a text the right to print, publish, and sell copies of an original work. Integrated circuit design protection safeguards the original three-dimensional pattern of layout design embodied in an electronic circuit. Finally, innovations can be protected through secrecy. Trade secrets can be

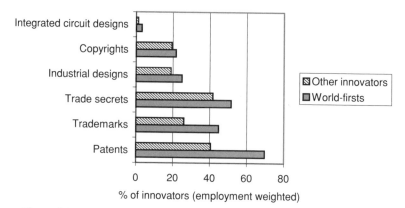

Figure 8.11. Usage of Intellectual Property Protection, by Novelty Type

licensed to others with the requirement that the recipient not divulge information about the secret. These agreements are enforced by the courts under common law.

Firms may choose other methods as substitutes or complements for those forms of intellectual property protection that depend upon regulatory or judicial oversight – what are referred to here as statutory rights. Careful design of products may make it difficult for others to copy the innovation. Being first in the market may give a firm a large enough lead that imitators cannot hope to produce a similar product at the same cost. Firms can bundle complementary characteristics such as services to reduce the chance that their customers will switch to the products of imitators.

Both world-firsts and other innovators use intellectual property rights extensively (see Figure 8.11). Patents are used most frequently, followed by trademarks, trade secrets, industrial designs, copyrights, and integrated circuit designs.

There are two reasons to expect that world-first innovators will make greater use of patents, trademarks, and other statutory forms of intellectual property protection. First, statutory rights all require a degree of novelty that world-firsts clearly possess. Secondly, world-firsts focus more on new product characteristics than do non-world-firsts. Intellectual property rights like patents are better suited for protection of products than they are for processes.

Confirming this hypothesis, world-first innovators do indeed use intellectual property rights more frequently then do other innovators. Over 78% of world-firsts use at least one patent, trademark, trade secret, industrial design, copyright, or integrated circuit design, while only 61% of other innovators use any of these intellectual property rights. Some 70% of world-first innovators use patents; only 41% of other innovators use them. Trademarks are used by some 45% of world-firsts and by 35% of other innovators. Trade secrets are used by 51% of world-firsts and by 42% of other innovators. Protection is offered by industrial designs for 25% of world-first innovators and by 19% of other innovators.

World-first innovators, then, are more likely to make use of statutory forms of intellectual property protection. Their innovations are sufficiently unique that making use of intellectual property protection is an important strategy for this group of firms. However, it must be noted that while other innovators are less likely to make use of statutory forms of intellectual property protection, they nevertheless make use of them. They use patents at about one-half the rate of world-firsts. Their relative use of trademarks and industrial designs is even higher. Whereas the innovations of non-world-firsts tend to be more concentrated on incremental process improvements, which do not lend themselves as frequently to protection via intellectual property rights, they nevertheless have a substantial novelty content. Once more this confirms the importance of this group of innovations as a whole.

8.14 IMPEDIMENTS TO INNOVATION

Innovation is hindered by many factors in addition to problems with intellectual property rights. These problems range from inadequate benefits to excessive costs. For example, benefits from innovation may be inadequately exploited if firms are unable to capitalize on new products because they cannot adequately market them. Or firms may choose not to innovate because they perceive the costs of capital required for the commercialization of an invention to be too high. Comparing impediments for world-first and other innovations provides a guide as to whether policy programs differ in terms of their usefulness to each of these two groups.

Most of the returns and costs associated with innovation are determined by the actions of individuals and firms operating in market systems. However, some areas receive special emphasis from public policies. These are areas where the market is sometimes said to have particular problems.

Markets for labour and for information are often seen to be imperfect because both skills and information have characteristics of a 'public good' – a good whose benefits are not perfectly appropriable. When goods are not appropriable, markets do not function well and goods are not provided in adequate quantities.

The labour market is often used as an example of a market that suffers from the problem of appropriability. Firms may not provide the optimal amount of training if they perceive that the benefits they receive from investments in skill training are likely to be lost because of labour turnover. Similarly, markets for information that support innovation and technological change are imperfect if the information, once produced, becomes freely available for all to use, without due compensation being paid to the firm that produces the information.

Public policy intervention in these areas is aimed at overcoming market imperfections. Public education programs are aimed at improving labour skills. Government trade missions provide market information for export markets. Standards and regulations provide information that consumers can use to evaluate the safety of products. Public monies are used to fund basic research, both at research institutes and at universities. Technical service programs are subsidized to help provide information on technology and technical advice on how to improve operations. Networks of firms are encouraged to facilitate the spread of information on technology.

Innovators find that each of the areas addressed by public policies presents them with impediments. Figure 8.12 plots the percentage of firms indicating that these areas posed serious problems – the percentage of firms that scored these problems as a 4 or a 5 on a 5-point Likert scale. The impediment most frequently cited by both world-first and other innovators is the lack of skilled labour. While the percentage of both groups of innovators that report that they experienced increases in skill requirements as a result of innovation is about the same, a higher percentage of world-first innovators felt that skill deficiencies offered impediments. Some 60% of world-first innovators report this to be a problem; only 45% of other innovators do the same. The difference between these two groups suggests that deficiencies in the type of skills required are more difficult to overcome in the case of world-firsts than for non-world-firsts.

The second most frequently mentioned problem by world-first innovators is a lack of market information (40%). A lack of good information about the prospective market for new products creates uncertainty, and this reduces the tendency to invest in the innovation process.

That the lack of worker skills and a dearth of market information rank first and second, respectively, is consistent with the results of the Growing Small and Medium-Sized Enterprise (GSME) Survey (Baldwin, Chandler, et al., 1994). This study found that the two government

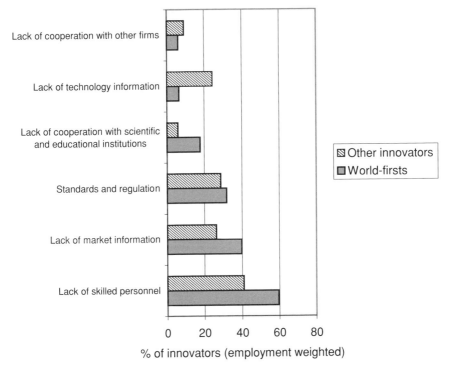

Figure 8.12. Impediments to Innovation, by Novelty Type

programs given the greatest importance were labour training and market information programs.

Both world-first and other innovators also find that standards and regulations are an important impediment, with 31% of the former and 32% of the latter indicating that this was a difficulty.

The pattern of problems experienced by other innovators is somewhat similar to that of world-first innovators, although each of the problems mentioned occurs less frequently. While it has been argued that both types of innovators contribute positively to economic growth, the intensity of the problems that have to be solved probably differs in each of these two groups. World-first innovators, as we have shown, are more likely to be innovating simultaneously on both the product and process side. They run into more problems and thus note these problems more often. But the similarity in the relative emphasis that is placed on the different impediments by each type of innovator suggests that these problems are not so much barriers to innovation activity as they are problems that accompany it. Innovations involving more novelty experience the same pattern of problems, but they experience them more frequently because the innovation process is more complex for innovations that are more novel.

There is one exception – where other innovators report impediments with greater frequency than do world-first innovators. Some 28% of other innovators report impediments due to a lack of technology information. Less than 7% of world-firsts report this to be a problem. This accords with the picture of the group of non-world-first innovators that has been drawn here. They tend to rely less frequently on internal ideas. They perceive a greater need for help in finding sources of external information about technologies that are appropriate for their operations. These firms, then, benefit more from outside technology-support programs.

8.15 WHAT HAPPENS TO WORKERS IN INNOVATIVE FIRMS?

Process innovations often reduce costs by decreasing the quantity of factors of production required per unit of output. This would reduce the demand for labour if output of the innovative firm remains constant after the innovation. However, innovative firms also indicate that their share of domestic and foreign markets often increases as a result of their introducing an innovation. This increase in market share may be the result of either the commercialization of product innovations that allow firms to change their output mix, or the introduction of process innovations leading to more efficient production methods and lower prices which, thereby, allow firms to aggressively compete for market share. Whether the demand for labour in a particular firm increases or decreases as a result of innovation will depend on which of these offsetting forces is largest.

The net effect of the two forces on the demand for labour is generally positive (see Table 8.7). Over 40% of all world-first innovative firms indicate that they increased demand; only 7% decreased demand. Other innovators are also more likely to increase than decrease their demand for workers.

It is hypothesized that the two groups of innovators will have different impacts on labour demand. While both indicate that their innovations increase their market share, the non-world-firsts probably operate more frequently at that stage of the product life cycle where process innovation is associated with a reduction in unit costs and an increase in labour productivity associated with the introduction of labour-saving technologies. In contrast, world-firsts are still at the stage where innovation focuses on

Table 8.7. *Impact of Innovation on Workers, by Novelty of Innovation (% of Innovators)*

	World-First Innovators	Other Innovators
Employment		
Increase in total workers	42 (5)	37 (2)
Decrease in total workers	7 (2)	18 (2)
Increase in non-production workers	22 (40)	22 (2)
Decrease in non-production workers	1 (1)	4 (1)
Increase in production workers	34 (5)	31 (2)
Decrease in production workers	9 (3)	19 (2)
Skill levels		
Increase in skill requirements	57 (5)	71 (2)
Decrease in skill requirements	1 (1)	2 (1)
No change in skill requirements	43 (5)	27 (2)

Notes: Larger firms (IPs) only. Standard errors are in parentheses.

new products and where process innovations do not yet involve radical labour savings.

As was hypothesized, differences between those non-world-firsts that are expanding and those contracting are smaller and less significant than for world-firsts.

Not all workers are affected in the same way by innovation. Nonproduction workers consist mainly of white-collar workers who are seen as having the skills necessary to benefit from the computer-based technological revolution. Production workers, on the other hand, are usually blue-collar workers, and their relative numbers have decreased recently compared to nonproduction workers (Berman, Bound, and Griliches, 1993; Baldwin and Rafiquzzaman, 1999).

The demand for labour in these two groups has been affected quite differently by innovations (Table 8.7). Innovators increase demand for nonproduction workers substantially more frequently than they decrease demand – both in the case of world-first innovators and other innovators. However, there is less of a difference in the percentage of firms indicating that innovation increased the demand for production workers and the percentage of firms that decreased the demand for production workers, especially in the case of other innovators. As was demonstrated, non-world-first innovators tend to place greater emphasis on automation and introduce fewer new products than do their counterparts. These tendencies are more likely to result in a decrease in the amount of production workers required.

New technologies affect not only the demand for labour but also the type of skills required. The changes that technology-driven process innovations have on the skill levels of workers are the subject of controversy. In some circles, the introduction of new technologies has been equated with de-skilling. Others have claimed that innovation and technological change are associated with increases in skill levels. New more flexible forms of production, it is argued, can only be achieved through a highly skilled workforce with greater conceptual skills than were previously required.

The differences in the demand for blue- and white-collar workers suggest that greater skill levels are associated with innovation. Indeed, this is reflected in the effect of innovation on skill levels (Table 8.7). Less than 2% of innovators, whether they be world-firsts or others, indicate that skill levels decreased as a result of innovation. Between 27% and 43% indicate no change in skill levels. Some 71% of non-world-first innovators indicate that skill levels increase; some 57% of world-first innovators require higher skill levels after innovation. Non-world-first firms may be more likely than world-firsts to decrease demand for production workers, but their overall skill levels are increasing, not decreasing. Thus, new technologies that are being implemented by non-world-firsts are having almost as great an impact as those being implemented by world-firsts. Once more the similarities between the two groups in terms of the effects of innovation strongly suggest that non-world-firsts are forging changes that are significant and worthy of study in their own right.

8.16 CONCLUSION

Innovation policies encourage both the development of new ideas and their widespread diffusion. The development of major inventions, from the stage of brand new ideas to commercial production, captures the imagination of most scientific writers. These inventions often have dramatic visible effects on the economic system. Transistors and other electronic components have created a computer-based revolution over the last forty years. Lasers have dramatically altered both communications and production systems.

Other types of innovations are more incremental in nature and receive less attention. In some cases, they involve the application of new products in new ways – such as the application of the laser to bar coding and

Conclusion 215

point-of-sale analysis. In other cases, they involve incremental changes in the production process. While each of several incremental changes has a relatively small effect when considered by itself, the cumulative effect of a sequence of incremental changes is often large.

This chapter focuses on both types of changes by using the taxonomy of world-first innovations and all other innovations. Both types potentially make significant, though different, contributions to economic growth. While world-firsts are important, the other category should not be ignored. Canada-firsts, for example, would include the production and development of products introduced into Canada for the first time by multinationals. Even if the resulting products serve only the domestic market, Canadian consumers benefit from new products, an improvement in product quality, or lower costs. Innovations that are neither world-firsts nor Canada-firsts can also have a substantial effect on economic performance. A firm that buys new machinery and equipment from a foreign supplier to duplicate the equipment that has already been purchased by its domestic competitors will fall into the third category. The benefits associated with a reduction in production costs that result from this action can also be substantial.

Quantitatively, non-world-first innovations are introduced more frequently than are world-firsts. Some 51% of innovators (weighted by employment) introduce non-world-firsts over a three-year span. Another 33% are responsible for Canada-firsts, and 16% are involved with world-firsts.

Innovators in all three groups report similar effects with about the same frequency. Each type of innovation improves product quality, extends the product line offered, reduces lead times, and improves working conditions as well as technological capabilities. Indeed, in all but the latter case, other innovators report these results as frequently as do world- and Canada-first innovators. More than 58% of innovators in each of these groups report that innovation improves profit margins or increases market share. Both types of innovations, then, contribute significantly to growth in the business population.

While the effects of innovation are relatively similar across innovation types, there are differences in the nature of the innovations and in the inputs used to support innovation. The innovation process is often described in terms of highly visible new products, such as steam engines, gasoline, airplanes, computer chips, and lasers. However, innovation in production processes, which is less visible than new product innovation, underpins the evolution of new industries that produce innovative products. The

Canadian evidence substantiates the importance of technological competence in process engineering. Process innovations are important in at least 80% (weighted by employment) of innovators – either because the innovation involves purely process changes or because it simultaneously involves new products and processes.

The main difference between world-first and all other innovators lies in the extent to which world-firsts stress both product and process innovations. World leaders show a special ability to manage both the product conception stage and changes in technology associated with the production process.

The characteristics of new products range from providing new functions to the use of new materials. Here, world-firsts are more likely to characterize their product innovations as being new in all dimensions.

Many sources of ideas are used for innovation. Internally, research and development, sales or marketing, management, and production all make a contribution. The emphasis given to each of these sources differs by novelty type. World-first innovators tend to rely more frequently on research and development divisions than on any of the other internal sources – sales, marketing, management, or production. In contrast, non-world-first innovators rely more evenly on all of these latter sources, with management being cited slightly more often.

Research and development is given relatively more emphasis by world-first than by non-world-first innovators. Some 88% of world-firsts indicate that they find research and development divisions to be a useful source of ideas for innovations as compared to 55% of non-world-firsts. Some 84% of world-firsts perform R&D on a continuous basis; some 57% of other innovators do so. The main difference, then, is that non-world-first innovators do less R&D and rely on a number of other sources for development of their innovations.

External sources of innovation ideas are also used extensively by all types of innovators, but they are used more frequently by non-world-first innovators. The most important outside sources for non-world-first innovators are customers, followed by suppliers and related firms. The most important external sources for world-firsts are suppliers, followed by customers and related firms. In both cases, innovation depends upon forward links from suppliers to their customers and backward linkages from customers to their suppliers. In this respect, both groups of innovators are similar. They do, however, differ in terms of the other sources from which innovation ideas are obtained. In particular, non-world-firsts rely heavily on competitors as an external source of ideas.

Process innovations involve the use of new technologies. New technologies consist of new production equipment, new production techniques, and new organizational structures. Internal sources of ideas for new technologies originate in the pure research group, in the experimental development division, and in production engineering. All three are used frequently by world-firsts. Non-world-firsts tend to rely primarily upon experimental development and production engineering, and the frequency with which this occurs is about the same as for world-firsts. Non-world-firsts differ from world-firsts primarily because they concentrate much less on pure research. Research *and* development may be important for non-world-firsts but it is the development component that receives the most emphasis in this group.

Both groups of innovators make heavy use of suppliers for new technologies. But with the exception of suppliers, external sources of technology differ for the two groups of innovators. World-firsts are more likely to use outside groups that complement their research and development facilities – related firms, industrial research firms, universities – than are non-world-first firms. Non-world-firsts are more likely to use the services of consultants, along with information garnered from publications.

It is important to understand not only what inputs are used by the innovation process but also how innovation affects the nature of the inputs used. Nowhere is this more important than in understanding the effect of innovation on the demand for labour. Innovative firms more frequently indicate that innovation had the effect of increasing their demand for labour, not decreasing it. However, innovation increases the demand for white-collar workers more than it does blue-collar workers. Innovators increase demand more frequently for nonproduction workers than they decrease it. The differences are less marked for production workers.

In this area, world-firsts are more likely to increase the demand for workers than are non-world-firsts, probably because the latter are at the stage of the product life cycle where process innovations have a more dramatic effect on unit costs and reduce labour requirements.

Innovators also stress that skill levels of workers increase as a result of the innovation. Some 71% of non-world-firsts indicate that skill levels increase; some 57% of world-first innovators require higher skill levels as a result of innovation. Both groups of innovators, then, serve to increase the need for a more highly skilled workforce.

Public policy is directed at supporting the innovation process in a number of different ways. Policies are aimed at reducing impediments in several areas where markets are seen to have imperfections. These encompass such areas as labour training, market information, regulations and standards, and technical services. Innovators confirm that they indeed experience impediments in each of these areas. The areas that give them the greatest difficulty are lack of skilled personnel, lack of market information, and government standards and regulations. In each of these cases, world-first innovators generally experience these problems more frequently than do non-world-first innovators. This suggests these problems do not block innovation as much as they accompany more intense innovative efforts.

Public policy also facilitates innovation by setting the framework in which innovators can protect the intellectual property that accompanies innovations. Both world-first and non-world-first innovators use intellectual property rights extensively. Over 78% of world-first innovators possess at least one patent, trademark, trade secret, industrial design, copyright, or integrated circuit design. The percentage of non-world-firsts that do so was lower (61%), but nonetheless significant.

In summary, the Canadian innovation system produces a variety of innovations that range in importance from world leaders to the incremental and not-so-incremental changes that result from the general diffusion of knowledge about new production techniques. These different innovations have similar beneficial effects in terms of improving a firm's market share or profit margin – though the causes of the changes are sometimes different. Innovation is more likely to result in foreign market share improvements for world-firsts and domestic market share improvements for Canada-first and other innovators. The different innovation types are the product of quite different innovation systems.

NINE

The Use of Intellectual Property Rights

9.1 INTRODUCTION

This chapter examines the extent to which innovative firms use and appreciate various forms of protection for the intellectual property that is developed as part of the innovation process. Innovation involves the development of new ideas that lead to new products or new processes. Investment in the development of new products and new processes will not be made unless the investment is profitable – unless the intellectual property that results from the investment has some private value. Unfortunately, in many cases, ideas can be easily duplicated or stolen. Without some form of protection for the knowledge assets developed by the investments in ideas that are required for innovation, innovation will not take place, or at least not in optimal quantities (See Arrow, 1962). When intellectual property is protected, an innovator is able to appropriate the benefits of innovation.

Appropriability is facilitated by various methods that are used to establish and protect intellectual property rights in knowledge assets. These property rights are protected by statutes, by common law, and by strategies that make it difficult for others to imitate or copy the innovation. In creating these rights, the political system has continuously wrestled with competing objectives–the creation of property rights that protect ideas versus the desire to diffuse information so as to facilitate the widest possible benefits from the innovation; the consequences of creating monopolies by the provision of protection for ideas against the desirability of having a competitive market structure producing goods and services.[1]

The theoretical literature on the optimal type of intellectual property regime[2] is more extensive than applied studies on the use of intellectual property. One of the reasons is that detailed data on intellectual property use are difficult to obtain. Empirical studies on the use of intellectual property in Canada by Firestone (1971), Séguin-Dulude and Desranleau (1989), and Consumer and Corporate Affairs Canada (1990) have had to rely on specially designed surveys. Because of their cost, surveys of this nature are done infrequently.

Other studies (Etemad and Séguin-Dulude, 1987; Hanel and St.-Pierre, 2002) make use of patent registrations, derived from either data on Canadian patent registrations that were maintained by Industry Canada (the patent database [PATDAT] file), or international data from the World Intellectual Property Organization (WIPO). These international data suffer from several problems if they are to be used to judge the innovativeness of different countries. First, the international data are not always comparable. For instance, Japan allows narrower criteria for patenting and, therefore, has a larger number of patents than do other countries. Second, it is difficult to judge the importance of patents from data on patent filings. Some countries may patent a large number of relatively unimportant new ideas, while other countries may create a smaller number of commercially important ideas.

The most significant problem with patent registration data is that alone, it provides very little information about whether firms are using intellectual property protection, in particular whether the patent is being worked. For example, Firestone (1971, p. 96) found that not more than 45% of Canadian patents were being worked. Moreover, patent data alone tell us little about the characteristics of the firms that are using intellectual property protection. It is difficult, though not impossible,[3] to link firm characteristics to the data on filings. Without having characteristics of those firms that use intellectual property rights, it is difficult to understand who is using them and under what circumstances. Recently, several surveys in the United States (Levin et al., 1987) and in Europe (Bussy, Kabla, and Lehoucq, 1994; Arundel, van de Paal, and Soete, 1995) have

---

[1] A discussion of these tradeoffs can be found in Taylor and Silbertson (1973).
[2] See Cohen and Levin (1989).
[3] See Hanel and St.-Pierre (2002) for a study that links profitability data to PATDAT patent data.

focused more broadly on how intellectual property protection complements the innovation process. But while these surveys expanded the information on use and on users' attitudes towards the efficacy of the various forms of intellectual property protection, they either did not link the information on use to other characteristics of the firm or had a relatively limited set of firm characteristics that could be related to intellectual property use.

This chapter builds on these previous studies and extends them. It focuses broadly on the use of intellectual property protection in Canada by examining the extent to which it is an integral part of the innovation process. It focuses on more traditional issues, such as the intensity of use – although it tries to obtain a comprehensive picture of the different instruments (both statutory and other forms of protection) that are employed to protect intellectual property assets. It also measures the degree to which firms perceive different instruments to be more or less effective in protecting intellectual property. Finally, the different forms of intellectual property protection that are used can be related to the innovation profile of the firm.

9.2 CANADA IN AN INTERNATIONAL CONTEXT

Before we discuss the way in which intellectual property protection is used in Canada, it is important to situate the Canadian experience with intellectual property protection relative to that of other OECD member countries. For this purpose, one form of intellectual property protection (patents) will be used for a comparison. Two measures of patent use will be employed. The first captures the extent to which the citizens of different countries potentially benefit from the consumption of goods and services that are protected by patents. The second is the extent to which the innovation systems of different countries make different use of patent protection.

Patents are one of the primary methods used to protect the investment made in ideas that produce innovations. Since patent applications are filed to provide protection for knowledge assets, the number of filings is related to the value of capital invested in the innovation process that are available to benefit the population of a country. The OECD publishes the number of patent applications in each member state – both the total applications and the number filed by just the residents of that country. Patents are filed by foreign residents to protect their right to exploit their own ideas either through trade or through direct production. Since

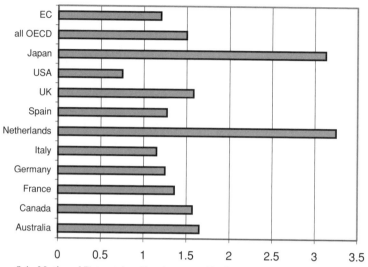

Figure 9.1. National Patent Applications per Capita, 1992. *Source:* WIPO, OECD

foreign patent applications provide as many, if not more, benefits to the inhabitants of the country in which they are filed as resident patent applications, total patents both by residents and nonresidents will be used here for cross-country comparisons.[4] In order to standardize for the number of people who are expected to benefit from patent filings, the national totals are divided by the population in each country.[5] The resulting ratios of 1992 patent filings per capita are presented in Figure 9.1 for Canada and all other OECD countries that have more than 15 million inhabitants. Canada has about 1.54 filings per 1000 inhabitants. The OECD total is 1.49 and the EC total is 1.17. Canada is behind Japan and the Netherlands but equal to or ahead of the other large members of the OECD.

Patent statistics may also be used to characterize the productiveness of different innovation regimes. For this purpose, only patents filed by residents are relevant. In order to reduce the problems referred to previously – differences in standards of patenting and in the importance of patents across countries – Patel and Pavitt (1991) have suggested that patent filings in a third country like the United States be used. This will

---

[4] Firestone (1971, p. 132) reports that over two-thirds of U.S. corporations that were found to own patents in his sample also possessed operating subsidiaries in Canada. Therefore, there is a strong presumption that these patents are being used in Canada.

[5] Alternately, gross national product (GNP) could be used. But such a measure is biased against more productive countries.

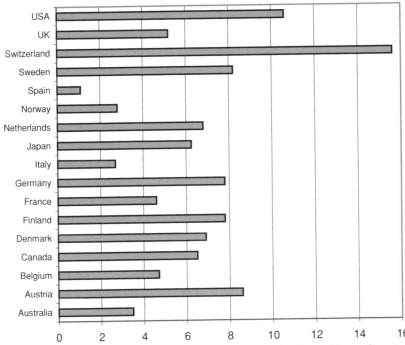

Figure 9.2. Patent Filings in U.S. Market per R&D Scientists in Home Country, 1992. *Source:* WIPO, OECD

impose a common standard and will cull out those patents that inventors do not feel possess enough commercial importance to warrant the expense of protecting them in the largest and wealthiest OECD market. In order to compare the relative productivity of different innovation systems, patents, which are a measure of output in that they are used to protect innovation assets, need to be standardized by a measure of input to the innovation process. The measure used for that purpose is the number of research and development scientists and engineers.[6] The resulting measure of the productivity of the innovation process for each country is presented in Figure 9.2. With 6.2 patents filed in the United States per research scientist, Canada compares favourably to most countries. It is ahead of Japan, France, Belgium, Italy and the U.K., but behind Germany, Sweden, Denmark, Finland, and Switzerland.

---

[6] The patent filings data are for 1993 and come from WIPO, Industrial Property Statistics, Part 1, 1993, 1995. The data for R&D scientists and engineers are mainly for 1990 and come from the OECD, 1994, Part 2, p. 18.

This comparison does not permit an evaluation of the relative innovativeness of Canada and the United States because it uses U.S. home market data. However, if Canada and U.S. experience in a third market (patent applications in the U.K., Germany, and France) are compared, then Canada does just as well (over 10 patent applications per 100 R&D scientists) if not better than the United States (less than 5 patent applications per R&D scientist).

9.3 FORMS OF INTELLECTUAL PROPERTY PROTECTION

The protection given to intellectual property takes different forms. Intellectual property rights can be grouped as works of identification (trademarks, appellations of origin), works of expression and information (copyright, industrial design, trade secrets), and works of function (patents, copyright on computer software, and trade secrets).

Companies may protect their knowledge assets by keeping them secret and may enforce the responsibilities of their employees not to divulge proprietary information through the courts. Unlike other areas of intellectual property rights in Canada, no distinct statute protects trade secrets. Trade secrets can be licensed to others with the requirement that the recipient not divulge information about the secret. Trade secrets violations are dealt with by the courts as unfair trade practices under common law. The owner of a trade secret is entitled to its exclusive use, at least until it is lost due to independent development by another company, reverse engineering, espionage, or an unauthorized disclosure. A trade secret is advantageous in that there is protection for an unlimited time if it remains undisclosed and if it can be exploited immediately without the time and expense of registration.

Another form of protection for knowledge assets is afforded by statute.[7] These statutes create and protect property rights in these assets. The most familiar type of statutory protection is given by patents. A patent gives to the inventor the exclusive right to exploit an original invention for a limited period in return for the public disclosure of information about the innovation. While patents are the most visible form of protection, there are a number of other types of protection that are provided by legislation, by administrative practices, or by regulation. Trademarks are devices or words legally registered as distinguishing a manufacturer's

---

[7] Throughout this chapter, trade secrets are grouped under statutory forms of protection since they are enforced by an arm of the state – the court system.

goods. The Industrial Designs Act protects the ornamental aspects of goods. Copyrights are a form of protection given by federal statute to an author for the right to print, publish, and sell copies of the original work. Although copyrights are normally thought of in the context of book publishing, their use extends also to the product and service sector. Copyrights are also used to protect computer software. To the extent that documentation is key to the understanding of the operation of a product, copyrights offer important forms of protection to goods.

Two forms of statutory protection are highly specific to particular industries. Plant breeders' rights protect seeds or other propagation material. Integrated circuit design protection safeguards the original three-dimensional pattern of layout design embodied in an electronic circuit.

Statutory forms of protection essentially enhance the degree of protection provided to knowledge assets. However, it must be recognized that some protection would exist without patents and other forms of statutory protection. Those who stress difficulties in the innovation process associated with a lack of appropriability often treat the process as one in which ideas easily flow from company to company. If this is the case, ideas can be readily stolen and intellectual property will have little value. However, it is argued that in many cases, the knowledge that is important to the innovation process is 'tacit'. It is not easily codified or communicated and depends on innate skills that are specific to particular firms. As such, appropriability exists for some innovations (the intellectual property therein is protected) even in the absence of statutory intellectual property rights. Protection for innovation in these instances is given by the very difficulty in copying new ideas, new products, or new processes. Even when new products are fully described in patent documents, the act of turning that knowledge into a new product or process is difficult and costly. Since new technology involves a mix of codified knowledge and implicit know-how that is difficult to transmit or to digest (Mowery and Rosenberg, 1989), intellectual property associated with innovation retains its value even in the face of attempts to copy it.

There are a number of methods that a firm can use to augment the protection that its knowledge assets are given by the inherent difficulty in copying ideas. These include being first to market, exclusive contracting, reputation and goodwill, and tie-ins to services.

Some innovations have characteristics that give them innate protection. These are forms of protection that do not originate in standard intellectual property legislation. They either result from natural characteristics of the product or are the result of specific strategies adopted by the firm. Process innovations can sometimes be hidden behind factory walls and the know-how required to make them work can be kept secret. On the other hand, product innovations by their nature circulate. While the secret of new products is difficult to conceal, product innovations may nevertheless be protected via several different strategies. For example, complexity of design can provide enough of an advantage for the innovative firm to permit research and development expenses to be recouped. Being first in the market can offer substantial protection if it engenders enough consumer loyalty or if it generates a cost advantage because of cumulative learning. In addition, firms can bundle complementary characteristics like service or quality with a good in order to reduce the chance that new competitors will capture substantial market share. These instruments offer a form of protection that derives from a firm's strategy and not from legislative authority.

9.4 USE OF INTELLECTUAL PROPERTY RIGHTS BY MANUFACTURING FIRMS OPERATING IN CANADA

The use of statutory forms of intellectual property protection provides a measure of the output of the innovation system – at least of those outputs that receive some form of administrative or legislative protection (see Griliches, 1990). The intensity of use at the level of the firm is investigated here for several forms of intellectual property protection – protection that is granted by statutory rights associated with copyrights, patents, industrial designs, trade secrets, trademarks, integrated circuit designs, and plant breeders' rights.

Overall, some 24% of Canadian manufacturing firms utilize at least one of these statutory forms of protection (see Table 9.1). Some 14% of firms own only one of these forms of protection. About 6% have two forms

Table 9.1. *Multiple Use of Statutory Forms of Intellectual Property Protection (% of Firms)*

Weighting Factor	Number of Intellectual Property Types (IPTs)				
	At least 1	1	2	3	4+
Company weighted	24	14	6	3	1
	(.8)	(.7)	(.5)	(.3)	(.2)
Employment weighted	50	21	12	7	11
	(1)	(.8)	(.7)	(.5)	(.6)

Note: Standard errors are in parentheses.

Table 9.2. *Usage of Individual Forms of Intellectual Property (% of Firms)*

Type	No. of IPTs in All Firms (Company Weighted)			No. of IPTs in All Firms (Employment Weighted)		
	Any	1 to 5	6+	Any	1 to 5	6+
Copyrights	4 (.4)	3	2	13 (.7)	6	8
Patents	7 (.5)	6	1	29 (.9)	16	13
Industrial designs	6 (.5)	4	2	15 (.7)	10	6
Trade secrets	8 (.6)	6	3	20 (.8)	11	9
Trademarks	11 (.6)	9	2	32 (1)	18	14
Integrated circuit designs (semiconductor chips)	1 (.2)	–	–	1 (.2)	1	–
Plant breeders' rights (plant variety rights)	–	–	–	1 (.2)	–	–
Other	1 (.2)	1	.1	2 (.3)	1	1

Notes: IPTs are intellectual property types. Standard errors are in parentheses.
– = none reported.

of statutory protection. Very few have more than this. Manufacturing firms do not tend to utilize multiple forms of statutory protection very frequently.

The employment-weighted use rates are more than double the company-weighted rates. Those firms that have some form of statutory protection account for 50% of total employment (Table 9.1). The large difference between the company-weighted and the employment-weighted results extends across each of the usage categories. Large firms, then, are much greater users of the various statutory forms of intellectual property protection.

Use varies substantially by type of intellectual property protection (see Table 9.2). Trademarks are the most popular form, with 11% of firms using at least one trademark. Statutory trade secrets and patents are second and third, with 8% and 7%, respectively. Industrial designs are used by 6% of firms and copyrights by 4% of firms. Less than 1% of firms report use of integrated circuit designs and plant breeders' rights.

Once again, the employment-weighted use rates are substantially higher then the company-weighted results – although the relative importance of the various categories is about the same in each case. Those firms that used trademarks made up only 11% of the population but they accounted for 32% of total employment. Firms that used patents made up only 7% of the population but they accounted for 29% of total employment.

The ranking of trademarks and patents derived from these use rates accords broadly with the relative size of the number of trademarks and patents that are registered annually in Canada. For example, in 1993, some 14,580 patents were granted, while 15,121 trademarks were registered in Canada.[8] On the other hand, Canadian manufacturing firms indicate a greater reliance on industrial designs than the figures for formal registration of industrial designs would suggest. In the case of industrial designs, only 1,638 deposits were registered in 1993 – a little more than 12% of the number of patents granted. Yet almost the same percentage of firms indicated they protected their intellectual property with industrial designs (6%) as indicated they used patents (7%). Respondents most likely took a broader definition of industrial designs than just those officially registered and included in their response those unique features that served to establish a valuable advantage for their product.

Patents offer statutory protection but require that information regarding the invention be placed in the public domain. A patent is a compromise between two offsetting objectives – that of protecting the rights of the innovator and that of disseminating information about the invention. On the one hand, patent protection grants appropriability for the invention and provides incentives for innovation. On the other hand, the information filed with a patent facilitates the spread of information and may aid the general process of innovation. More importantly, the establishment and enforcement of property rights facilitate trade in intellectual property. Without well-defined property rights, markets do not function efficiently.

Although patents offer protection for a firm, the protection may not always be very strong. Competitors can patent around an invention when there are many known means to achieve an effect equivalent to the patented one. Patents also suffer from difficulties in enforcement (Von Hippel, 1988, p. 52).

Secrecy offers an alternative to patents as a way to protect an innovation. An innovator who possesses a trade secret can prevent the disclosure of the secret through fraudulent or dishonest means. Other firms can be licensed to use the secret and bound to keep the information secret. The disadvantage of the secrecy route is that the holder cannot prevent imitation if that imitation is independently discovered, acquired legally, or reverse

---

[8] The relative usage rates do not, however, closely correspond to the relative usage rates of patents and trademarks by residents that are reported in the WIPO statistics. Here trademarks registered to Canadian residents outnumber patents granted to Canadian residents by a factor of 7 to 1.

engineered. Thus, trade secrets are most effective for process innovations where the process can be hidden behind factory walls or with products that incorporate various barriers that prevent reverse engineering.

Trade secrets and patents need not be regarded as being strict substitutes. They can be used together. If an innovation involves both a process and product change, as is often the case (see Chapter 3), the process innovation may be protected via a trade secret while the product innovation may receive protection from a patent.[9]

In light of these considerations, it is noteworthy that use rates in the Canadian manufacturing sector for patents (7%) and trade secrets (8%) are about the same (Table 9.2). Trade secrets are used just as much as the patent process to safeguard innovations. Trade secrets are somewhat less important than patents when the employment-weighted estimates are used – though not much so.

Most firms do not make use of a particular form of protection more than once. Most firms indicate that they only have 1 to 5 instances of a property right. As such, the relative rankings (company weighted) in this category of usage are much the same as for overall use. However, for higher use categories, the ranking of trade secrets (company weighted) rises to first place. Trade secrets are relatively more important for owners of multiple assets. On an employment-weighted basis, trademarks increases in relative importance and surpasses trade secrets. The large multiple users, then, are relatively heavy users of the trademark system.

9.5 EFFECTIVENESS OF INTELLECTUAL PROPERTY PROTECTION

9.5.1 Overall Evaluations

Information on intellectual property use provides one indicator of the efficacy of different forms of statutory intellectual property protection. If a form of protection is not used, it serves little purpose.

An alternate measure is provided by evaluations given by respondents about the effectiveness of the various forms of intellectual property protection in preventing their innovation from being duplicated (see Table 9.3). The innovation survey asks firms to rank the seven forms of protection enumerated in Table 9.3 on a scale of 1 to 5, where 1 is 'not very

---

[9] Patents protecting a new production process may also protect the product produced by the new process.

Table 9.3. *Effectiveness of Intellectual Property Protection (Mean Score on a Scale of 1–5)*

Intellectual Property Rights Associated with	Average Score				
	All Firms 1	Users of Any Statutory Right 2	Nonusers of Any Statutory Right 3	Users of Specific Statutory Right 4	Nonusers of Specific Statutory Right 5
Statutory					
Copyrights	1.6 (.04)	1.9 (.06)	1.3 (.05)	2.8 (.11)	1.4 (.04)
Patents	1.9 (.05)	2.3 (.06)	1.4 (.06)	3.0 (.07)	1.5 (.05)
Industrial designs*	1.6 (.04)	2.0 (.06)	1.3 (.05)	2.5 (.09)	1.4 (.04)
Trade secrets	2.1 (.05)	2.6 (.07)	1.6 (.07)	3.2 (.08)	1.6 (.05)
Trademarks	2.0 (.05)	2.6 (.06)	1.4 (.05)	3.1 (.07)	1.5 (.05)
Integrated circuit designs	1.3 (.04)	1.4 (.07)	1.2 (.04)	3.2 (.24)	1.2 (.04)
Plant breeders' rights	1.2 (.03)	1.2 (.04)	1.1 (.03)	2.3 (.36)	1.2 (.03)
Other	1.4 (.06)	1.8 (.12)	1.2 (.06)	3.3 (.42)	1.3 (.06)
Other strategies:					
Complexity of product design	2.6 (.05)	3.2 (.06)	2.2 (.08)		
Being first in market	3.2 (.05)	3.4 (.06)	3.0 (.09)		
Other	2.3 (.14)	2.7 (.12)	2.0 (.18)		

Notes: Scored as 1: not at all effective; 2: somewhat effective; 3: effective; 4: very effective; 5: extremely effective. Standard errors are in parentheses.
* This category probably involves protection granted both under the Industrial Designs Act and more innate forms of design protection.

effective', 2 is 'somewhat effective', 3 is 'effective', 4 is 'very effective', and 5 is 'extremely effective'. The average scores given to copyrights, patents, industrial designs, trade secrets, trademarks, integrated circuit designs, and plant breeders' rights are given in column 1 of Table 9.3. In order to provide a comparison to each of these, the average score given to other forms of protection – the complexity of product design, being first in the market, and other strategies – is also included in Table 9.3. The average score presented in the first column is derived from just those giving a positive score for that category (that is, it omits nonresponses). The second and third columns contain the average scores for firms divided into those who gave a positive score and indicated that they used any of the statutory forms of protection in question 5.1 and those who gave a positive score to effectiveness but did not indicate that they possessed any of the intellectual property rights. This divides the population into what might be referred to as general users of intellectual property rights versus those not possessing any of the mentioned property rights. The fourth and fifth columns contain the average score for the firms that indicated they used the particular form of protection in question and those that did not – what might be referred to as specific users and others. Thus, the average score for all firms evaluating copyrights is 1.6; for those using any of the forms of statutory protection (copyrights, patents, trademarks, etc), it is 1.9; for those indicating that they possessed a copyright, it is 2.8.

When all respondents are considered (Table 9.3, column 1), most of the statutory forms of protection receive low average scores. None of the statutory forms of protection is deemed to be very effective by the population at large. The most effective are trademarks and trade secrets, which are also the two forms of statutory protection that are most heavily used. Nevertheless, at 2.1, they receive only average scores that are slightly more than 'somewhat effective', the second lowest ranking.

Innate protection derived from the complexity of product design and being first in the market receives the highest average scores – 2.6 and 3.2, respectively. Both are scored significantly above the forms of statutory protection. Thus, innate, not statutory, forms of protection are valued more highly by the population at large.

When the sample of firms is restricted to those using any one of the forms of statutory intellectual property protection (general users), the average scores increase slightly for each of the forms of intellectual property protection (Table 9.3, column 2). The scores also increase for the strategies involving complexity and being first in the market, and these still receive the highest average scores.

When the sample is further restricted to include just those firms using the forms of intellectual property protection in question (specific users), scores increase somewhat more (Table 9.3, column 4). For example, those who possess patents score this form of intellectual property as 3.0 or 'effective'; those who have no patents give it a score of 1.5 – less than 'somewhat effective'. The average score given to trade secrets by all firms is only 2.1 – 'somewhat effective' – but users of trade secrets give them an average score of 3.2 – 'effective' – while nonusers rank it only as 1.6 – between 'not at all effective' and 'somewhat effective'. This difference, between those who use the particular form of protection in question and those who do not, can be found in almost all the categories. Innovators who use intellectual property protection rank this protection well above those who do not. The low average scores that the population gives to intellectual property rights is due to the large number of nonusers who do not regard property rights as effective.

For users of intellectual property, trade secrets and trademarks are still at the top in terms of ranking, though now they receive an average score slightly above 3, indicating that users regard both of them as 'effective'. Most of the other forms of protection rank between 2.3 and 3.3. Patents receive a score of 3.0 and are only slightly behind trade secrets.

In the case of nonusers, trademarks, trade secrets, industrial designs, and patents all receive scores between 1.4 and 1.6. The reason, then, for the low overall score given to statutory protection lies in the evaluation of nonusers. Users believe that they are basically effective.

The previous section examined the value placed on various forms of intellectual property protection by examining central tendencies of the distribution of their scores. An alternate method of evaluating differences across the various forms of protection is to compare the entire distributions for different categories of firms. Distributions show that the group who do not use intellectual property protection have a very different distribution than the group who make use of it.

The differences in the distributions of user and nonuser patent scores are presented in Figure 9.3. The scores of the effectiveness of intellectual property protection for each group range from 1, not at all effective, to 5, extremely effective. Firms' responses across the scoring categories are depicted as percentage distributions.

The vast majority of nonusers rated patent protection as either not effective, a score of 1, or only somewhat effective, a score of 2. The

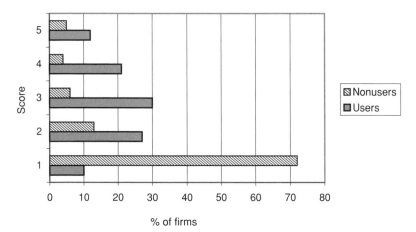

Figure 9.3. Distribution of Scores for Users and Nonusers of Patents

distribution of scores for specific users of patents is somewhat more symmetric. The majority of firms using patent protection score it as either 3, effective, or 4, very effective, or 5, extremely effective. This difference holds true for each of the five major forms of statutory protection – although copyrights and industrial designs are viewed as somewhat less effective forms of protection than patents, trade secrets, and trademarks, even among firms who make use of them.

9.5.2 Canada/United States Comparisons

The Canadian experience that patents are not valued as much as alternate nonstatutory protection has a parallel elsewhere. Mansfield (1986) asked some 100 firms in 12 two-digit industries how many innovations would not have been developed in the absence of patent protection. His findings were that except in pharmaceuticals and chemicals, patents were not judged to be essential for innovation. Patents were described as essential in about 10% to 20% of commercially introduced inventions in petroleum, machinery, and metal products – less in other industries. If patents have little value, alternate forms of protection for innovation must solve the appropriability problem.

Research by Levin et al. (1987), which uses a different strategy, confirms that other forms of protection are more important than patents. Some 650 individuals – high-level R&D managers – representing firms in 130 narrowly defined lines of business were asked to evaluate the effectiveness of patents, secrecy, lead time, moving down the learning curve, and sales or service efforts as a means of protecting the competitive advantages of new products or processes. A 7-point scale (in contrast to the 5-point scale employed here) was used to rank each from 'not at all effective' to 'very effective'. The mean results are reproduced in Table 9.4.

Other means than patents were found to be just as or more effective than patents. For process innovations, lead time receives a mean score of 5.11(standard error = .05) and secrecy 4.31(.07), but patents as a means to prevent duplication receives only a score of 3.52(.06). In the case of product innovations, patents are given a higher score relative to secrecy, but lead time and sales or service efforts still outrank patents.

When only firms that use statutory forms of intellectual property protection are considered (a sample closer to the R&D managers used in the U.S. study), the Canadian results compare closely to those of Levin et al. (1987) for the United States. In the Canadian case, being first in the market receives the highest score; in the United States, it is lead time

Table 9.4. *Effectiveness of Intellectual Property Protection in the United States (Mean Score on a Scale of 1–7)*

	Overall Sample Means	
Method of Appropriation	Processes	Products
Patents to prevent duplication	3.52	4.33
Patents to secure royalty income	3.31	3.75
Secrecy	4.31	3.57
Lead time	5.11	5.41
Moving quickly down the learning curve	5.02	5.09
Sales or service efforts.	4.55	5.59

Note: Scale: 1 = not at all effective; 7 = very effective.
Source: Levin, Klevorick, Nelson, and Winter, 1987, p. 794. Reprinted with permission by the Brookings Institution Press.

that is first for process innovations and sales or service efforts that is first for product innovations. In both countries, patents trail these firm-based strategies. In both cases, patents receive an average score less than the median point on the scoring scale.[10]

9.6 LARGE VERSUS SMALL FIRMS

Innovation to some is synonymous with large firms. Schumpeter stressed the seeming advantage of large firms in the innovation process (Scherer, 1992). If large firms are more innovative, they might also be expected to make more use of statutory forms of intellectual property protection. Even if they are not more innovative, they may use intellectual property protection more frequently if there are substantial cost barriers involved with intellectual property protection that only they can overcome.

In order to study differences in the use of intellectual property protection by size of firm, firms were divided into four size classes – those with fewer than 20 employees, 20–99 employees, 100–499 employees, and more than 500 employees – and use rates were calculated for each group (see Table 9.5).

There is a substantial difference between the percentage of small and large firms that possess any of the specific intellectual properties. Only 20% of the smallest group use any one of the statutory forms of intellectual property protection, while more than 50% of each of the two largest

---

[10] The median of the U.S. scale is 4. It is 3 for the Canadian survey.

Table 9.5. *Multiple Use of Intellectual Property Protection, by Size Class (% of Firms)*

Firm Size	Number of Intellectual Property Types				
	Any	1	2	3	4+
0–20	20 (1)	12 (1)	5 (1)	2 (.4)	1 (.3)
21–100	27 (2)	17 (1)	6 (1)	3 (1)	1 (.4)
101–500	52 (3)	23 (2)	17 (2)	7 (1)	4 (1)
500+	64 (3)	29 (3)	13 (2)	10 (2)	12 (2)

Note: Standard errors are in parentheses.

Table 9.6. *Usage of Individual Forms of Intellectual Property Protection, by Size Class (% of Firms)*

Type	Size Class (Employees)			
	0–20	21–100	101–500	500+
Copyrights	4(.6)	4(.8)	12(2)	13(2)
Patents	5(.7)	8(1)	22(2)	38(3)
Industrial designs	5(.6)	7(1)	15(2)	17(2)
Trade secrets	8(.9)	5(1)	18(2)	24(3)
Trademarks	7(.8)	16(1)	29(2)	41(3)
Integrated circuit designs (semiconductor chips)	.5(.2)	1(.4)	.4(.3)	2(1)
Plant breeders' rights (plant variety rights)	0	0	2(1)	1(.5)
Other	2(.4)	1(.4)	.3(.3)	2(1)

Note: Standard errors are in parentheses.

groups do so. The largest groups are also relatively more likely to be multiple users. They are four to eight times more likely to use three or four forms of protection than are the smallest group.

Large firms are also more likely to make use of each of the specific forms of intellectual property protection (see Table 9.6). Only 5% of the smallest group are likely to avail themselves of patents, while 38% of the largest group possess at least one patent. Only 7% of the smallest group possess a trademark, while 41% of the largest group have at least one trademark. There is less of a difference in the use of industrial designs and trade secrets – though even here the differences are still significant. Small firms place relatively greater emphasis on trade secrets, compared to the emphasis they place on other forms of protection. In contrast, large

Table 9.7. *Effectiveness of Intellectual Property Protection, by Size Class (Mean Score on a Scale of 1–5)*

	Size Class (Employees)			
Type of Protection	0–20	20–99	100–499	500+
Patents	1.8(.1)	1.8(.1)	2.7(.1)	2.9(.1)
Trade secrets	2.0(.1)	2.1(.1)	2.4(.1)	2.6(.1)
Trademarks	1.9(.1)	2.1(.1)	2.5(.1)	2.8(.1)
Complexity	2.5(.1)	2.7(.1)	2.7(.1)	3.0(.1)
Being first in market	3.3(.1)	3.0(.1)	3.3(.1)	3.2(.1)

Notes: Scored as 1: not at all effective; 2: somewhat effective; 3: effective; 4: very effective; 5: extremely effective. Standard errors are in parentheses.

firms make relatively greater use of patent protection. As firms progress from small to large, they continue to focus more on trademarks than any other statutory instrument, but they reduce the emphasis on trade secrets and increase the emphasis on patents.

The valuation placed on the effectiveness of the forms of intellectual property protection also differs by size class. The average scores attributed by firms in each of the four size classes are presented in Table 9.7 for patents, trade secrets, trademarks, complexity of product design, and being first in the market.

The average score given to the statutory forms of intellectual property protection – patents, trade secrets, trademarks – for large firms is above that of small firms for all the statutory forms of protection. The largest difference occurs for patents, the smallest for secrecy. The differences between the largest and smallest group are statistically significant (at the 1% level) for patents and trademarks. They are weakly significant for trade secrets (5% level). As firms grow, they move from giving trade secrets their highest score to placing their highest value on patents. This change in valuation also accords with the differences in the relative patterns of use across size classes.

The difference across size classes in the scores attributed to the effectiveness of alternative strategies for protecting investments in intellectual property are smaller than for the statutory forms of intellectual property protection. In the case of both complexity of design and being first in the market, both small and large firms assign a higher value to these strategies than to the statutory forms of protection, and this value does not differ much across size classes.

9.7 DIFFERENCES IN THE USE OF INTELLECTUAL PROPERTY PROTECTION BY INNOVATIVE AND NON-INNOVATIVE FIRMS

9.7.1 Distinguishing Innovative and Non-innovative Firms

The innovation survey is meant to provide measures of innovation that can be used to examine differences in strategies being followed by firms. This section makes use of these measures to investigate the difference in the use of intellectual property protection by innovative and non-innovative firms.

For the purposes of this chapter, firms were defined as innovative using a combination of several questions in the Canadian innovation survey. Firms are classified as being innovative if they introduced or were in the process of introducing a product or process innovation during the period 1989 to 1991 (question 3.1), if they listed product or process innovations between 1989 and 1991 (question 3.2 and question 4.1), *or* if they reported sales in 1991 resulting from a major product innovation between 1989 and 1991 (question 1.4). The sample that is used for the comparisons of innovative versus non-innovative firms is restricted to those firms that answered the question in the first section of the survey – dealing with the percentage of sales accounted for by innovative products – that allows us to distinguish between innovative and non-innovative firms.[11]

9.7.2 Intellectual Property Use and Innovativeness

If innovation is closely associated with the use of intellectual property protection then there should be large differences in the percentage of each group that make use of protection.

Differences between innovative and non-innovative firms with regards to their exploitation of intellectual property rights and other means of preventing imitation of their products or processes are presented in Figure 9.4. Non-innovative firms make some use of intellectual property protection either because they may have innovated some years ago or because they may have purchased an intellectual asset from another firm

---

[11] The other questions used for the classification of innovative firms did not offer any problem, since nonresponse was interpreted as non-innovative. However, the question on the percentage of sales derived from innovative products has a lower response rate because of the inherent difficulty of answering the question, and no imputation for nonresponse was made. Therefore, nonrespondents to this question are excluded when comparisons are made of innovative and non-innovative firms.

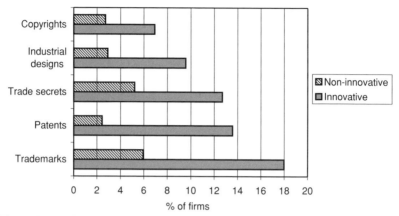

Figure 9.4. Incidence of Use of Intellectual Property in Innovative and Non-innovative Firms

and may now be exploiting that asset. However, innovative firms make significantly greater use of each form of intellectual property protection. The greatest difference occurs for patents and trademarks. Some 18% of innovative firms possess trademarks and only 6% of non-innovative firms do so.

The effectiveness (on a scale of 1 to 5) of the various forms of intellectual property protection and other protective strategies is presented in Figure 9.5 for all innovative and non-innovative firms rating the particular factor. Innovative firms generally place a higher weight on both statutory forms of intellectual property protection and the innate strategies – being first in the market and complexity of product design. However, the differences in scores between the two sets of firms are greatest for the latter strategies. Despite the fact that innovative firms tend to attribute greater value to all of the proposed methods, they are distinguished from non-innovative firms primarily in terms of the value that they attribute to the latter strategies. By itself, innovativeness is not as good a predictor of perceived effectiveness as it is a predictor of use of intellectual property protection.

The score that firms give to the various forms of intellectual property protection differs between firms using a particular form and those not using the particular form. Since innovative firms are also more likely to use a particular form of protection, the differences in the scores outlined in Figure 9.5 may be due mainly to different tendencies of firms to make use of a particular form of protection.

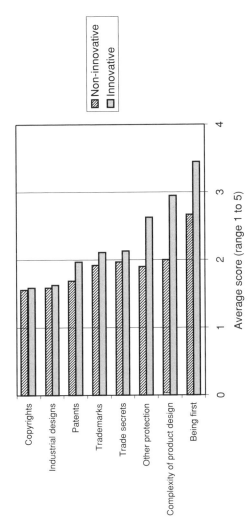

Figure 9.5. Perceived Effectiveness of Intellectual Property Protection and Other Strategies in Innovative and Non-innovative Firms

To investigate this possibility, firms are divided into four groups – based on whether or not they are innovative and whether or not they make use of one of the statutory forms of intellectual property protection. The mean scores for innovative and non-innovative firms for those that possess a specific form of protection and for those that do not make use of one of the statutory forms of protection are presented in Table 9.8. When this is done, the importance of being innovative is much reduced. Innovative firms generally place more value on intellectual property than do non-innovative firms, both for the group that use statutory forms of intellectual property protection and for those that do not, but the differences are not significant. There is, however, a significant difference in the scores given to the alternative strategies – that is, being first in the market.

The comparison of firms that made or did not make use of intellectual property protection presented in Table 9.8 suggests that innovative firms fall into two groups. One group utilizes statutory forms of intellectual property protection and perceives them to be valuable, but perceives other strategies, such as being first in the market and the complexity of product design, to be equally or more effective in preventing imitation of their new products and processes. The second group of innovative firms does not perceive statutory forms of intellectual property protection to be effective methods of preventing imitation and, consequently, does not make use of these methods. The second group of firms believes the other strategies – related to being first in the market – are the only effective means of reaping the benefits of innovation activity.

9.7.3 Innovation Differences Across Size Classes

Intellectual property protection is used to protect the fruits of innovation. Since so much emphasis has been placed on differences in the abilities of large and small firms to innovate, it is important to examine the extent to which differences in intellectual property use are linked to differences in tendencies to innovate. To do so, the percentage of firms that were innovative, the percentage that performed R&D continuously, and the percentage that protected themselves with statutory forms of rights are compared across size classes (see Table 9.9).

Two definitions were used to define innovation. The first is whether a firm introduced a product or process innovation. Second, innovative firms were defined as those that reported 1991 sales from a major product innovation introduced between 1989 and 1991. These firms make up a

Table 9.8. *Effectiveness of Alternative Means of Protecting New Products and Processes from Imitation: For Innovators and Non-innovators, Users and Nonusers of Intellectual Property Protection (IPP) (Mean Score on a Scale of 1–5)*

	Firms That Used IPP			Firms That Did Not Use IPP		
	Innovative	Non-innovative	Sign of difference	Innovative	Non-innovative	Sign of difference
Intellectual property protection associated with:						
Copyrights	2.85	2.77	+	1.33	1.39	−
Patents	3.02	3.03	−	1.51	1.47	+
Industrial designs	2.50	2.63	−	1.40	1.39	+
Trade secrets	3.25	2.97	+	1.64	1.58	+
Trademarks	3.11	3.05	+	1.50	1.45	+
Integrated circuit designs	3.34	2.75	+	1.28	1.15	+
Plant breeders' rights	2.26	N.R.		1.14	1.15	−
Other strategies						
Complexity of product design	3.21	2.47	+*	2.64	1.65	+*
Being first in the market	3.61	2.93	+*	3.26	2.49	+*
Other	3.12	1.91	+**	2.16	1.89	+

Note: The first three columns examine the difference for firms that used any intellectual property right; the second three columns compare scores for firms that did not use specific intellectual property rights (i.e., patents).
* Significant difference at 1% level.
** Significant difference at 5% level.
N.R. = not reported.

Table 9.9. *A Comparison of Innovation Intensity, R&D, and Intellectual Property Use, by Size Class (% of Firms)*

% of Firms	Size Class (Employees)			
	0–20	21–100	101–500	500+
Introducing a product or process innovation	30 (1)	39 (2)	42 (3)	62 (3)
With sales from a major product innovation	19 (1)	9 (2)	36 (3)	37 (3)
Performing ongoing R&D	19 (1)	36 (2)	43 (3)	58 (3)
Using any form of statutory intellectual property protection	20 (1)	28 (2)	52 (3)	64 (3)
Using patents	5 (1)	8 (1)	22 (2)	38 (3)

Note: Standard errors in parentheses.

subset of those used in the previous section of this chapter.[12] The percentage of firms that are innovative varies substantially by size class. For the three largest size classes, the percentage of firms that perform R&D on a continuous basis is about the same as the percentage of firms that introduced a new product or process, but above the percentage that reported sales from a major product innovation – because the definition of innovation being used here excludes process innovations. But the percentage of firms making use of any form of statutory protection is quite similar to the percentage of firms that perform R&D continuously.

Differences across size classes in the use of intellectual property protection closely mirror other differences in both the use of inputs (R&D) to the innovation process and the production of outputs. While only 20% of the smallest group of firms possess intellectual property, compared to 64% for the largest group of firms, small firms are also less likely to have introduced a major product innovation or to be doing research and development. Only about 20% of the smallest group of firms report an ongoing research and development program, while 58% of large firms do the same.

Although the percentage of firms that perform R&D or possess intellectual property rights is quite similar to the percentage of firms that report sales during 1992 from a major product innovation, this similarity cannot be used to infer that innovative firms are all performing R&D or that all those performing R&D are innovative. Nor would we expect either condition to occur. Not all new products need R&D; not all R&D

---

[12] This variant must be used because it is the only innovation question that was answered by small firms that were sent the R&D and the intellectual property questions.

Table 9.10. *A Comparison of R&D and Intellectual Property Use for Innovators, by Size Class (Indexed to Firms Reporting Sales from Major Product Innovation)*

% of Firms	Size Class (Employees)			
	0–20	21–100	101–500	500+
With sales from a major product innovation	100	100	100	100
Performing ongoing R&D	40 (3)	58 (4)	51 (4)	78 (4)
Using any form of statutory intellectual property protection	32 (3)	43 (4)	71 (4)	81 (4)
Using Patents	8 (2)	16 (3)	33 (4)	61 (5)

Note: Standard errors are in parentheses.

is successful. Nevertheless, it is interesting to examine the percentage of innovative firms that do perform R&D or use intellectual property. These percentages are tabulated just for those firms reporting sales from a major product innovation (see Table 9.10).

In the largest size class, almost 80% of product innovators perform R&D on a continuous basis. The same percentage makes use of some form of statutory intellectual property protection. About 61% possess at least one patent. The percentage of innovators doing R&D on a continuous basis falls to between 51% and 58% for the middle size classes and to only 40% for the smallest size classes. The smallest innovators, then, tend to make less use of the R&D process. The use of statutory intellectual property also falls off for the smaller size classes.

In conclusion, the tendency of smaller firms to make less use of intellectual property does not stem entirely from differences in tendencies to innovate. Within the group that innovates, a smaller percentage performs R&D and also a smaller percentage uses intellectual property.

The limited financial resources of small firms provide one explanation of this pattern. As indicated in Chapter 7, the majority of small firms do not perform regular R&D and rely more on management and production personnel. As such, they are less likely to produce the type of original innovation that is protected by patents. Moreover, the limited use that small firms make of intellectual property, according to Lerner (1994), arises because small firms are deterred by litigation costs. This view is supported by the recent findings of Cohen, Nelson, and Walsh (2000), who found a positive correlation between the size of firm and the cost of defending patents in court in the United States as a reason not to apply for a patent.

9.7.4 Intellectual Property Protection and the Characteristics of Innovations

A firm's attitude toward intellectual property will depend upon the nature of its innovation strategies – whether it is a leader or a follower. A firm may adopt different intellectual property strategies for different types of innovations. If this is so, a firm-based response to the importance of intellectual property protection may hide important differences in the use of intellectual property for different types of innovations.

In order to investigate how intellectual property protection varies for different types of innovations, we use the information that the innovation survey collected on whether *the most important innovation* was protected by intellectual property rights – by copyrights, patents, industrial designs, trademarks, secrecy agreements, integrated circuit designs, and plant breeders' rights, or through other means. In the next three subsections, methods of protection used by large firms (IPs) for their most important innovation are tabulated by novelty and type of innovation.

9.7.4.1 Intellectual Property Protection for Major Innovations

Some 33% of large firms with a major product or process innovation use at least one of the forms of intellectual property protection listed in Canada (see Table 9.11). Over 22% do so in the United States, 9% in Europe, and 7% in the Pacific Rim. The statutory system of intellectual property

Table 9.11. *Usage of Intellectual Property Protection for Major Innovation of Innovator, by Region (% of Innovators)*

	Canada	USA.	Europe	Pacific Rim	Other	All Regions
Copyrights	3	2	.1	0	–	3
Patents	14	12	5	4	1	14
Industrial designs	6	3	1	1	1	6
Trademarks	10	5	2	1	1	10
Secrecy agreements	11	6	2	2	1	12
Integrated circuit designs	1	1	–	–	1	1
Plant breeders' rights	–	–	0	0	0	1
Other	3	2	1	1	1	4
All property rights	33	22	9	7	4	35

Notes: Larger firms (IPs) only. Standard errors range from .2 for the estimates less than 5 to around 1 for the estimates that are greater than 10.
– = nonexistent.

protection is used by about one of every three innovative firms for their major innovation.

Some 14% of large firms introducing a major innovation made use of patent protection in Canada or elsewhere (Table 9.11).[13] This means that fewer than 1 in 5 innovations that are classified by the respondents as major innovations makes use of the patent process.

While the use of most forms of intellectual property protection abroad is less important than in Canada, this is not the case for patents. Almost as many of the firms indicate that they protect their innovations by taking out patents for their innovations in the United States (12%) as do so in Canada (14%). Patents are taken out in Europe at less than one-half the rate they are used in the United States. The Pacific Rim follows at about one-third the U.S. rate. Cross-border protection is important for innovations of considerable magnitude, such as those being investigated here.

The next most important manner of protection used for major innovations is secrecy agreements.[14] Some 11% of large firms with a major innovation indicate that secrecy agreements in Canada made up part of the strategy used to protect their innovation. Secrecy agreements are relatively less important in other countries than are patents.[15] Some 12% of innovations are protected by patents in the United States, but only 6% via secrecy agreements.

The third most important method of protection for a major innovation in Canada (10%) involves the use of trademarks. Trademarks are more important than industrial designs (6%) and copyrights (3%).

In summary, information on the use of intellectual property protection for major innovations confirms the general importance that is attached to protection at the firm level. But it also shows some differences. When the protection afforded to the major innovation of the company is examined, the importance of both patents and trade secrets increases, while it falls for trademarks.

---

[13] These are the figures for large firms (IPs) that come from a question that asks whether the major innovation made use of intellectual property, while the questions in Table 9.2 for all firms ask whether intellectual property is used at all.

[14] The reader should note that secrecy agreements used here are more proactive than trade secrets. A trade secret is simply a knowledge asset that is protected by nondisclosure; secrecy agreements are pacts that are made with other firms.

[15] There is no international coordination of trade secrecy agreements comparable to the international conventions governing patents, trademarks or copyrights. The legal costs of defending secrecy agreements is in some countries even larger than patent litigations costs.

When the use of intellectual property is tabulated by different characteristics of firms' major innovations, even more marked differences from the earlier profile emerge. Two characteristics are used here. The first is whether the innovation was a product or process. The second is whether the innovation was a world-first or otherwise.

9.7.4.2 Products Versus Processes

Previous work (Firestone, 1971; Levin et al., 1987; and most recently Cohen et al., 2000) indicates that products are better suited to patent protection than are processes. Processes by their very nature can be better protected by trade secrets because they can be kept behind closed doors. Dividing major innovations into those involving new products versus those involving new processes allows us to test this hypothesis in a Canadian context.

For this purpose, firms were divided on the basis of whether their major innovation involved a product and a process change, only a product innovation that did not involve any process changes, or only a process innovation. The type of protection for the most important innovation was then investigated.

Major innovations involving combinations of products and processes are more likely to make use of a statutory form of intellectual property protection than are those that only involve new products or only new processes (see Table 9.12). Some 61% of firms indicating that they

Table 9.12. *Usage of Intellectual Property Protection for Major Product and Process Innovations (% of Innovators)*

	Combination of Product/Process	*Only* Product Innovation Without Change in Man. Technology	*Only* Process Innovation Without Product Change
Copyrights	6 (1)	4 (2)	1 (1)
Patents	23 (2)	28 (5)	10 (3)
Industrial designs	12 (2)	8 (3)	7 (3)
Trademarks	20 (2)	10 (3)	20 (4)
Secrecy agreements	18 (2)	20 (4)	7 (3)
Integrated circuit designs	2 (1)	2 (2)	2 (1)
Plant breeders' rights	1 (1)	0.3 (1)	1 (1)
Other	5 (1)	7 (3)	2 (1)
Any	56 (3)	61 (5)	34 (5)

Notes: Larger firms (IPs) only. These firms are generally larger than 20 employees. Standard errors are in parentheses.

Table 9.13. *Usage of Intellectual Property Protection, by Region and by Product or Process (% of Firms with a Major Innovation That Reported Making Use of Intellectual Property Rights for This Innovation)*

	Canada	USA	Europe	Pacific Rim	All
Panel A: Combination of product process innovation					
Copyrights	11 (2)	6 (2)	1	.1	11 (2)
Patents	39 (3)	35 (3)	17	12	41 (4)
Industrial designs	20 (3)	10 (2)	2	1	21 (3)
Trademarks	31 (3)	14 (2)	5	4	32 (3)
Secrecy agreements	34 (3)	20 (3)	8	8	35 (3)
Integrated circuit designs	3 (1)	2 (1)	.2	0	3 (1)
Plant breeders' rights	1 (1)	1 (2)	0	0	1 (1)
Other	7 (2)	3 (2)	1	1	9 (2)
Any	96 (1)	60 (4)	26 (3)	19 (3)	100
Panel B: *Only* product innovation with no change in man. technology					
Copyrights	7 (4)	2 (2)	0	0	7 (4)
Patents	43 (7)	38 (7)	13	10	45 (7)
Industrial designs	12 (5)	7 (4)	4	4	12 (5)
Trademarks	32 (6)	19 (5)	.4	1	32 (6)
Secrecy agreements	16 (5)	5 (3)	4	4	17 (5)
Integrated circuit designs	4 (3)	4 (3)	3	3	4 (4)
Plant breeders' rights	.4 (1)	0	0	0	.4 (1)
Other	7 (4)	5 (3)	4	5	11 (4)
Any	95 (3)	68 (6)	24 (6)	22 (6)	100
Panel C: *Only* process innovation without product innovation					
Copyrights	0	0	0	0	4 (4)
Patents	26 (8)	25 (8)	8	8	30 (8)
Industrial designs	16 (7)	0	0	0	19 (7)
Trademarks	15 (6)	7 (5)	7	7	19 (7)
Secrecy agreements	48 (9)	36 (9)	6	3	59 (9)
Integrated circuit design	1 (1)	0	0	0	5 (4)
Plant breeders' rights	0	0	0	0	4 (4)
Other	5 (4)	5 (4)	0	0	5 (4)
Any	89 (6)	73 (8)	21 (7)	18 (7)	100

Notes: Larger firms (IPs) only. Standard errors are in parentheses.

introduced only product innovations make use of formal statutory protection either in Canada or abroad. Some 34% of firms with only process innovations are likely to protect themselves with intellectual property rights.

The relative tendency to make use of the intellectual property protection system in Canada in contrast to protection abroad also varies between product and process innovations. This is outlined in Table 9.13 where the distribution of intellectual property use is depicted just for those firms that made use of any intellectual property right.

The use of the various forms of protection varies considerably between product and process innovations. Of those large firms introducing a product innovation with no change in manufacturing technology, some 43% make use of patents to protect it in Canada. Trademarks are second (32%). Secrecy agreements come third at 16%. By way of contrast, for those large firms introducing process innovations with no change in products, secrecy agreements are used most (48%) and patents are used by only some 26%. Not surprisingly, for cases where the innovation involved both a product and process change, patents and secrecy agreements are adopted more equally, 39% and 34%, respectively.

In all three cases, patent protection is generally taken out in U.S. markets about as frequently as in Canadian markets. On the other hand, other instruments tend to be used less frequently in the United States. In particular, secrecy agreements tend to be used less frequently in U.S. than in Canadian markets.

9.7.4.3 World-First Versus Other Types of Innovations

Not all innovations are equally significant. In order to examine how intellectual property protection varies by degree of novelty, firms are divided into two groups – those that indicated their most important innovation was a world-first, and those that indicated it was a Canadian-first or 'other'. Only 16% of innovations are classified as being world-firsts, some 33% are firsts for Canada, and the remaining 51% are essentially improvements of existing products or processes.[16]

Innovations that are world-firsts are most likely to involve the need for intellectual property protection. Firms that are imitators either purchase patents or invent around existing patents. In the latter case, they have less of an incentive to register their new designs or processes for two reasons. Registration will not be successful unless they have created a new product or new process with sufficient originality to pass the patent examiners. Or the very act of registration may provide information to the original innovator that would allow it to challenge the validity of the innovation in court. As a result, innovations that are world-firsts are hypothesized to make greater use of intellectual property laws.

---

[16] Only some 9% of firms indicating they had a major innovation did not answer this question.

Table 9.14. *World-First/Non-World-First Usage of Intellectual Property Protection for Major Innovation, by Region (% of Innovators)*

	Canada	USA	Europe	Pacific Rim	All
Panel A: World-first					
Copyrights	10 (3)	7 (3)	–	–	12 (3)
Patents	39 (5)	36 (5)	21 (4)	17 (4)	42 (5)
Industrial designs	12 (3)	8 (3)	7 (3)	4 (2)	14 (3)
Trademarks	29 (4)	18 (4)	9 (3)	6 (2)	31 (5)
Secrecy agreements	30 (5)	23 (4)	9 (3)	10 (3)	33 (5)
Integrated circuit design	1 (1)	1 (1)	0	0	3 (3)
Plant breeders' rights	0	0	0	0	2 (1)
Other	5 (2)	3 (2)	0	–	5 (2)
Any	75 (4)	60 (5)	29 (4)	24 (4)	78 (4)
Panel B: Non-world-first					
Copyrights	2 (1)	1 (1)	–	0	2 (1)
Patents	11 (1)	9 (1)	3 (1)	2	11 (1)
Industrial designs	5 (1)	2 (1)	–	–	5 (1)
Trademarks	8 (1)	3 (1)	1 (1)	1	8 (1)
Secrecy agreements	9 (1)	4 (1)	2 (1)	1	9 (1)
Integrated circuit design	1 (1)	1 (1)	–	–	1
Plant breeders' rights	–	0.2	0	0	0.5
Other	2 (1)	1 (1)	1 (1)	1	4
Any	28 (2)	18 (2)	6 (1)	5 (1)	30 (2)

Notes: Larger firms (IPs) only. Standard errors are in parentheses.

This is confirmed by the differences in the use of intellectual property laws found in these two groups (see Table 9.14). Over three-quarters of world-firsts make use of some form of statutory protection, compared to less than one-third for the other group. As expected, the gap is largest between the two groups for patent protection. Some 42% of world-firsts make use of patent protection. Only 11% of the rest make use of patents. Some 31% of world-firsts make use of trademarks; only 8% of the other group do so. Some 33% of world-firsts use secrecy agreements; only 9% of the other innovations do so.

World-firsts are also more likely to be protected in other countries. The ratio of patents taken out in Europe and the Pacific Rim relative to their use in Canada is higher than for other innovations.

9.8 INDUSTRY DIFFERENCES

The use of statutory forms of intellectual property protection might be expected to vary across industries for several reasons.

First, the scientific climate that is conducive to the discovery of new product entities is not the same everywhere. The superior science base for some industries means that more scientific discoveries are made in these sectors than elsewhere.

Second, industries differ in the extent to which statutory forms of protection for intellectual property provide the most efficacious method of protecting innovations. Patents and other forms of intellectual property protection are not equally useful across industries. Patents, it has been stressed, require clear standards for definition and for defence against infringements. Research by Taylor and Silbertson (1973) and Mansfield (1986) suggests that pharmaceuticals and chemicals, followed by mechanical engineering, benefit most from the patent system. Chemical entities are relatively easy to define and, therefore, to protect; mechanical inventions satisfy the same preconditions of discreteness and identifiability (Levin et al., 1987).

In order to allow for inherent differences in the scientific climate in each industry, industries are grouped here into three sectors based on the intensity of innovative activity in each – core, secondary and 'other'.

Canadian patent statistics reflect these differences among the three sectors. These statistics are kept both on an industry of manufacture basis (the industry where the patent is taken out) and on an industry of use basis (the industry that will use the product or process covered by the patent). The cumulative totals of all patents granted between 1972 and 1987 are presented in Table 9.15, for both industry of manufacture and for industry of use.

The core group of industries is responsible for the highest intensity of patent creation (industry of manufacture), with electrical and electronic products leading the way, but with machinery a close second and chemical products following third. The secondary group produces goods covered by fewer patents than the core group. In the secondary group, transportation and metal fabrication industries take out the most patents for their innovations. Industries in the other group of industries generally use patents less. The most active industry here is paper products. It resembles nonmetallic minerals, primary metals, and rubber products, all of which are in the secondary group.

The core group tends to produce more patentable innovations than it uses. The ratio of patents used to those manufactured is .6 for this group. The secondary group makes use of about the same number of patents as it creates – with a ratio of use to manufacture of 1:1. The other sector uses 2.7 times what it makes. Canada's experience in these three

Table 9.15. *Industry Patterns of Patent Use*

Industry Sector	Cumulative Patents Granted, 1972–87		
	Industry of Use	Industry of Manufacture	Industry of Use/Industry of Manufacture
Core sector	104628	173349	0.6
Refined petroleum and coal	2000	626	
Electrical and electronic products	45651	58320	
Chemicals and pharmaceuticals	29010	44694	
Machinery	21244	52344	
Scientific instruments	6723	17365	
Secondary sector	41483	37429	1.1
Transportation	16602	12021	
Nonmetallic mineral products	2861	2454	
Primary metal	4676	2322	
Fabricated metal	8347	14751	
Plastics	6472	4657	
Rubber	1525	1224	
Other manufacturing	16352	6066	2.7
Beverage	545	15	
Wood	1177	459	
Printing	2630	332	
Clothing	725	129	
Food	3367	964	
Primary textiles	1740	803	
Furniture and fixtures	659	901	
Textile products	2171	957	
Paper	2874	1347	
Leather	464	159	
Other	1438	4157	

Source: Special tabulations of the Department of Consumer and Corporate Affairs (1988).

sectors – core, secondary, and other – conforms to the same pattern that has been observed in the U.K. and the United States.

9.8.1 Industry Use of Intellectual Property Protection

Data on the actual usage of intellectual property protection taken from the innovation survey also show that there is a considerable variation across the core groups in the intensity of use of patents, trademarks, trade secrets, and industrial designs (see Table 9.16). The most frequent users of patent

Table 9.16. *Usage of Individual Forms of Intellectual Property Protection, by Individual Industry (% of Firms)*

Industry Sector	Patents	Industrial Designs	Trademarks	Trade Secrets	Any of Previous
Core sector	16 (2)	8 (1)	17 (2)	11 (1)	32 (2)
Electrical & electronic products	13 (3)	20 (2)	17 (3)	12 (3)	30 (4)
Chemicals and pharmaceuticals	20 (3)	6 (2)	30 (4)	14 (3)	38 (4)
Machinery	19 (3)	9 (2)	13 (3)	8 (2)	32 (4)
Refined petroleum and coal	10 (5)	7 (4)	19 (7)	19 (7)	26 (7)
Secondary sector	11 (1)	9 (1)	8 (1)	11 (1)	24 (2)
Fabricated metal	12 (2)	10 (2)	6 (2)	8 (2)	21 (3)
Nonmetallic mineral products	6 (2)	3 (1)	8 (2)	13 (3)	21 (3)
Primary metal	10 (4)	7 (3)	10 (4)	9 (4)	21 (5)
Plastic	16 (4)	10 (3)	15 (4)	12 (3)	28 (4)
Rubber	18 (7)	19 (7)	20 (7)	19 (7)	44 (8)
Transportation equipment	12 (3)	11 (2)	11 (2)	13 (2)	30 (4)
Other sector	3 (1)	3 (1)	11 (1)	7 (1)	17 (1)
Beverage	1 (2)	1 (2)	33 (7)	1 (2)	35 (7)
Clothing	2 (1)	1 (1)	8 (3)	5 (2)	12 (3)
Food	3 (1)	3 (1)	23 (3)	13 (2)	29 (3)
Furniture and fixtures	2 (1)	8 (3)	6 (2)	5 (2)	15 (4)
Leather	6 (3)	4 (3)	10 (5)	9 (4)	17 (6)
Other manufacturing	5 (2)	5 (2)	11 (3)	11 (3)	26 (4)
Paper	5 (2)	8 (3)	18 (4)	13 (4)	27 (5)
Printing and publishing	1 (1)	1 (1)	8 (2)	3 (1)	11 (2)
Primary textiles	4 (3)	12 (6)	10 (5)	2 (2)	22 (7)
Textiles products	8 (3)	5 (3)	11 (4)	12 (4)	19 (4)
Wood	1 (2)	3 (1)	5 (2)	3 (1)	9 (2)

Note: Standard errors are in parentheses.

protection are machinery, rubber, plastics, and chemicals. Industries in the core group have the greatest intensity (16%), those in the secondary group are next (11%), and other industries come last (3%). These differences have the same sign as the number of patents registered (applications) but not the same magnitude. Patent registrations in the core sector are five times those in the secondary sector (Table 9.16).

There are a number of reasons for these differences. First, patent registrations obtained from administrative data do not have to be made by firms operating in Canada, as was the case for the innovation survey.

Second, patent registrations in administrative data measure total output and not output per firm as is calculated here.

Industrial design use exhibits an intersectoral pattern similar to that of patents. Industries in the core and secondary industries have the highest intensity of use – 8% and 9%, respectively. Only 3% of firms in 'other' industries make use of industrial designs. The industries that most frequently use industrial designs are rubber, plastics, primary textiles, transportation equipment, fabricated metal, and electrical.

The pattern evident with patents and industrial designs disappears for trademarks. Industries in the core group are most likely to use trademarks (17%), but it is the other set of industries that comes second (11%). The highest-using industries – beverages, chemicals, refined petroleum, food, and rubber – are scattered across all three groups.

The pattern for trade secrets falls somewhat between the two other models. The core and the secondary groups are the heaviest users of trade secrets – around 11% each – while the third group follows more closely behind (7%) than in the case of patents or industrial designs. The most frequent industry users come from all three groups – refined petroleum, rubber, and paper. The least frequent users are found in the other industry group.

Thus, electrical, chemicals, petroleum, plastics, and rubber are particularly intensive users of several different forms of intellectual property patents, trademarks, industrial design, and trade secrets. Food and paper are particularly strong users of trademarks and trade secrets. Beverages use trademarks intensely. Primary textiles relies on industrial designs and trademarks. Wood and clothing are infrequent users of almost all forms of intellectual property protection.

The uses of the various forms of intellectual property protection are related. Cross-industry correlation is high between patents and design (.65), and between patent use and trade secrets (.60). In contrast, trademarks have low correlation with patents (.27), industrial designs (.02), and trade secrets (.44). The highest correlation is between design and trade secret use (.77).

9.8.2 Industry Effectiveness of Intellectual Property Protection

Industries may differ with regard to their evaluation of the efficacy of intellectual property protection for two reasons. First, firms may have views regarding the efficacy of the various forms of protection that do not depend upon whether they use the protection, but which differ by industry because conditions in some industries make it easier to protect appropriability using patents. In this case, average scores per sector would be similar for most firms in an industry but would differ across industries. Second, firms may not differ in the extent to which they value intellectual property protection if they use it, but they may differ in the extent to which they use protection. The latter would occur if different scientific environments lead some industries to offer more scope for innovation than other industries do. In this case, the evaluation of the form of intellectual property protection will differ across users and nonusers in an industry but would be relatively similar across industries for users. Industries would differ with regards to average scores of users and nonusers taken together, primarily because the proportion of users and nonusers varies across industries.

To examine whether one or both of these explanations is at work, the average scores given to different forms of intellectual property protection are tabulated in Table 9.17 by industry sector. Three different methods are used. The average score for all firms is given in panel A. It will differ across industries either because the percentage of firms using intellectual property protection varies across industries or because the valuation of

Table 9.17. *Effectiveness of Individual Forms of Property Protection, by Industry Sector (Mean Score on a Scale of 1–5)*

Sector	Copyrights	Patents	Industrial Designs	Trade Secrets	Complexity	Lead Time	Other
A) All firms							
Core	1.8	2.3	1.9	2.2	3.0	3.3	2.9
Secondary	1.5	2.0	1.7	2.2	2.7	3.2	2.1
Other	1.6	1.6	1.4	1.9	2.4	3.2	2.3
B) Firms with any intellectual property							
Core	2.0	2.6	2.2	2.5	3.2	3.3	3.2
Secondary	1.6	2.4	1.9	2.7	3.0	3.4	2.9
Other	2.1	2.1	1.9	2.6	2.9	3.6	2.1
C) Firms with specific forms of intellectual property							
Core	2.8	3.0	2.1	3.3			
Secondary	2.4	3.0	2.4	3.4			
Other	3.0	3.2	2.3	2.9			

Notes: The scores in panel B for complexity, lead time, or other are taken for the same sample used for the main instruments – firms that indicated that they used one of the copyrights, patents, industrial designs, or trade secrets. In panel C, the sample in each column is taken as the firms using the instrument in that column. Since there is no counterpart to use of complexity or lead time in question 5.1, no data are included for these methods in panel C. Standard errors are generally less than .1.

those who use it differs. The average score of all firms that actually use any form of intellectual property protection is given in panel B. The average score of all firms that use the particular property right being evaluated is given in panel C. If the scores differ by sector in panel A but not in panel B or C, firms differ not so much in terms of their evaluation of the efficacy of protection as in their tendency to be innovative and to make use of the protection.

The score given to the various forms of protection by all firms in panel A tends to be highest in the core sector and declines in the secondary and 'other' sectors. Innate forms of protection – such as complexity, lead time, and other means – receive higher scores in all sectors, compared to the statutory means of protection.

When only firms that possess any intellectual property right are used for tabulations (Table 9.17, panel B), the gradation between the core, secondary, and other sectors remains for patents, but the differences disappear elsewhere. It is still the case that complexity, lead time, and other natural strategies are given higher scores than copyrights, patents, industrial designs, and trade secrets.

By contrast, the average score given to the efficacy of patents by just the users of the patents (Table 9.17, panel C) tends to vary much less than in panel A or panel B. This suggests that the average differences across industries for patents for all firms do not reflect inherent differences in the efficacy of patents; rather, they reflect differences in inherent opportunities for innovation. This confirms findings in the United States. Cockburn and Griliches (1988) attempted to explain cross-industry differences in patent/R&D ratios using the U.S. data on the effectiveness of patents, but found that the latter had little effect on the number of patents taken out per dollar expenditure on R&D – primarily because the measures of effectiveness have very little cross-industry variation (Griliches, 1990).

9.9 MULTIVARIATE ANALYSIS OF INTELLECTUAL PROPERTY USE AND EFFECTIVENESS

Previous sections of this chapter have examined how two different aspects that measure the importance of intellectual property protection vary by the characteristic of reporting firms. The first aspect involves use; the second consists of evaluations of the usefulness of different instruments in protecting innovative ideas. Both use and self-evaluations were generally found to be higher for larger firms, for foreign-owned firms, for more innovative firms, and for firms in certain core industries that are responsible for the largest proportion of innovations.

These findings are all based on simple two-way comparisons of the use of different forms of intellectual property or on evaluations of these forms of protection – tabulations by size or by ownership or by innovativeness. This section uses multivariate analysis to ask whether each of these characteristics still matters once the others have been taken into account. It may be, for instance, that foreign-owned firms make greater use of different forms of intellectual property protection simply because they are larger. Being able to correct for the effect that firm size has on intellectual property use permits conclusions to be drawn about whether foreign ownership matters once the effect of size has been taken into account.

In order to disentangle the various effects, a multivariate analysis is used to examine the connection between use or the efficacy scores attached to intellectual property protection and various firm characteristics. Two models are employed for each of the five major statutory forms of intellectual property protection – copyrights, patents, industrial designs, trade secrets, and trademarks. The first is a binary variable that measures use – 0 if the form of intellectual property protection in question is not used, 1 if there is at least one use made of it. For this variable, a probit regression is employed. The second variable used is the score attached to the form of intellectual property protection. For this ordinary least squares (OLS) regression analysis is used. In both cases, weighted regressions (using company weights) are employed.

The regressors are size of firm (SIZE), nationality (FOREIGN), innovativeness (INNOVATE), and three industry classifications – core (CORE), secondary (SECONDARY), other manufacturing industries and all other industries (OTHER).[17] SIZE is defined as 1 if the firm has more than 200 employees and 0 otherwise. FOREIGN is defined as 1 if the firm is foreign controlled and 0 otherwise. INNOVATIVE is defined as 1 if the firm has introduced an innovation and 0 otherwise. The definitions employed are those used in this chapter. The definition of innovativeness is the comprehensive definition that was used previously. Each of the industry sectors is also represented with an industry variable. The regression uses both the larger firms (IPs) and the smaller firms (NIPs).

---

[17] Although only manufacturing establishments were sampled, some of the owning enterprises fell outside the manufacturing sector.

Table 9.18. *Regression Coefficients for Utilization of Intellectual Property Protection*

Variable	Patents	Copyrights	Designs	Trade Secrets	Trademarks	Any of Previous
SIZE	0.74	0.39	0.51	0.48	0.68	0.82
	(0.0001)	(0.03)	(0.003)	(0.03)	(0.0001)	(0.0001)
INNOVATE	0.82	0.43	0.55	0.45	0.63	0.75
	(0.0001)	(0.0001)	(0.0001)	(0.0001)	(0.0001)	(0.0001)
FOREIGN	0.24	0.33	0.06	−0.06	0.46	0.30
	(0.12)	(0.06)	(0.74)	(0.72)	(0.009)	(0.02)
CORE/SECONDARY	0.76	−0.09	0.42	0.20	−0.08	0.17
	(0.0001)	(0.33)	(0.0001)	(0.01)	(0.27)	(0.006)
Log likelihood	−495	−415	−476	−649	−754	−1150
N	2368	2368	2368	2368	2368	2368

Note: Number in parentheses is the probability value of the null hypothesis that 'the coefficient is zero' is true.

The results presented in Table 9.18 are for usage. The omitted category here is a small, domestically owned, non-innovative firm located outside the core and secondary sectors. Table 9.19 includes results for the score attached to the effectiveness of intellectual property protection. The omitted category is a small, domestically owned, non-innovative firm located outside the manufacturing sector.

Looking first at use, it is evident that the innovativeness of a firm has a significant (at the 1% level) impact everywhere. Size is also significant across all the categories. Nationality of ownership is not significant everywhere. It is insignificant in three areas – in the case of patents, trade secrets, and industrial designs. It is only highly significant for trademarks. This suggests that foreign firms cross geographic boundaries to exploit knowledge-based assets relating to hard-to-transfer marketing brand name assets (Caves, 1982). Foreign-owned firms are therefore not more likely to make use of patents, trade secrets, or industrial designs once other characteristics of these firms, such as size and innovation potential, are taken into account. It may be that firms that are using trademarks are making their innovations elsewhere and, therefore, trademarks is catching the innovation differences that are not being captured in the survey. Firms in the core and secondary sectors are also more likely to have higher use rates for patents, industrial designs, and trade secrets.

Several variants of the regressions reported in Table 9.18 were also estimated. The first utilized a continuous size variable, employment in the firm, as opposed to the binary variable reported here. It was included to

Table 9.19. Regression Coefficients for Effectiveness Score Attached to Intellectual Property Protection

Variable	Patents	Copyrights	Designs	Trade Secrets	Trademarks	Complex	Being First
SIZE	0.49	0.14	0.32	0.13	0.43	−0.09	−0.14
	(0.0001)	(0.19)	(0.001)	(0.29)	(0.001)	(0.48)	(0.23)
INNOVATE	0.22	0.02	0.01	0.15	0.21	0.90	0.78
	(0.016)	(0.78)	(0.88)	(0.14)	(0.03)	(0.0001)	(0.0001)
FOREIGN	0.53	0.33	0.36	0.49	0.39	0.47	−0.007
	(0.003)	(0.07)	(0.04)	(0.02)	(0.03)	(0.02)	(0.97)
CORE	0.48	0.31	0.43	0.18	0.22	0.69	0.61
	(0.007)	(0.08)	(0.014)	(0.38)	(0.30)	(0.002)	(0.003)
SECONDARY	0.55	0.27	0.44	0.28	0.17	0.48	0.53
	(0.003)	(0.13)	(0.01)	(0.17)	(0.41)	(0.035)	(0.01)
OTHER MANUFACTURING	−0.06	0.31	0.18	−0.02	0.43	0.25	0.52
	(0.73)	(0.07)	(0.29)	(0.91)	(0.02)	(0.24)	(0.007)
R^2	0.11	0.007	0.04	0.03	0.05	0.13	0.07
F	18.18	1.81	6.31	3.20	6.51	19.48	12.53
Prob$>F$	0.001	.095	0.001	.004	0.0001	0.0001	0.001
N	804	673	645	704	825	694	806

Note: Number in parentheses is the probability value of the null hypothesis that 'the coefficient is zero' is true.

see whether nationality might partially be capturing size. The inclusion of the continuous variable has no effect on the significance of the nationality of ownership variable, except in the case of patents, where its inclusion makes the nationality variable positive and significant. The continuous size variable is significant (at the 5% level) for patents and trade secrets, but not for industrial designs.

The second variant includes the score given to each intellectual property right on the grounds that firms may acquire intellectual property only where they feel it is useful. In each case, the score was associated with a highly significant and positive coefficient on use. But its inclusion does not remove the other regressors as significant explanatory variables, with a few minor exceptions.

In the case of the score given to the different categories (Table 9.19), the same general patterns exist, but they are not quite as significant across all categories.[18] For patents, all variables are significant. Firms give a higher rating to patents if they are larger, more innovative, foreign owned, and in the core or secondary sector. Both trademark and industrial design scores are also generally related to all the explanatory categories. By way of contrast, the scores given to copyrights and trade secrets are not as significantly related to size or innovativeness. However, both copyrights and trade secrets are given a higher evaluation by firms that are foreign controlled.

The same regression that was performed for the various forms of statutory intellectual property protection was performed for the two innate or natural forms of protection – being first in the market and having a complex product design. In both cases, almost all coefficients, with the exception of size and nationality, are significant and positive. Smaller firms, then, see the nonstatutory forms of intellectual property protection to be as effective as do large firms.

Variants of the multivariate analysis for the scores were also estimated with a continuous employment-size variable and with the usage included as separate regressors. As was the case for the use equations, neither of these variants changed the reported results substantially.

Thus, the multivariate results for the score equation show that each of the characteristics that was previously examined – size, nationality,

---

[18] There are fewer observations available for estimation of Table 9.19 than 9.18. Therefore the results in Table 9.18 were rerun to test to see whether there was anything unusual about the sample that responded to the efficacy question. There was not, since results similar to those reported in Table 9.18 were generated with the smaller sample.

innovativeness, industry location – is significant in its own right. This is particularly the case for the scores given to patents, trademarks, and industrial design.

The multivariate results are not quite so clear-cut when it comes to the use of intellectual property protection. Here, too, size and innovativeness generally are significant; however, nationality is not as frequently significant as it is in the case of the score attached to the efficacy of the intellectual property right in question. In particular, foreign-owned firms are not significantly more likely to possess a patent than are domestically owned firms. On the other hand, they are more likely to attach a higher value to patents as a method of protection. This may occur because foreign-owned firms hold a patent elsewhere than in Canada or because the size variable is capturing most of the foreign-ownership effect. The latter is likely the case, since the inclusion of the continuous-size variable results in the nationality variable gaining significance in the equation that estimates use.

Both the innate or natural forms of protection are perceived to be more effective in firms that are innovative and in the core or secondary sectors. However, size is not significantly related to the scores given to these forms of protection. Large firms may be more inclined to give a higher score to statutory forms of protection like patents, industrial design, and trade secrets; but they are not different from small firms in their evaluation of strategies like complexity of design and leadership. Both groups find them equally important.

In summary, innovative firms are more likely to make use of each of the statutory forms of intellectual property – patents, copyrights, industrial designs, trade secrets, and trademarks. They also assign a higher score to patents and trademarks, as well as to being first and to having a more complex form of product. Both size and nationality are also frequently related to either use or the score given to the effectiveness of intellectual property. Both size and nationality are probably capturing other aspects of innovativeness that the innovation variable used here does not reflect – the importance of the innovation, its originality, or its use of proprietary knowledge.

9.10 CONCLUSION

This chapter sets the use of intellectual property protection in the context of innovation activity. Several broad forms of protection for intellectual property are supported by the state – patents, copyrights, trademarks, industrial designs, and trade secrets. These range from patents, which are registered by an administrative system and enforced by the courts, to trade secrets, which are supported through the legal system. Only about one-quarter of the population of manufacturing enterprises, both large and small, make use of at least one of these forms of protection. Only about 7% specifically use patents.

The importance of these forms of protection for intellectual property differs between small and large firms. While only about one-quarter of manufacturing firms use one of these forms of protection, these firms account for over 50% of employment. The difference in these two measures of use – company weighted versus employment coverage – results from very large differences in the extent to which small and large firms avail themselves of the statutory forms of protection. Over 60% of large firms protect themselves with any one of the statutory rights investigated here; less than 30% of those with fewer than 100 employees do so. Part, but not all, of this difference is accounted for by different tendencies to innovate. But even when these differences are taken into account, small innovative firms are seen to make use of the formal forms of protection less frequently than do large firms. Of those large firms reporting sales from a product innovation, almost 80% possess one of the statutory rights; about 40% of those with fewer than 100 employees who have recently introduced a major product innovation do so.

Being innovative is a primary determinant of the use of intellectual property protection. There are substantial differences in the use of trademarks, patents, trade secrets, industrial designs, and copyrights between those who had just innovated in the three preceding years and those who had not. But this demonstrates as much about those firms who indicated that they were not innovating as it does about the effectiveness of the intellectual property regime. It is, of course, possible that a significant portion of the group of non-innovators had innovated in the not-too-distant past and still would have been availing themselves of intellectual property protection. However, while some of this group possessed some forms of intellectual property protection, the percentage was small. For example, the proportion of firms that were innovative over the three-year period preceding the survey and that possessed a patent was about 14%, but only 2% of firms that had not innovated over these three years did so. The size of this difference suggests that the non-innovating population possesses relatively little of the knowledge-based capital that would be based on past activities. This, in turn, indicates that firms in the innovative group have not supplanted a previous set of innovators; rather, they have been the leaders in the development of intellectual property for some time.

While being innovative is a prerequisite for the need for protection, not all forms of statutory protection are sought equally by innovative firms. When the effect of being innovative is separated from the effect of size, nationality, and industry, innovativeness has its largest effect on the use of patents and trademarks. However, large and significant effects are also found on the use of industrial designs, trade secrets, and copyrights. Innovative firms, then, concentrate on patents but also use a wide range of other statutory forms of protection.

Although many innovators make some use of statutory intellectual property protection, there are a substantial group who do not. There are a number of reasons why firms do not seek to protect their intellectual property from being copied by using such statutory forms of protection as patents. First, while each of the innovating firms has developed new products or processes, not all of the ideas imbedded in these innovations are unique enough to be patentable. About 16% of innovations in large firms are world-firsts; only 33% are firsts for Canada. This study finds almost 80% of world-first innovators protect themselves with a form of statutory protection either in Canada or abroad. It is, however, the case that less than half use patents. Second, processes lend themselves better to protection through secrecy than do products. Pure process innovation is only about half as likely to make use of patents as are pure product innovations, and it is more likely to focus on trade secrets than on patents.

In addition to providing a broad overview of the extent to which intellectual property rights are used, the study also examines the extent to which participants value the system. Previous work has asked whether forms of protection, such as patents, are essential to the innovation process. Mansfield (1986), using a sample of 100 firms, asked whether patent protection was essential to innovation and found that most firms felt it was not – with the exception of those in chemicals and pharmaceuticals. Levin et al. (1987) had R&D managers score the effectiveness of patents as a means of protecting an innovation and found that the average score received was 'less than effective'. Firms gave a higher ranking to alternative protection strategies like being first in the market or having a complex product design.

This chapter confirms the findings of Levin et al. (1987). The population tends to value alternate strategies more highly than the statutory forms of protection. Moreover, the population as a whole ranks such strategies as patent protection as being less than 'effective'. However, these rankings depend very much on the characteristics of a firm. If a firm is innovative, large, and foreign owned, and is in one of those industries that tend to produce more innovations, the score given to the statutory forms of protection like patents increases greatly. On average, users of patents find them effective; so too do large foreign firms.

Mansfield (1986) observed that although firms may not have thought the patent system was essential to innovation, they nevertheless took out patents on most patentable products. Thus, measures of use tend to give a different picture of the importance of the patent system than do firms' own evaluations. This is not the case here. The firms that tend to give statutory forms of protection like patents a lower score are those that do not make use of patents. Those who have undertaken the effort to acquire a patent tend to give them a passing grade.

The findings do, however, confirm that statutory protection is only one of the methods used by firms to defend their intellectual property. Innate forms of protection are seen by almost every subgroup – large, foreign, innovative – as being equally if not more important. Substantial forms of other protection exist that are seen as being effective in protecting knowledge assets.

It is interesting to note that there is much less variance across size classes in the perception of the effectiveness of these alternative strategies than there is for the statutory forms of protection. Small firms do not feel that protection for intellectual assets that they might develop during innovation is lacking. They feel that innate strategies, such as being first, are effective while patents are not. In the multivariate analysis that holds other characteristics like innovativeness and industry of location constant, the size of firms has a strong positive effect on the evaluation given to patents and trademarks, but it is negative for the strategies of being first or of having a more complex product.

Differential usage patterns between small and large firms also reflect these differential opinions on effectiveness. Small firms use trade secrets more frequently relative to patents than do large firms. When other differences between small and large firms, such as differences in innovativeness, nationality, and industry of location, are considered, size has the greatest impact on patent and trademark use.

We confirm the findings of others (Taylor and Silbertson, 1973; Levin et al., 1987) that the industry environment affects the use that is made of intellectual property. Cross-industry differences in intellectual property usage in Canada are closely related to differences in innovativeness that have been described by Robson et al. (1988). There are a core set of industries – chemicals, pharmaceuticals, refined petroleum, electrical products, and machinery – that tend to produce a large number of inventions used downstream in other industries as inputs or as capital equipment. These industries make greater use of almost all forms of statutory protection than do other industries. This is particularly true of patents and of trademarks. This relationship also is found in the multivariate analysis after other characteristics of the firm are considered. Thus, the industrial structure of a country very much determines the use that will be made of its intellectual property system.

Finally, it should be noted that firms in different sectors take a very different view of the effectiveness of both the statutory and the innate forms of protection. Even after allowing for differences in size, nationality, and innovativeness of the firm, being in the core sector substantially increases the score given to patents, copyrights, and design – though not to trademarks and trade secrets. The industry environment conditions a firm's view of the effectiveness of the intellectual property system. Some have suggested that differentials in the evaluations given to patents across industries may simply reflect the fact that patents are easier to defend in some industries because of the specificity of the product (Levin et al., 1987). However, there is something else at work here, since this effect is found across almost all of the statutory forms of protection. If a firm is in an industry that is generally innovative, it develops an attitude that engenders the use of protection – probably because a learning process is required before a firm understands how to exploit and protect its advantage. This study has reported several pieces of evidence to support this view. Invariably, users of intellectual property manifested a different view of the effectiveness of the various forms of protection. These consistent differences suggest that intellectual property use – like any other strategy – involves acquired skills that only develop in use. As firms innovate, they learn which strategies best protect their knowledge assets. The study also suggests that these skills, in that they are associated with size, are part of the growth experience and tend to increase as a firm successfully masters a range of strategies and grows.

TEN

Multinationals and the Canadian Innovation Process

10.1 INTRODUCTION

As multinational corporations (MNCs) have evolved over the last 50 years, the framework that has been used to analyze their role in the Canadian innovation system has changed.

Early theories use oligopoly models that portray multinationals as entering the Canadian market to control strategic Canadian natural resources (Aitken, 1961) or avoid tariffs (Eastman and Stykolt, 1967). Caves (1971, 1982) extends this framework to argue that foreign investment by a firm just as often stems from the exploitation of a key asset that is indivisible and not easily transferred from one firm to another except through foreign direct investment (FDI). This asset might involve marketing competencies (brand recognition) or scientific knowledge arising from research and development expenditure. Asymmetric knowledge and an imperfect contractual environment associated with the characteristics of these assets means that the exploitation of this key asset is most efficiently done via the extension of the firm across international boundaries via direct investment.

The alternative to foreign investment is to export or license a foreign producer. Licensing is more likely to occur when the rent-yielding capability of the asset lies in a one-time innovation of a technique or product. In these cases, the information on which the asset rests is more easily transferred intact via an arm's-length sale to an unrelated foreign firm. In most other cases, either the information cannot be transferred without simultaneously providing entrepreneurial manpower, or the uncertainty about the value of the knowledge in the foreign market will preclude agreement about the terms of a licensing agreement with an unrelated third party.

In this framework, foreign investment allows the host or recipient country to benefit from the specialized competency of the multinational firm because it facilitates the transfer of technical know-how. But in this model, as with the previous one, the MNC was not seen to develop local innovation capabilities.[1]

This model can be described as resembling a hub and spoke system – with the key home-country asset being transferred in a single direction with little development of local capabilities at the end of each artery. Local subsidiaries were regarded as branch plants, with the capacity to exploit the key asset but little capability to develop assets that could be, in turn, transferred and exploited in the worldwide operations of the parent company. The benefits of foreign investment were perceived to arise from the direct exploitation of the transferred asset and via indirect spillovers that affected domestic competitors (Dunning, 1958; Globerman, 1979; Saunders, 1980).

The theory of the multinational that is based on the exploitation of its proprietary technological advantage can be viewed as the analogy of the linear model of innovation. Both were derived from observation and based on factual evidence. But both were so focused on a single representation of the facts that they tended to hide other salient processes at work or to blind observers to changing trends. The linear model of innovation tended to focus narrowly on the importance of R&D and to underrate the effect of interactions with customers, suppliers, and local scientific or technical human resources on innovation. New technology and institutions have made these linkages more important and have changed the nature of the model that has come to be used as a framework. In the same way, changes in technology and the world-trading environment have led to changes in the type of operations characteristic of multinationals.

The growing availability, importance, and adoption of information technologies have reduced the need for centralization since the new information technologies have made it easier to coordinate dispersed activities. New transportation systems and communications technologies allow for better coordination of the plants of large geographically diversified firms. New organizational structures for the production and distribution process, such as 'just in time' that rely on the new information technologies, shorten

---

[1] A summary of this conclusion can be found in Safarian (1973) and Britton and Gilmour (1978).

the time of delivery and make new types of organizations possible. Flexible design and manufacturing based on information technologies increase the capacity of firms to respond at short notice to orders of customers. New communications technologies allow scientific personnel to work at geographically dispersed locations but still have their activities coordinated.

These changes in the ways of managing manufacturing, marketing, and R&D activities have been associated with a change in the type of activities of foreign affiliates, particularly with regards to their R&D. Foreign subsidiaries increasingly compete with their sister companies for worldwide product mandates. Recent case studies (Eden, 1993) suggest that those foreign subsidiaries in Canada that are taking an entrepreneurial approach to subsidiary management are increasingly receiving international responsibilities. As part of these initiatives, research and development activity is seen to be a key support activity and, therefore, has taken on a new role in the subsidiaries of multinationals. Activities for which Canadian subsidiaries of MNCs are responsible must be globally competitive, and competitiveness requires an advanced scientific and technological foundation (Birkinshaw, 1995).

In addition to these technological and organizational developments, changes in international commercial policies and trading relations have also affected the localization and organization of R&D and innovation activities in MNCs. These include the creation of regional trading blocs (Canada, the United States, and Mexico; the European Union) and the inclusion of intellectual property rights under the jurisdiction of the World Trade Organization. The decline in trade barriers that have been caused by these changes has made it easier for the MNC to take advantage of different national capabilities. Multinationals, which originally were highly focused on their country of origin, have begun to become more 'multinational', that is, stateless organizations. The MNC is increasingly becoming a complex network of interactions among subsidiaries, the parent company, and their local environments. The central hub-spoke relationship is being replaced in some instances by a matrix type of organizational structure.

These changes have been accompanied by new theoretical perspectives that reflect the increasing internationalization of R&D activities, which has seen multinationals based outside the United States entering the United States to take advantage of local R&D capabilities.[2] In

---

[2] Foreign-owned companies accounted for 18% of total company-funded R&D in the United States in 1995 (Serapio, 1997).

contrast with the traditional theorizing about the MNC that stressed the home-base asset-exploiting motive for FDI, the recent theories explicitly take into account the contribution to an MNC's technology and expertise from interactions with and spillovers from the human resources and institutional environment in the host country.

Foreign-owned firms are now portrayed as needing to conduct R&D abroad for more than one reason. The first is to exploit foreign firms' specific capabilities in the foreign environment. The second is to augment a firm's knowledge base by tapping potential spillovers from existing local resources (local firms, higher education, publicly funded or executed R&D, and the intellectual property regime). This extension of the simple hub and spoke model to a more comprehensive matrix model of the firm can be found earlier in Teece's (1986) model of complementary assets, which argues that firms make FDIs in R&D performers in order to secure critical assets from abroad that are complimentary to the key assets held by the investing firm, or to provide complimentary assets that are essential to the success of the investing firm's international operations (Serapio, 1997).

As intriguing as these new perspectives are, there has been little empirical evidence regarding their applicability to the Canadian economy. The objective of this chapter is to ask how the multinational fits into the Canadian innovation system and to ascertain which of the alternative models best describes the activities of MNCs in the early 1990s. We recognize that economic structures change slowly and, therefore, we ask whether a combination of both old and new models is needed to explain MNC activity.

The two opposing models outlined above have different predictions. The hub and spoke model suggests that the R&D functions in Canadian subsidiaries should be relatively truncated. The mature matrix model suggests that we should expect to find that multinationals are more likely to have developed a relatively robust local R&D capacity. If local assets complement foreign assets held by a multinational, we might expect that MNCs actually are more, not less, likely to engage in local R&D than are domestic firms.

Therefore, this chapter examines whether the nationality of firm ownership affects the organization of R&D activity, the sources of innovative ideas, and the incidence and effects of innovation. Throughout, we compare the multinational firm operating in Canada to Canadian, domestically owned firms. By comparing the performance of the MNC to domestic corporations, we allow for the fact that technological opportunities and

Introduction 269

other structural characteristics will condition the nature of the innovation system in Canada.

We compare the multinational firm first to all Canadian firms in order to see whether there is any indication that foreign firms are disadvantaged relative to those based in Canada. But we also examine differences between foreign multinationals and Canadian firms with a foreign orientation – those with operations abroad or with export sales.[3] The latter allows us to ask whether foreign subsidiaries are disadvantaged relative to Canadian firms that successfully operate in foreign markets, what we shall refer to as domestic multinationals for the sake of brevity.

The first section deals with similarities and differences in the organization of R&D activity between Canadian- and foreign-owned firms. It asks if foreign-controlled firms are dependent on R&D from their parents or are just as active with respect to R&D as their Canadian-owned counterparts.

The second part of the chapter compares the foreign and domestic groups with respect to their sources of ideas for innovation. We examine the patterns of internal and external sources of innovative ideas in both domestic and foreign-controlled firms. In doing so, we focus on two issues. The first is the extent to which the foreign-owned firm chooses to use an extensive set of information sources and, thus, shows that it is making as much use of local information sources as are domestic firms. The second is the extent to which it makes such heavy use of its own affiliated partners that it neglects developing a local R&D capacity.

Throughout, we also investigate the extent to which technological opportunities are an important determinant of the knowledge-sourcing pattern. While this chapter focuses primarily on the issue of nationality, we continue to develop the themes that are present in other chapters – that size of firm is closely related to the innovation pattern and that technological opportunities differ across industries and often serve to condition interfirm differences. In this vein, we ask whether once we have accounted for differences in the proportion of domestic and foreign-controlled firms performing R&D by sector, there are still inherent differences by nationality type.

A comparison of the extent and impact of innovation activity of locally and foreign-owned firms concludes the chapter. There has been a long-standing interest in the productivity difference between

---

[3] We combined firms with foreign operations with those having exports because they exhibited a similar profile.

Table 10.1. *Composition of the Survey Sample, by Nationality and by Size Class (% of Manufacturing Firms)*

Employment Size Class	Country of Majority Ownership					
	Canadian	Foreign	USA	Europe	Pacific Rim	Other
All	84	17	10	5	1	1
0–20	97	4	2	2	0	0
21–100	90	10	7	3	1	0
101–500	72	28	17	8	2	1
501–2000	63	37	22	14	2	0
>2000	50	51	26	22	2	2

Note: Larger firms (IPs) only.

Canadian- and foreign-owned firms (Safarian, 1973). Recently, Globerman, Ries, and Vetinsky (1994) argued that most of the differences in labour productivity are accounted for by differences in capital intensity. But that still begs the question as to why the latter occur. Are foreign-controlled firms better able to incorporate more advanced techniques into their production process? We know that they are more likely to adopt advanced manufacturing technologies (Baldwin and Diverty, 1995; Baldwin and Sabourin, 1998; Baldwin, Rama, and Sabourin, 1999). Is this because they are more adept at innovating? In the concluding section of the chapter, we tentatively address this issue. We ask how the innovation regime that is developed in the first section of the chapter is translated into innovative activity. We also ask whether foreign and domestic firms make different use of intellectual property protection.[4]

10.2 CHARACTERISTICS OF CANADIAN AND FOREIGN-OWNED FIRMS: SIZE AND INDUSTRY DIFFERENCES

Foreign-owned firms are neither the same average size nor located in the same industries as domestically owned firms. Firms with a majority of Canadian shareholders (Canadian owned) are generally smaller than those belonging to foreign shareholders (see Table 10.1). The distribution of foreign-owned firms according to the country of origin of the majority

---

[4] Because of the complexity of the tables in this chapter, standard errors are not included. They may be found at www.statcan.ca in Baldwin and Hanel (2000).

shareholders is dominated by the United States, followed by European and Japanese firms.

In the group of firms with fewer than 100 employees (small firms), over 90% are Canadian owned. In the medium-sized group (100 to 500 employees), about 70% are Canadian owned. In the largest firm class (over 2,000 employees), only 50% are Canadian owned. Since the proportion of foreign-owned firms regularly increases with size, we can expect that many attributes of large size are likely to be associated with foreign ownership, and vice versa. Therefore, we shall make most comparisons here between foreign and domestic firms within size classes.

Since foreign ownership is really only significant for firms of medium and larger size, our comparison in this chapter will concentrate on the subsample of the survey – those coming only from that part of the sample, that consists of the firms referred to as the integrated portion of the Business Register (IPs). These are firms that generally possess more than 20 employees. This group excludes most of those that we have referred to in previous chapters as microfirms – those with fewer than 20 employees. We shall refer to the former as 'larger' firms in this chapter.

Foreign ownership in the manufacturing sector also varies considerably by industry. In order to investigate these differences, we use a core/secondary/'other' taxonomy that is used elsewhere in this book. The importance of foreign ownership is measured by the percentage of shipments made by plants under foreign control.

In 1973, foreign ownership was highest in the core sector (76%), followed by the secondary sector at 60% and lowest in the tertiary other sector (32%). Up to the early 1980s, the importance of foreign ownership declined in the other sector to 25% in 1986 and then increased to 33% in the early 1990s. The secondary manufacturing sector also follows this pattern, falling from 60% in 1973 to 51% by 1981 and then gradually increasing back to 60% by the early 1990s. On the other hand, the importance of foreign ownership in the core sector declines steadily over the period from over 76% in 1973 to only 63% by 1993.

The high percentages of foreign ownership in the core and secondary sectors confirm the advantages of foreign investment where proprietary assets exist. Corporations with proprietary technology have three ways to appropriate benefits from their technology in foreign markets. They can license their technology to foreign firms, they can export new products from the home base, or they can extend their production to foreign countries

Table 10.2. *Distribution of Firms, by Nationality and Industrial Sector (% of Firms)*

Sector	Canadian	Foreign
All	84	17
Core	69	31
Secondary	80	20
Other	91	9

Note: Larger firms (IPs) only.

through direct investment. As the industry data show, the last alternative is frequently chosen in those industries where there is proprietary technology or related soft assets, such as brand name, marketing, and distribution networks.[5]

Because of differences in the importance of foreign ownership, we will compare domestic and foreign firms within industries to correct for those industry characteristics that otherwise would be related to aggregated differences between foreign and domestic firms. For this purpose, we once again make use of the industry taxonomy that was developed by Robson, Townsend, and Pavitt (1988), dividing industries into three – the core, secondary, and other sectors.

The distinction that we have drawn in previous chapters among the core, secondary, and other sectors is based on the degree and type of innovation produced in each one. These differences are related to the science base (either R&D or technological) of the sectors. We should, therefore, expect to find significant differences in foreign penetration across these sectors because some are more likely than others to utilize the type of scientific assets that multinationals specialize in transferring from one country to another. The differences in foreign ownership of the core, secondary, and other sector are listed in Table 10.2. The core sector has the highest percentage of foreign firms (31%). The secondary sector follows with 20%. The lowest proportion (9%) is found in industries belonging to the other sector.

The share of foreign-owned firms increases with the size of firm in each sector. In the large-firm size categories, foreign affiliates outnumber domestically owned firms in both the core and secondary sectors.

---

[5] See Dunning (1992, 1993).

10.3 INCIDENCE AND ORGANIZATION OF R&D

Since innovative inputs like R&D are essential to innovative outputs and the former have received more attention than the latter, we focus first on whether foreign affiliates perform R&D in Canada more or less frequently than do locally owned firms. If foreign affiliates operate as truncated branch operations, they would be expected to be less inclined to set up R&D facilities.

Technology-policy discussions on this matter in Canada have traditionally been characterized by two opposing views on the consequences of truncation. On the one side are those who believe that owing to extensive foreign ownership, Canadian firms have easy access to new technology developed abroad and that R&D performed in Canada is therefore not very relevant (Globerman, 1979). On the other side are those who defend the need for a stronger Canadian R&D involvement (Britton and Gilmour, 1978).

The debate on whether deficiencies exist in multinational R&D performance has generally occurred without detailed statistics on R&D incidence. The present survey provides information on the similarities and differences between domestically and foreign-owned firms with respect to organization of R&D and innovation in Canada. It also allows us to divide domestically owned firms into those with an international orientation – that operate in foreign markets, either because they possess foreign operations (production or R&D facilities) or because of any foreign sales – and those that are purely domestic in scope.

When we examine the differences in the probability of foreign and domestic firms conducting R&D (see Table 10.3), it is apparent that foreign-owned firms generally perform R&D more often than domestic ones, taken as a whole. Since larger firms perform ongoing R&D more often than the smaller ones, the employment-weighted results give higher proportions for both groups but generally do not reverse the pattern revealed by the company-weighted results. When domestic firms are broken into two groups, the domestic firms with an international orientation look very much like the foreign subsidiaries, with no significant difference between the two groups.[6]

There is also a significant difference in the way foreign- and Canadian-owned firms approach R&D activity. Only a minority of pure domestic

---

[6] Domestic multinationals are slightly lower using company-weighted estimates but slightly higher using employment-weighted estimates.

Table 10.3. *Incidence and Delivery Mechanism for R&D, by Nationality (% of Firms)*

		Population: % Conducting R&D	Of R&D Performers				
Ownership	Weight		% Conducting Ongoing R&D	% with Separate R&D Department	% with R&D in Other Departments	% That Contracts Out R&D	% That Collaborates on R&D
Foreign	Company	89	53	44	55	20	35
All Canadian	Company	77	43	28	64	23	21
Canadian with international operations	Company	86	45	30	61	24	22
Canadian with domestic operations only	Company	67	39	25	70	23	19
Foreign	Employment	90	75	30	49	66	58
All Canadian	Employment	88	49	34	56	41	38
Canadian with international operations	Employment	95	51	44	53	33	43
Canadian with domestic operations only	Employment	71	40	32	68	36	20

Note: Standard errors are less than 4 for the company-weighted estimates and less than 6 for the employment-weighted estimates.

firms conducting any R&D engage in ongoing R&D (39%). They account for less than half of employment (40%) in the Canadian-owned segment of manufacturing. In contrast, about half of foreign firms (53%) perform ongoing research and development, and they employ three-quarters of workers in their segment of industry. Foreign-controlled firms are also more likely to perform ongoing R&D than the internationally oriented domestic firms, but the differences are not statistically significant. This suggests that foreign firms operating in Canada are not truncated, at least on the basis of comparison with their domestic counterparts.

Foreign-owned R&D performers are also more likely to conduct R&D in a separate department than are both types of Canadian-owned firms, which tend to perform R&D activities more often in other departments. There are relatively minor differences in the extent to which foreign-owned and domestic-owned firms contract R&D out. However, the openness of foreign subsidiaries to external sources of expertise is reflected in a more frequent participation in R&D collaboration agreements. Multinationals take advantage of expertise from networks more frequently than do domestic firms. Moreover, they even do this more intensively than Canadian firms operating in world markets.

Since Canadian firms are generally smaller than their foreign counterparts, it is possible that the less intensive engagement of Canadian-owned firms in R&D is due to their smaller size. A breakdown of the type of R&D by size of firm shows that the proportion of smaller foreign- and domestically owned firms that perform ongoing R&D (see Figure 10.1)

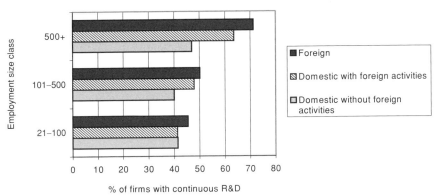

Figure 10.1. Foreign Versus Domestic Firms with R&D Performed on a Regular Basis, by Size Class

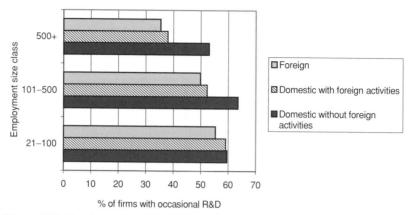

Figure 10.2. Foreign Versus Domestic Firms with R&D Performed Occasionally, by Size Class

and occasional R&D (see Figure 10.2) is rather similar over the smaller size classes.

The most notable difference between the two groups occurs in the largest firm category (firms employing more than 500 employees). Only 47% of pure domestic Canadian firms and 64% of domestic multinationals in this size group have an ongoing R&D program, compared to 71% of foreign-owned firms. On the other hand, domestic firms are more likely to conduct R&D on an occasional basis. This holds true for both domestic firms with and without international activities – although there is much less of a difference between internationally oriented and foreign subsidiaries in the larger two size classes.

Our results, then, show that foreign firms are more likely to perform R&D activity, especially compared to purely domestic firms.[7] But these findings bear only on the incidence and not the intensity of R&D activity. However, Holbrook and Squires (1996) have examined intensity for those firms reporting R&D expenditure and report that after controlling for size and industry, the largest Canadian-owned firms have a lower R&D intensity (as measured by the value of R&D expenditures/sales) than do foreign-owned firms. Together, these two sets of findings suggest that in

---

[7] To check whether foreign-owned firms might have been referring to their parents' R&D facilities in the R&D question, rather than to R&D capacity in Canada, we compared the percentage of foreign firms reporting that they performed R&D in a separate unit to the percentage of foreign firms reporting separate R&D units in Canada (the latter came from a different question on the location of production, assembly, and R&D units). The two percentages were the same.

the large firm size class, it is the domestic, not the foreign, sector that has an underdeveloped R&D capacity.

10.3.1 Sector Differences

Technological opportunities are not evenly distributed over all industrial sectors, and neither are foreign-owned firms. Since firms that operate in industries exploiting the latest scientific discoveries are naturally more research intensive, it may be that the location of foreign-owned firms explains the striking contrast in R&D involvement between the largest firms of the two ownership groups.

A tabulation of R&D involvement and organization by industry sector and by nationality of ownership in Table 10.4 reveals that, generally, a higher proportion of foreign- than domestically owned firms perform ongoing R&D in each sector. In the core sector, foreign firms and domestic firms with an international orientation do not differ significantly; but in the other sector, foreign firms exceed both domestic groups in their likelihood of conducting ongoing R&D. Two-thirds of Canadian firms in the other sector do not have a separate R&D unit and perform research and development activities in other departments. This suggests that multinationals are bringing specialized technological assets to all industries, not just to those industries that have been characterized as the most advanced technologically.[8] It also means that differences across these sectors are greater for domestic than for foreign participants.

In related work (Baldwin and Gellatly, 1999a and 1999b), we have argued that the classification of industries into high and low tech is problematic because there are high-tech firms in every industry. The results here reinforce this point by indicating that foreign firms in all three sectors resemble one another in terms of their innovative capabilities.

The ever-increasing speed and cost of innovation is inducing firms to collaborate in order to spread the risk and cost of precompetitive research. These collaborative agreements are most frequent in the R&D-intensive core sector, but foreign-owned firms collaborate in all sectors more frequently than do domestic ones, and they collaborate in the other sector just about as much as they do in the core sector. Once again, it appears that foreign firms in the other sector are exploiting technical capabilities that are firm- rather than industry-specific.

---

[8] See Baldwin and Caves (1991) for a similar finding about differences in the effect of foreign and domestic mergers.

Table 10.4. *Delivery Mechanism for R&D, by Industrial Sector and Nationality (% of Firms)*

Sector/Firm Type	% of All Firms Conducting Ongoing R&D	% of R&D Performers			
		With Separate R&D Department	With R&D in Other Departments	Contracting Out R&D	Collaborating on R&D
Core					
Foreign	62	50	57	20	40
Canadian with foreign operations	64	53	48	21	32
Canadian without foreign operations	39	38	68	30	20
Secondary					
Foreign	47	33	61	20	30
Canadian with foreign operations	44	25	66	21	19
Canadian without foreign operations	37	23	64	21	19
Other					
Foreign	50	51	46	19	35
Canadian with foreign operations	37	24	62	27	20
Canadian without foreign operations	40	22	74	22	19

Note: Standard errors range from 4 to 8 for foreign, 3 to 5 for Canadian, and 3 to 5 for international Canadian.

10.3.2 Probabilistic Models of R&D Organization

In the previous two sections of this chapter, we have shown that both firm size and industry affect the nature of R&D activity, providing a magnitude of the effect of each factor when considered by itself. In order to investigate the joint effect of size and industry and the importance of other factors like the competitive environment on the incidence and organization of R&D, we use multivariate probability (logit) analysis here.

Each statistical model evaluates the correlates associated with the probability that a particular outcome occurs. In the first instance, we evaluate which variables are associated with the probability that a firm performs research and development. Here, the dependent variable takes a value of 1 if the firm performs R&D in any form; it takes a value of 0 if it does not perform R&D at all. The next four models estimate the probability that an R&D-performing firm adopts a particular organization of its research and development activities – that it performs ongoing R&D, has a separate R&D department, performs ongoing R&D in a separate department, performs R&D only through contract research, performs R&D itself and contracts it out, and performs R&D and collaborates on R&D with other firms.

We are interested in examining how firm size, sector of operation, and ownership jointly affect each of these facets of the R&D process. While the hypothesis that we test changes from model to model, the set of explanatory variables is, at the beginning, identical for all models. In addition, we test the effect of competition on the choice of an R&D system because of the importance that has been attached to the variant of the Schumpeterian hypothesis, which argues that it is not just size that matters but also the competitive environment.

The following set of explanatory variables is used:

(1) A binary variable identifying the origin of a firm's ownership: DOMESTIC = 1 if the firm is domestically owned and has neither foreign sales nor production facilities and 0 otherwise.
(2) A binary variable identifying the origin of a firm's ownership: MULTDOM = 1 if the firm is domestically owned and has foreign sales or production facilities and 0 otherwise.
(3) The size of the firm is represented by four binary variables representing four separate size classes – firms from 1 to 20 employees (SIZE1); firms from 21 to 100 employees (SIZE2); firms from 101 to 500 employees (SIZE3); firms with more than 500 employees (SIZE4).

Each takes on a value of 1 if the firm belongs to that category and 0 otherwise.

(4) A set of binary variables to define the three sectors. These are CORE, SECONDARY, and OTHER. Each takes on a value of 1 if the firm belongs to that sector and 0 otherwise.

(5) The degree of competition that a firm faces is captured with two sets of variables. Firms in the survey were asked to identify the extent to which competition came from different regions – Canada, the United States, Europe, and the Pacific Rim. In each case, firms were asked to score the degree of competition on a scale of 1 to 5, with 0 being the least intensive and 5 the most intensive. Two variables were used here (DOMCOMP and FORCOMP). The first variable measures the degree of domestic competition and the second the degree of foreign competition from the United States.

The various models that were estimated are:

For the population of all firms –
- Model 1   Probability that a firm performs R&D versus no R&D
- Model 2   Probability that a firm performs ongoing R&D

For the population of firms conducting R&D –
- Model 2a Probability that an R&D performer does ongoing rather than occasional R&D
- Model 3   Probability that an R&D performer has a separate R&D department
- Model 3a Probability that an R&D performer does ongoing R&D and has a separate R&D department
- Model 4   Probability that an R&D performer contracts R&D and does no R&D himself
- Model 4a Probability that an R&D performer performs R&D in-house as well as contracting it out
- Model 5   Probability that an R&D performer has R&D collaboration agreements

Coefficients of the estimated logit function associated with a particular variable are presented in Tables 10.5 and 10.7. A positive sign indicates that the variable is associated with more activity. The reference firm, against which each effect is estimated, is in the core sector, is a foreign firm, and has 0 to 19 employees. The coefficients presented in Tables 10.5

Table 10.5. *Multivariate Analysis: Performing R&D (All Firms)*

Dependent Variables	Any R&D: Model 1		Ongoing R&D: Model 2a	
	Coefficients	$(P > \lvert t \rvert)$	Coefficients	$(P > \lvert t \rvert)$
Intercept	1.23	.005	−0.84	.023
SIZE2	0.53	.031	0.75	.015
SIZE3	1.00	.002	0.96	.004
SIZE4	1.21	.007	1.69	.000
DOMESTIC	−0.62	.041	−0.17	.482
MULTDOM	0.07	.810	0.06	.777
DOMCOMP	0.57	.030	−0.18	.401
FORCOMP	0.61	.004	0.35	.054
SECONDARY SECTOR	−0.90	.006	−0.63	.002
OTHER SECTOR	−1.39	.000	−0.91	.000
Log likelihood				
No. observations	1320		1320	
F	10.07		8.03	
P > F	.0000		.0000	

P = probability value.

and 10.7 allow us to evaluate the qualitative importance of each variable, but do not provide an effective mechanism to evaluate the magnitude of the effect of each. This can be done by calculating the probability values attached to each stratum.[9] Table 10.6 contains the probability that some form of R&D is performed for each of the strata represented by the dependent variables – for example, for foreign in contrast to the domestic categories. In our discussion, attention is focused on the influence of the ownership of the firm.

The probability that a firm performs any form of R&D is higher for firms that are larger, for firms that are in the core rather than the secondary and other sectors, and for firms that feel that they face more competition from both domestic and foreign sources (Table 10.5). There are substantial differences in terms of the probability of R&D occurrence across size classes (Table 10.6). The smallest size class has only a 72% chance of performing R&D, while the largest has an 89% chance. The magnitude of the differences across sectors is about the same, ranging from 76% in the tertiary 'other' sector to 93% in the core sector. The effect of competition is gauged by the difference between the probability of experiencing the mean value of competition plus or minus one standard deviation from

---

[9] The probabilities are evaluated at the mean values of all explanatory variables.

Table 10.6. *Estimated Probability of Performing R&D*

Variable	Probability(%)	
	Any R&D	Ongoing R&D
SIZE1	72	22
SIZE2	81	37
SIZE3	87	42
SIZE4	89	60
DOMESTIC	76	38
MULTDOM	86	38
FOREIGN	86	38
CORE SECTOR	93	54
SECONDARY SECTOR	84	39
OTHER SECTOR	76	32
DOMCOMP + 1 st.dev.	85	38
DOMCOMP − 1 st.dev.	79	38
FORCOMP + 1 st.dev.	86	42
FORCOMP − 1 st.dev.	78	34

St.dev. = standard deviation.

the mean. It varies from 79% to 85% for domestic competition and from 78% to 86% for foreign competition.

Of note is the fact that the incidence of R&D is significantly higher for foreign-owned firms than for domestically owned firms that are purely domestically oriented (Table 10.5). But there is no significant difference between foreign and internationally oriented domestic firms. The probability of conducting R&D for purely domestic firms is only 76%, whereas it is 86% for foreign multinationals. Thus, foreign-controlled firms are at least as likely to conduct R&D in Canada as are domestically controlled firms, even after their larger size and industry location are both taken into account. Nationality also has the same qualitative effect on whether ongoing R&D is conducted, but the difference is not significant.

Once firms conduct R&D, there are fewer variables that are significant in determining how they do it, whether it is ongoing, whether it is in a separate R&D department, whether it is ongoing with a separate R&D department, whether it is contracted out, or whether collaboration occurs (Table 10.7). In almost all cases, larger firms are more likely to engage in these activities (the one exception being the case of contracting out and not doing R&D in-house). But the degree of competition and the nationality of ownership rarely have a significant effect on the type of R&D

Table 10.7. Probability Models: Type of Delivery Mechanism for R&D

Variable	Ongoing Model 2a	Separate R&D Department Model 3	Ongoing and Separate R&D Department Model 3a	No In-House R&D and Contracts Out Model 4	R&D In-House and Contracts Out Model 4a	R&D In-House and Collaborative Model 5
Intercept	−0.22	−0.10	0.90	−2.00*	−2.21*	−1.17*
SIZE2	0.64***	0.32	0.01	−0.23	0.29	0.53
SIZE3	0.75**	0.80**	0.43	−0.44	0.64	0.80**
SIZE4	1.52*	1.49*	1.44**	−0.57	1.09***	1.65*
DOMESTIC	0.01	−0.19	−0.45	−0.45	0.63***	−0.15
MULTDOM	0.34	−0.16	−0.21	0.91	0.37	−0.24
DOMCOMP	−0.44	−0.50**	−0.11	0.23	−0.38	−0.68**
FORCOMP	0.18	−0.03	−0.13	−0.62**	−0.04	0.37***
SECONDARY SECTOR	−0.51**	−0.89*	−0.98*	0.67***	−0.62**	−0.35
OTHER SECTOR	−0.68*	−0.89*	−0.87*	0.77***	−0.47	−0.34
Log Likelihood	−730.4	−638.1	−329.2	−396.3	−369.1	−565.1
No. observations	1094	1094	510	1094	1094	1094

Notes: Significance level: * = 1%; ** = 5%; *** = 10%. Larger firms (IPs) only. These generally have more than 20 employees.

process that is undertaken. A notable exception is that purely domestic firms are more likely to contract out R&D.

10.3.3 R&D Collaboration Partnerships

R&D activity is increasingly performed in partnerships, alliances, and joint ventures. These forms of collaborative research efforts attempt to reduce the cost, to speed up the innovation and the product cycle, and to share the risk involved in discovering, inventing, and innovating (Niosi, 1995a).

When all R&D performers in the survey report their partnerships, foreign affiliates report that they have collaborative partners almost twice as often as do pure domestic firms (Table 10.3). Their likelihood of collaboration is higher than for domestic multinationals as well. This difference could, however, be unduly influenced by the large number of small firms included in the Canadian-owned subsample. Even after controlling for the size of the firms, foreign-owned firms report R&D partnerships more frequently than do Canadian ones in all size categories, though the differences are only marked in the largest size group (see Figure 10.3).

In order to separate out the different influences on R&D collaboration, we estimated the probability of a firm having a collaboration agreement using a logit regression (model 5 in Table 10.7). The results show that even after the size of the firm, the extent of foreign competition, and the industry sector are taken into account, foreign affiliates are more likely to

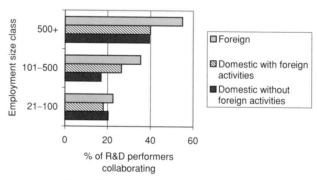

Figure 10.3. Foreign Versus Domestic R&D Performers with R&D Collaborative Agreements, by Size Class

participate in R&D partnership than are Canadian-owned firms, though these differences are not significant.

Facing foreign rather than domestic competitors also increases the probability of R&D collaboration. The relationship among the ownership structure, competitive environment, and the size of the firm on the probability of an R&D partnership provides evidence that the conduct of firms in a given industry sector is very much affected by the structural characteristics of the industry.

R&D collaboration is not a substitute for a firm's own research and development activity but a complement to it (Cohen and Levinthal, 1989). Firms conduct their own research in order to develop in-house expertise that enables them to participate in and absorb the research output of cooperative research, to develop the in-house expertise so as to contribute to cooperative projects, and to monitor the development of science and technology in a particular field (Mowery and Rosenberg, 1989). We should, therefore, expect to find that those firms that are more likely to conduct R&D are also more likely to engage in collaborative ventures.

The findings of the present survey support this view. The fact that foreign affiliates have a higher propensity to conduct R&D and to participate in partnerships than do purely Canadian companies are two related aspects of their innovative strategy. In addition to the capabilities that arise from their conducting R&D in Canada, foreign affiliates bring technological and other intangible assets of their parent multinational enterprise to an R&D partnership. Their 'entry ticket' to cooperative ventures is thus provided by their 'family' connection and the 'membership fee' by their ongoing R&D (Kumar, 1995).

Foreign innovators establish research partnerships just as frequently as do domestic innovators, not only abroad but also in Canada, for example, with customers, universities, and research institutions. In Figure 10.4, we present the percentage of large innovators reporting a partnership with a particular type of partner (customers, suppliers, etc.). In addition, we subdivide this information by the nationality of the partner (domestic and foreign). The results show that foreign firms have collaborative agreements with foreign partners and with domestic partners more frequently than do Canadian firms. There are considerable differences in terms of partnerships with customers located abroad – 29% and 22% for foreign and domestic firms, respectively. This pattern also exists for R&D institutions and Canadian universities. In contrast, domestic firms are

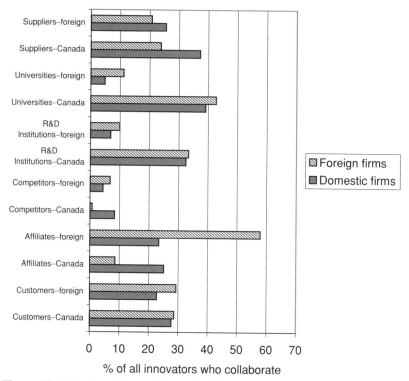

Figure 10.4. Foreign- Versus Domestic-Firm Use of Different Collaboration Partners

more likely to have domestic collaborative agreements with domestic suppliers and domestic competitors.

In summary, we conclude that multinationals operating in Canada are not operating as truncated firms without a research and development capacity. Although they are larger and tend to be more concentrated in the high-tech sectors, they are just as likely to possess an R&D capacity and to develop separate R&D facilities as are their domestic counterparts. Canadian firms with an international orientation approach foreign firms with regard to their likelihood of performing R&D, but they lag behind with respect to their tendency to conduct ongoing R&D. Another important conclusion is that foreign-owned firms do not collaborate less frequently than do domestic firms with Canadian partners (customers, R&D institutions, universities and colleges, and other partners). Thus, as far as R&D collaboration is concerned, there is no evidence in support of the once-popular argument (Britton and Gilmour, 1978, 1980) that foreign-owned firms do not develop links in Canada and are thus responsible for a truncated pattern of corporate behaviour in Canadian manufacturing.

10.4 SOURCES OF NEW IDEAS AND INSPIRATION FOR INNOVATION

Interactions between foreign affiliates in Canada and their parent and sister companies provide a fertile ground for innovative ideas and inspiration. Increasingly, the R&D activity of foreign affiliates serves a double purpose. On the one hand, it serves to adapt technologies developed by other firms in their group of companies to local conditions. On the other hand, it acts as a corporate antenna for capturing new ideas and technological spillovers for potential use in their MNCs' innovation activity.

Yet R&D provides only one of the sources of ideas for innovation. As the chapter on sources of innovative ideas stressed, R&D is neither necessary nor sufficient for innovation. In some sectors, other sources like management partially substitute for it. In some cases, it is combined with external sources to facilitate the ingestion of outside ideas.

This section examines the importance that is attributed to both inside sources of ideas for innovation, including R&D, and outside sources, such as affiliates. It builds on the previous section by asking not only whether the multinational relies on these parent and sister firms but also whether there is any indication that this reliance occurs to the detriment of the R&D capacity of the firm.

This section also provides an indirect measure of the importance of the R&D unit. The previous section provided evidence only on the incidence of R&D, although its description of the type of organization associated with the establishment of an R&D facility provided a measure of the capacity of the firm. Firms that perform ongoing rather than only occasional R&D are perceived by themselves to be more competitive (see Chapter 5). In this regard, multinationals were more likely to be conducting R&D but also more likely to be conducting ongoing R&D, which entails a greater commitment to the innovation process. In this section, we therefore ask whether there is evidence suggesting that the local R&D units do not really contribute to the local innovation process of multinationals. We do so by comparing the importance attributed to R&D and to other sources. If the multinational attributes little importance to its R&D unit as a source of ideas, especially compared to the importance it attributes to information coming from its parent company, then we shall infer that the local R&D operation is little used.

The personal interactions and exchanges among member firms of the multinational enterprise provide a potentially favourable environment for innovation. The rigid, hierarchical organizational structure that is still typical of some multinational enterprises may, however, prevent this potential advantage from being fully realized. Thus, it remains an empirical question whether most foreign affiliates in Canada make use of this potential advantage over their locally owned counterparts.

Since Canadian firms are on average smaller than foreign affiliates, observed differences in the source of innovative ideas between the two groups may simply reflect differences in the organization of smaller and larger firms. There is some evidence of this in the type of internal sources that are used by foreign and domestic firms. R&D personnel is the most frequently mentioned internal source of innovative ideas in foreign-owned firms, whereas Canadian-owned firms rely more on management for inspiration (see Table 10.8). This accords with earlier reports that ideas from management are substituted for ideas from R&D units in smaller size classes and in the 'other' sector. It also confirms that differences in the incidence of R&D are largest in the other sector (Table 10.4).

Of equal interest is whether there are differences between Canadian and foreign firms once we allow for industry effects. There do not appear to be. Once industry differences are accounted for (Table 10.8), there are few consistent differences between the foreign and the domestic groups for management, sales, and production as internal sources of innovative ideas. But it is the case that foreign firms report a significantly greater tendency to use R&D facilities in the other sector, which is generally a user of technology created in the core and secondary sectors.

This difference in the emphasis given to R&D also exists if we compare foreign firms to those domestic firms with an international orientation (see Figure 10.5). Foreign firms are more likely to cite internal R&D sources than are domestic multinationals, and the difference is largest in the other sector.

In conclusion, the evidence on sources of ideas corroborates our previous findings that foreign firms were more likely to have an R&D unit, compared to domestic firms. They are also considerably more likely to make use of their R&D unit as a source of ideas for innovation.

10.4.1 How Important as Sources of Innovative Ideas Are Foreign Parents and Sister Companies of Foreign-Owned Firms?

A principal advantage of the presence of foreign-owned firms in Canada is their access to the knowledge and technology pool of their parent

Table 10.8. *Internal Sources of Innovative Ideas, by Nationality and Industrial Sector (% of Innovators)*

Sector	All		Core		Secondary		Other	
	Canadian	Foreign	Canadian	Foreign	Canadian	Foreign	Canadian	Foreign
Management	54	49	47	40	53	60	58	48
R&D	41	55	60	68	46	42	26	53
Sales & Marketing	44	41	48	40	41	45	42	34
Production	37	34	21	21	50	44	38	43
Other	3	4	1	2	3	6	5	5

Notes: Larger firms (IPs) only. Standard errors range from 3 to 6 for Canadian and 6 to 10 for Foreign.

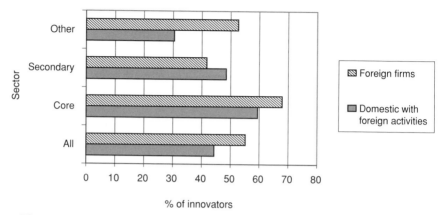

Figure 10.5. Internal R&D Used for Sources of Ideas by Foreign and Domestic Multinationals, by Industry Sector

and sister companies abroad (Globerman, 1979). Confirming this, related firms were mentioned most frequently as an external source of innovation ideas by 40% of foreign-owned firms (see Table 10.9). In comparison, only 9% of Canadian-owned firms reported a related firm as their source of innovation ideas. The lesser importance attributed to related firms extends to both groups of domestic firms – those with and without an international orientation (see Figure 10.6). Access to the knowledge, information, and contacts provided by related firms constitutes the principal difference between the external sources of innovative ideas of locally and foreign-owned firms.

This finding does not mean that the connection with related firms leaves foreign firms with underdeveloped facilities in other areas. Local initiatives of foreign subsidiaries that develop their own path-dependent process of learning-by-doing and innovation are as important as are intra-company linkages. Even though foreign-owned firms often reported their parent or sister firms as the source of inspiration for innovation, their own R&D was even more important. It is reported as a source of innovation ideas by 55% of foreign-owned firms – significantly more than the percentage reporting inspiration from their parent or sister companies (40%). This difference is even greater in the core sector. R&D is mentioned as a source of ideas by more than two out of three foreign-owned firms in the core sector, significantly more frequently than ideas from the parent or sister companies (42%). A breakdown by firm-size category (not presented here) indicates that in the core sector, only smaller foreign-owned firms rely on their parents for ideas more frequently than on R&D. Those

Table 10.9. *Main External Sources of Innovative Ideas, by Industrial Sector and Nationality (% of Innovators)*

Source	All		Core		Secondary		Other	
	Canadian	Foreign	Canadian	Foreign	Canadian	Foreign	Canadian	Foreign
Suppliers	29	27	22	28	25	17	35	44
Clients & customers	48	38	51	37	51	43	45	29
Related firms	9	40	10	42	8	37	9	40
Competitors	30	20	26	18	35	23	29	15
Government regulations & standards	9	6	16	5	12	5	3	10
Consultants	14	11	14	15	22	7	9	9
Trade fairs	19	13	10	16	13	10	27	12
Professional publications	13	11	12	18	8	5	17	5

Notes: Larger firms (IPs) only. Standard errors range from 2 to 3 for All Canadian and from 3 to 6 for All Foreign.

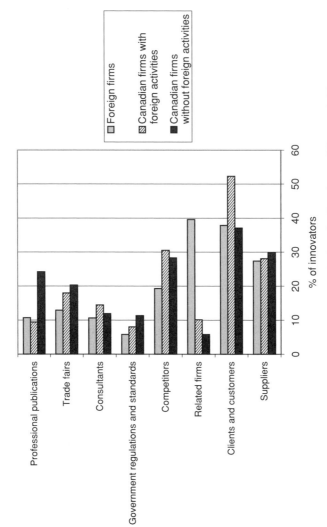

Figure 10.6. Use of External Sources of Ideas, by Foreign and Domestic Firms

Table 10.10. *Main Sources of Internal and External Innovation Ideas Reported by Foreign Affiliates, Broken Down by Whether Innovation Ideas Came from Related Firms (% of Innovators)*

Source	Idea from related firms	
	Did Not Use	Used
Management	37 (7)	67 (7)
R&D	59 (7)	49 (9)
Production	29 (6)	40 (9)
Sales & marketing	45 (7)	34 (8)
Suppliers	20 (6)	39 (9)
Clients and customers	47 (7)	24 (7)

Notes: Column 1 regroups technologically independent, column 2 technologically dependent related firms. Larger firms (IPs) only. Standard errors are in parentheses.

employing more than 100 persons report that their own R&D is a more frequent source of innovative ideas than are their parent or sister firms.

Foreign-controlled firms also use information from their customers about as frequently as they make use of ideas from their parents. The creation and maintenance of a dense set of local linkages lessen the parent firms' influence over the subsidiary (Goshal and Bartlett, 1990). Thus, foreign firms draw their ideas for innovation from a variety of sources and not exclusively from their parents' corporate headquarters.

In order to investigate differences across different types of foreign affiliates, these firms were separated into those that used innovation ideas received from their parent or sister company and those that did not. These two types will be referred to here as dependent and independent foreign affiliates (see Table 10.10, columns 2 and 1, respectively).

Dependent affiliates implement ideas received from parent companies and suppliers. These are channelled and adapted to Canadian conditions more often by affiliates' management and R&D departments than by their suppliers and sales and marketing services. Technologically dependent affiliates are also less receptive to innovative ideas from their Canadian customers. Their information sourcing fits the image of the more traditional 'branch plant' operation.

The second group of technologically more independent foreign affiliates rely relatively more on their own R&D, in that it is the most important source of ideas. The second most important groups providing innovative ideas are clients and customers, and sales and marketing departments.

Management is relatively unimportant. The independent affiliates, therefore, not only have a more developed R&D department but are also closer to their customers, which are the source of ideas for innovations in one out of every two of these firms.

10.4.2 Internal and External Sources of Technology

Innovation, even when it involves new products, often requires new technological expertise. Firms find it in their own production departments, with the help of suppliers, clients, and related firms, with the help of consultants, or by using other private and public sources of technical information.

Foreign-owned subsidiaries are part of a multinational corporation that can offer its corporate members many technical services that the mostly unaffiliated Canadian-owned firms must either provide for themselves, contract out, or find in the public domain. This differences is confirmed by the finding that 'related firms' are an important source of technology for foreign-owned firms, but not for domestic firms (see Table 10.11). Domestically owned firms generally do not have the same option of drawing on technical assistance from affiliated firms. As a result, they rely more on their suppliers and customers. Despite the importance of the technological contribution provided by related firms to their subsidiaries in Canada, foreign-owned firms still develop local capabilities as sources for ideas in this area. They use their own research about as often (33%), and experimental development more frequently (51%), than technology from related firms (32%).

Notwithstanding differences in the access that foreign firms have to their parents' technology, the most significant single difference between foreign and domestic firms is that the former rely more often on production engineering.

In summary, foreign affiliates, far from relying exclusively on their parent and sister companies, tap a larger variety of internal and external sources in Canada than do most of their Canadian counterparts in their quest for technology.

10.4.3 Transfer of Technology

One of the raisons d'être of multinational firms is their advantage in internalizing activities in areas where market transactions are inefficient and costly. Among the most important of these activities is the transfer of technology. Multinational firms prefer to exploit their technological

Table 10.11. *Principal Internal and External Sources of Technology, by Nationality and Industrial Sector*
(% of Innovators)

External Source	All		Core		Secondary		Other	
	Canadian	Foreign	Canadian	Foreign	Canadian	Foreign	Canadian	Foreign
Research	31	33	57	49	28	18	17	24
Experimental development	53	51	68	54	58	53	40	40
Production engineering	48	65	40	53	59	85	46	57
Related firm	11	32	8	37	11	32	13	23
Unrelated firm	14	20	9	20	6	16	22	24
Consultants and service firms	17	11	12	10	20	10	18	16
Publications	9	9	17	14	1	7	8	1
Trade fairs & conferences	14	14	10	18	11	13	18	6
Government labs	4	4	4	5	2	7	4	1
University labs	3	8	5	10	2	7	3	6
Customer firm	16	9	9	6	21	7	17	17
Supplier firm	34	29	23	26	25	33	45	31

Note: Standard errors range from 1 to 3 for All Canadian and 2 to 6 for All Foreign.

Table 10.12. *Type of Transfer for Technology and Associated License Restrictions (% of Firms)*

Type of Transfer	Ownership	
	Canadian	Foreign
Continuous transfer	30 (6)	52 (10)
One-time transfer	27 (6)	19 (7)
Cross-licensing	2 (1)	5 (4)
Right to manufacture	34 (6)	33 (10)
Right to sell	33 (6)	18 (8)
Right to use in manufacturing	31 (6)	24 (9)
Right to use patents	20 (5)	21 (9)
Right to use industrial designs	16 (5)	13 (8)
Right to use trademarks	13 (4)	14 (9)
Right to use trade secrets	19 (5)	21 (10)
Other rights	9 (4)	10 (7)

Notes: Larger firms (IPs) only. Standard errors are in parentheses.

lead through direct foreign investment, as compared to other forms of technology transfer such as licensing (Caves, 1971).

Easy and efficient transfer of technology between parent and sister companies is one of the principal advantages of foreign ownership (Teece, 1977). Ease of technology transfer is accomplished by multinationals through a continuous transfer of technology, rather than by costly one-time contracts with all their attendant negotiation costs. For several reasons, technology transfer is more difficult when arm's-length contracts have to be negotiated. First, asymmetric information leads to high negotiation and monitoring costs. Secondly, one-time contracts must take into account the need to transfer knowledge about improvements in technology, which are often an important part of the technological process. Without codicils that allow for these improvements, the technology that is transferred rapidly becomes obsolete.

The survey results on the nature of technology transfer confirm that substantial differences exist between foreign- and domestically controlled firms with regard to technological transfer (see Table 10.12). Foreign-owned firms have an advantage over locally owned firms in that they more frequently arrange a continuous transfer of technology, listed respectively by 52% and 30% of respondents from each group.

Newer technologies are constantly being improved, making it difficult to negotiate arms-length contracts for the transfer of technology. Domestically owned firms have more difficulties in arranging a continuous technology transfer and use one-time transfer more often than do foreign subsidiaries (27% and 19%, respectively). These differences confer a competitive edge to foreign affiliates in sectors where the rate of technological change is fastest.

Since domestically owned firms use one-time contracts more frequently for technology transfer, their contracts are more often subject to conditions that need to be spelled out in detail in these types of contracts; however, these differences are not very significant. Technology transfer contracts signed by domestically owned firms more frequently contain clauses conferring a specific right to sell and to use in manufacture. The difference between the two ownership groups is negligible in other areas. About one-third of contracts in each case restrict manufacturing or sales of transferred technology to a specified territory.

The survey also offers an insight into differences in the channels through which innovations are spread to the rest of the economy. The survey investigated whether and in which form an innovation was sold (that is, transferred) to other firms. Domestic- and foreign-owned firms transfer or sell their innovations by different means. The large foreign-owned firms transfer their innovations primarily through sales of capital goods, whereas domestic firms do so through sales of intermediate products.

Since new or improved capital goods are among the most important vehicles of technological change, the large foreign affiliates generate what to their customers are process innovations in downstream industries. This is also corroborated by the fact that a larger proportion of foreign firms introduce new production techniques (68.3%) than do locally owned firms (57%). Since foreign-owned firms tend to introduce world-first innovations more frequently than do locally owned firms, they are also more likely than domestic firms to transfer intellectual property.

10.5 IMPEDIMENTS TO INNOVATIONS

Because of the extent of foreign ownership in the Canadian manufacturing sector, the previous sections of this chapter have focused on the type of innovation system possessed by the Canadian affiliates of multinationals. The survey finds that foreign-owned firms have an advantage in that they rely heavily on foreign affiliates, but they also have developed local innovative capacities in the form of R&D units, as well as local supply and customer networks.

Foreign innovation rates are influenced by more than the capacity of firms' R&D units; they also may be affected by other problems – the

Table 10.13. *Impediments to Innovation, by Industrial Sector and Nationality (% of Firms)*

Sector	All		Core		Secondary		Other	
	Canadian	Foreign	Canadian	Foreign	Canadian	Foreign	Canadian	Foreign
Lack of skilled personnel	49	41	40	38	43	41	57	46
Lack of Information on technology	30	34	30	37	33	39	28	20
Lack of Information on markets	28	31	36	43	27	31	25	7
Lack of External technical Services	16	12	20	7	9	17	18	13
Barriers to interfirm Cooperation	15	15	17	11	9	25	17	3
Barriers to university Cooperation	6	4	6	7	5	3	7	0
Government standards & regulations	21	30	27	23	18	28	21	51
Other	22	17	20	18	26	19	22	11

Note: Standard errors are less than 4 for All Canadian and less than 6 for All Foreign.

extent to which local labour markets do not provide the type of skills required, the effectiveness of technology information networks, or the extent to which government regulations present obstacles.

Obstacles reported by both domestic and foreign-owned firms are remarkably similar (see Table 10.13). The most frequent one is the lack of skilled personnel, but here, foreign firms are slightly less likely to have this problem than are domestic firms, probably because they can draw on their affiliates for help. Elsewhere, differences are not significant.

An industrial breakdown reveals some sector-specific differences. Foreign firms operating in the secondary sector have more difficulties than do their Canadian counterparts, particularly with regards to interfirm cooperation and government standards.

The situation is reversed in the 'other' sector, where Canadian-owned firms report impediments to innovation more frequently, except for government standards and regulations. Complaints of foreign affiliates regarding government standards and regulations were particularly frequent (51%) in the other sector.

10.6 DO CANADIAN-OWNED FIRMS INNOVATE MORE OR LESS THAN FOREIGN AFFILIATES?

The impact of foreign ownership on the structure and performance of Canadian industry has often been at the centre of controversy. Our overview of R&D activities, sources of innovation, and the method used to transfer technology suggests that foreign-owned firms are more likely than domestically owned firms to build a stock of proprietary knowledge, which is one of the essential conditions for successful innovation strategy.

These differences are reflected in innovation activity. Foreign firms exploit their superior innovation potential and are more likely to introduce innovations than are their domestic counterparts. The foreign-owned rate of innovation is 52%, while the rate for domestically owned firms with no foreign orientation is just 29%, and for domestically owned firms that are internationally oriented it is 47%.[10] Thus, domestic firms with an international orientation are not very different in terms of their innovativeness than are foreign multinationals. Innovation gaps are largest between the group of firms with an international orientation (both foreign and domestic) and those who serve only domestic markets.

---

[10] This sample, as all others in this chapter, uses just the integrated group of firms (Ips), which are generally larger than 20 employees.

Table 10.14. *Percentage of Firms That Introduced, or Were in the Process of Introducing, an Innovation During the Period 1989–91, by Nationality and Size Class*

	Employment Size Class			
Ownership	21–100	101–500	501+	All
Canadian without foreign operations or sales	27 (4)	33 (7)	39 (13)	29 (3)
Canadian with foreign operations or sales	45 (5)	48 (6)	64 (8)	47 (3)
Foreign	56 (8)	39 (5)	72 (6)	52 (4)

Note: Standard errors are in parentheses.

Table 10.15. *Innovation Intensity, by Nationality, Size Class, and Industrial Sector (% of Firms)*

Sector	Ownership	Employment Size Class			
		21–100	101–500	500+	All
Core	Canadian without foreign operations	40	–	–	52
	Canadian with foreign operations	62	69	75	66
	Foreign	67	45	77	60
Secondary	Canadian without foreign operations	27	35	–	28
	Canadian with foreign operations	40	45	72	45
	Foreign	64	41	70	57
Other	Canadian without foreign operations	25	24	41	26
	Canadian with foreign operations	41	42	51	42
	Foreign	18	30	69	36

Note: Standard errors are generally less than 6 for the All category.

The difference in the innovation rate varies across size categories, sectors, and industry groups. Foreign-owned firms innovate more frequently than do all domestically owned firms in the smallest and largest size categories, but not in the medium-size category (100 to 500 employees), where Canadian-owned multinationals have an edge over their foreign counterparts (see Table 10.14).

When the proportion of innovating firms is broken down by industrial sector and by size of firm (see Table 10.15), the previously revealed relationship between the size of firm, nationality, and innovation incidence remains broadly unchanged. The purely domestic firms generally innovate less than do both types of multinationals across all size classes.

There are no consistent differences between the two types of multinationals (see Figure 10.7). In the core sector, domestic multinationals are

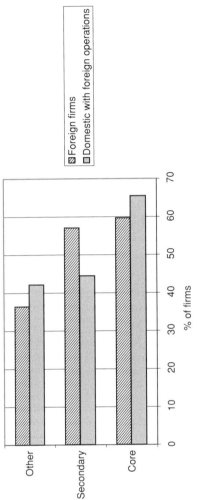

Figure 10.7. Innovation Rate for Foreign and Domestic Multinationals, by Industry Sector

more innovative overall but not in every size class; in the secondary sector, the domestic multinationals are less innovative overall, but this stems primarily from differences in the smallest size class. In the two largest size classes, there is little difference between the two groups.

10.6.1 Do Foreign-Owned Firms Introduce Process Innovations More Often Than the Canadian-Owned Firms?

A distinguishing feature of the innovation activity of foreign-owned firms in Canada that was noted in the 1980 Economic Council of Canada (EEC) innovation survey was the tendency of foreign-owned firms to introduce process innovations more often than Canadian-owned companies did (De Melto, 1980). However, the ECC survey covered only a limited number of manufacturing sectors. The present survey provides information on whether this result holds across the manufacturing sector, not just for process innovations but also for product innovations that are made without a change in process technology and product innovations that are accompanied by changes in technology (see Table 10.16).

In comparison with their domestically owned counterparts, foreign-owned firms are relatively more likely to introduce pure process innovations than product innovations (both without and with a change in manufacturing technology). While foreign-owned firms introduced all

Table 10.16. *Product Versus Process Innovation Intensity, by Nationality and Size Class (% of Innovators)*

Size	Ownership	Product Innovation Without Change in Man. Technology	Process Innovation Without Product Change	Product Innovation with Change in Man. Technology
All	Canadian	39 (3)	49 (3)	53 (3)
	Foreign	46 (6)	61 (6)	58 (6)
21–100	Canadian	37 (4)	47 (5)	53 (5)
	Foreign	41 (13)	60 (12)	70 (12)
101–500	Canadian	38 (6)	47 (6)	55 (6)
	Foreign	42 (9)	64 (9)	52 (10)
501–2000	Canadian	51 (9)	63 (9)	64 (10)
	Foreign	41 (9)	58 (9)	70 (9)
2000+	Canadian	38 (10)	83 (8)	58 (10)
	Foreign	78 (8)	57 (15)	64 (14)

Notes: Larger firms (IPs) only. Standard errors are in parentheses.

three types of innovations more frequently than the locally owned firms did, the largest differences between foreign- and domestically owned firms occur for process innovations. The tendency of foreign-owned firms to introduce process innovations that do not involve product change somewhat more often than Canadian-owned companies can also be found within each industry sector.

There are, however, some differences in this pattern when the relationship is broken down by the size of the firm (data not reported here). Small and medium size foreign-owned firms introduce process innovations more frequently than do comparable Canadian-owned firms. The opposite is true for the two largest size categories, where Canadian-owned firms introduce pure process innovations more frequently than do foreign-owned ones. A possible explanation of this pattern is that Canadian and foreign-owned firms of similar size in terms of Canadian employment (or sales) are not really comparable. A foreign affiliate employing, say, fewer than 500 persons in Canada may be related to a parent company abroad that is several times its Canadian size. If so, the affiliate's performance in Canada reflects in many respects the structure, conduct, and performance of a much larger parent firm and cannot be readily compared to a Canadian-owned company of a similar size. As evidence of this incomparability, results of the survey show that size affects the introduction of process innovation more for Canadian-controlled firms than for foreign-controlled affiliates. Smaller foreign affiliates can viably introduce pure process innovations because their innovation cost is spread over the volume of production of the whole multinational enterprise.

The ownership-specific innovation features (not presented here in tabular form) emphasize and illustrate the difference between the two groups. On the one hand, close to one-third of domestically owned firms specialize in product innovations with functional novelties. On the other hand, almost two-thirds of foreign affiliates list the major characteristic of their innovation as involving an improvement in production techniques.

10.6.2 Originality of Innovations

Another significant difference between domestically and foreign-owned firms is that the latter introduce a higher proportion of world-first and Canada-first innovations (see Table 10.17). This difference applies to both purely domestic firms and multinational domestic firms (see Figure 10.8). The imitative 'other' innovations are introduced more frequently by Canadian-owned firms.

Table 10.17. *Originality of Innovations, by Nationality and Industrial Sector (% of Innovators)*

Sector	All		Core		Secondary		Other	
	Canadian	Foreign	Canadian	Foreign	Canadian	Foreign	Canadian	Foreign
Company weighted								
World-first	14	25	23	33	11	22	11	12
Canada-first	31	42	35	39	37	49	26	34
Neither	55	34	43	28	53	29	63	53
Employment weighted								
World-first	20	38	32	50	20	43	12	13
Canada-first	39	29	47	37	40	14	34	36
Neither	42	33	22	14	41	44	54	51

Notes: Larger firms (IPs) only. Standard errors are 3 for the All Canadian category and 6 for the All Foreign category.

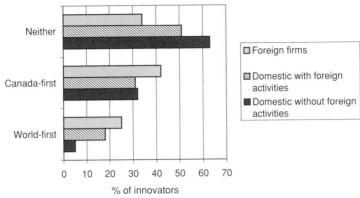

Figure 10.8. Originality of Innovations for Domestic Versus Foreign Multinationals

This pattern is partly related to the sectoral distribution of foreign ownership and the average size of foreign affiliates. Since foreign affiliates are concentrated in the core sector and are on average larger than their Canadian-owned counterparts, it is to be expected that they will display characteristics of the core sector and of large firms, that is, a heavier involvement in R&D and a correspondingly higher proportion of original innovations.

However, there are other factors at work. A comparison by sector shows that foreign-owned firms tend to introduce more original innovations in all sectors. About one in three affiliates in the core sector introduced a world-first innovation, compared to fewer than one in four Canadian-owned firms. There are also major differences in the secondary sector. Only in the tertiary other sector is the rate of world-first innovations that are created by both ownership groups about equal.[11] Moreover, these results are caused not just by larger firm size – as a comparison of the company- and employment-weighted results in Table 10.17 shows.

According to the conventional wisdom, foreign-owned subsidiaries introduce new technology only after it has been first introduced in the country of origin of the parent company. Accordingly, one would expect foreign affiliates to be in the forefront of the introduction of Canada-first innovations, thus confirming their role as the prime means of transferring foreign technology to Canadian industry. This is the case: Foreign affiliates report

---

[11] Even though these results concern only the most profitable innovation per firm, they would not be reversed by taking into account the number of innovations introduced by each firm. Foreign-owned firms introduced on average more innovations per firm than did Canadian-owned ones.

a higher proportion of those innovations than do locally owned companies in all sectors. However, it should be noted that this superiority disappears when innovation rates, weighted by employment, are compared. These are higher for domestically controlled than for foreign firms. Thus, it is the smaller foreign firms operating in Canada who are most likely to be bringing in technology that has already been introduced abroad. A similar comparison of the company- and employment-weighted estimates for world-firsts indicates that it is the larger firms that are introducing the most innovative products simultaneously in Canada and abroad. This provides evidence of the increasing interdependence of the largest foreign affiliates operating in Canada. One effect of the worldwide improvements in communication and transportation infrastructures is the development of increasing linkages among actors, both within and across national boundaries. Worldwide companies are increasingly striving to combine advantages of development and application of 'central innovations'.

Finally, it should be noted that imitative innovations, which are neither world-first nor Canada-first, are introduced in all sectors more frequently by Canadian-owned firms. The contribution of domestically owned firms to diffusion of technological change within Canada is negatively correlated with R&D intensity. It is least pronounced in the core sector and most notable in the other sector.

In summary, the evidence unequivocally supports the hypothesis that foreign-owned firms contribute significantly to technological progress in Canadian industry. A further breakdown of the data by size of firm could reveal that part of the lead of foreign-owned firms over Canadian-owned firms is related to their larger size and other activities like R&D. This issue is investigated at greater length in Chapter 14. However, a finding that other factors account for foreign versus domestic differences would not invalidate the conclusion that foreign-owned firms make a greater contribution to industrial innovation in Canada than their domestically owned counterparts that have no international orientation.

10.6.3 The Number of Innovations Introduced

To this point, we have concentrated our analysis on differences in the incidence of innovation by using information concerning the single most profitable innovation introduced by the firm in the 1989–91 period. In order to assess the quantitative importance of innovative activity, we can also examine how many major innovations were introduced by innovators during the period 1989–1991. The results (see Table 10.18) show that

Table 10.18. *Number of Innovations Introduced and in Progress for Product Versus Process Innovations, by Nationality*

Ownership	Product Innovation Without Change in Man. Technology		Product Innovation with Change in Man. Technology		Process Innovation Without Product Change	
	Company Weighted	Employment Weighted	Company Weighted	Employment Weighted	Company Weighted	Employment Weighted
Introduced						
Canadian owned	3	5	2	4	2	3
Foreign owned	5	6	3	3	2	2
Innovations in progress						
Canadian owned	2	3	2	2	1	2
Foreign owned	3	5	2	2	2	2

Notes: Larger firms (IPs) only. Standard errors for Canadian range from 0.2 to 0.5, and from 0.2 to 2.0 for Foreign.

foreign-owned innovators were more prolific in creating product innovations not involving any change in manufacturing technology (5 per firm) than were Canadian-owned firms (3 per firm).

This pattern indicates that foreign affiliates have an advantage in introducing new and improved products developed by their parents. However, in the other two innovation categories there are few differences, at least in the company-weighted estimates, and domestic firms do slightly better in the employment-weighted estimates. Thus, the large domestically owned firms outperform foreign affiliates in the other two categories. Differences between foreign-owned and domestically owned innovators for the number of innovations in progress – presented in the lower section of Table 10.18 – generally mirror the results for the number of innovations introduced.

A multiple regression of the number of innovations on the most important determinants reveals that the number of innovations increases with employment size of the firm. After controlling for the effect of size and ownership, firms operating in the tertiary other sector introduce fewer innovations than in the two remaining sectors. When the effect of the size of firm and the sector are both taken into account, Canadian-owned innovators do not introduce fewer innovations than do foreign owned innovators.

10.7 USE OF INTELLECTUAL PROPERTY RIGHTS

Firms vary in their innovation strategy. Some emphasize research and development and appropriability. Others pursue an imitative strategy. Intellectual property is protected and valued to different degrees by firms with different innovation strategies.

Foreign- and domestically owned firms might be expected to differ considerably in terms of their use of these forms of property protection. Foreign-owned firms are larger and tend to be located within high-tech sectors. More importantly, one of the advantages possessed by multinationals is their superior technological skills, including their ability to transfer these skills across national boundaries (Caves, 1982). They might, therefore, be expected to make greater use of intellectual property protection than domestically owned firms do.

Since foreign-owned firms are larger than domestically owned firms, differences between the two will be partly the result of nationality and partly the result of size. In order to account for size differences, the population was divided into large and small firms. For the purpose of this

Table 10.19. *Multiple Use of Intellectual Property Protection, by Nationality and Size Group (% of Firms)*

Number of Intellectual Property Types	Firm Type			
	Small Canadian	Small Foreign	Large Canadian	Large Foreign
Any	22 (1.3)	36 (5.6)	57 (5.0)	65 (4.7)
1	13 (1.0)	21 (5.0)	28 (4.1)	26 (4.6)
2	6 (0.7)	7 (2.3)	18 (5.8)	20 (3.8)
3	2 (0.4)	6 (1.8)	6 (1.6)	9 (2.1)
4+	1 (0.3)	1 (0.7)	4 (1.3)	10 (3.0)

Notes: Small and large firms are defined here as less than and more than 200 employees, respectively. Standard errors are in parentheses.

Table 10.20. *Usage of Individual Forms of Intellectual Protection, by Nationality and Size Class (% of Firms)*

Form of Intellectual Property Protection	Firm Type			
	Small Canadian	Small Foreign	Large Canadian	Large Foreign
Copyrights	4 (1)	9 (3)	11 (3)	15 (3)
Patents	6 (1)	12 (3)	22 (3)	39 (5)
Industrial designs	5 (1)	7 (2)	14 (6)	21 (4)
Trade Secrets	8 (1)	9 (2)	19 (6)	20 (4)
Trademarks	10 (1)	21 (5)	31 (4)	42 (5)

Notes: Small and large firms are defined here as less and more than 200 employees, respectively. Standard errors are in parentheses.

exercise, small firms are defined as those with fewer than 200 employees, large firms as those with more than 200 employees. Comparisons of domestic and foreign-owned firms within a size class allow the effect of nationality to be examined, holding size class constant.

Differences in the use of intellectual property protection are found both across size classes and across nationality groups (see Table 10.19). These differences are reflected in higher foreign use rates for almost all categories of statutory property protection (see Table 10.20). Innovations in foreign firms are more likely to use copyrights, patents, trademarks, industrial designs, and trade secrets. One of the largest differences occurs for patents, where 39% of large foreign-owned firms but only 22% of large domestically owned firms possess this form of intellectual property protection.

Table 10.21. *Effectiveness of Intellectual Property Protection, by Nationality (Mean Score on a Scale of 1–5)*

Intellectual Property Protection Associated with	Foreign			Canadian		
	All 1	Users* 2	Nonusers 3	All 4	Users* 5	Nonusers 6
Statutory protection						
Copyrights	2.0	2.8	1.4	1.6	2.8	1.3
Patents	2.7	3.4	1.7	1.8	3.0	1.4
Industrial designs	2.1	2.9	1.5	1.6	2.5	1.3
Trade secrets	2.7	3.2	2.5	2.0	3.2	1.5
Trademarks	2.6	3.1	1.6	2.0	3.1	1.4
Other strategies						
Complexity of product designs	3.1	3.2	3.1	2.6	3.0	2.2
Being first in the market	3.1	3.4	2.7	3.2	3.4	3.0
Other	2.5	3.5	1.3	2.3	2.6	2.1

Note: Standard errors are less than .1 for All Canadian and less than .2 for All Foreign.
* Users are defined as those having the particular property right being scored.

Not only do domestically owned firms use intellectual protection less frequently than do foreign firms, but they also value it less. When asked to score the efficacy of the different forms of intellectual property, foreign-owned firms rank every form of protection higher than do domestically owned firms (see Table 10.21, column 1 versus 4).

This may be the result of inherent differences in the attitude taken by domestically owned and foreign-owned firms towards intellectual property protection. Inherent differences in attitudes are tested by examining whether domestically owned and foreign-owned firms that use statutory intellectual property rights view the effectiveness of these rights differently. When these two groups are compared (Table 10.21, columns 2 and 5), the differences are reduced and sometimes eliminated.

The differences between the scores given by foreign-owned and domestically owned populations to the efficacy of intellectual property protection thus come partly from the larger percentage of Canadian firms who do not make use of the form of protection and partly from the lower scores that these firms give to the efficacy of protection in most areas. This nonusing group tends not to be innovative, tends to utilize imitator strategies, and thus finds intellectual property law to be relatively unimportant. But in the case of patents, the fact that this difference persists even when users are examined suggests a difference in the extent to which Canadian firms manage their intellectual property assets. This interpretation is born out by the fact that Canadian firms tend to have about the same attitude to the 'innate' forms of protection – whether complexity or being first in the market are compared.

10.8 EFFECTS OF INNOVATION

That innovative activity, innovation intensity, and the tendency to use intellectual property protection are higher in foreign-owned firms than in domestically owned firms may be either benefit or cost related. Foreign firms may be better able to reap particular types of benefits, or domestically owned firms may experience more barriers to innovation.

The effects of innovation range from improved profits, quality, and supplier and customer interactions to increased market share and to reductions in costs. A comparison of the frequency of occurrence of these effects between foreign and domestic firms (see Table 10.22) indicates that there are few major differences between the two groups, though there are some exceptions. Canadian firms are more likely to note that innovation increased domestic market share and improved customer relations.

Table 10.22. *Effects of Innovation, by Nationality (% of Innovators)*

Effects of Innovation	Ownership	
	Canadian	Foreign
Improvements		
Improved profit margin	63 (3)	61 (6)
Improved quality of products	61 (3)	58 (6)
Improved technological capabilities	51 (3)	49 (6)
Improved customer interactions	75 (3)	64 (6)
Improved suppliers interactions	27 (3)	19 (4)
Improved working conditions	31 (3)	27 (5)
Extended product range	56 (3)	58 (6)
Increased share		
In domestic market	69 (3)	57 (6)
In foreign markets	37 (3)	48 (6)
Cost reduction		
Reduced lead times	33 (3)	28 (5)
Reduced labour requirements	32 (3)	33 (5)
Reduced design costs	15 (3)	6 (2)
Reduced material requirements	19 (3)	19 (4)
Reduced capital requirements	6 (2)	4 (2)
Regulations		
Environmental regulations	17 (2)	17 (4)
Health and safety regulations	18 (2)	16 (4)

Note: Standard errors are in parentheses.

Foreign-owned firms are more likely to associate innovation with increased foreign market share. But elsewhere, about the same percentage of each group noted each benefit.

A breakdown by sector (not reported here in tabular form) finds additional differences between the groups, though they are not large. Since foreign-owned firms are more likely to introduce process innovations than are their Canadian counterparts, it follows that their innovations in the core sector tend to increase plant specialization and the scale of production, to reorganize work flows, and to increase production flexibility more often than in Canadian firms. On the other hand, domestic firms are more likely to introduce product innovations fulfilling new functions and supplying new functional parts that require improved interaction with customers and suppliers. Canadian-owned firms report the former two effects significantly more often (73% and 28%) than do their foreign-owned counterparts (50% and 14%).

10.8.1 Do Foreign Affiliates Export More or Less Than Canadian-Owned Firms?

The differences in innovation regimes that are outlined in this chapter suggest that foreign and domestic firms may differ in their penetration of foreign markets. Innovation is costly. It might, therefore, be expected that firms that can spread their costs across both domestic and foreign sales will have a greater ability to innovate. It is also the case that operations in foreign markets receive greater competitive stimuli to innovate than do firms operating solely in domestic markets. While not all domestically controlled firms operate only in domestic markets, the group as a whole is less likely to be doing so than are multinationals.

The importance of foreign markets to the innovation process is investigated further by examining differences in the extent to which an innovation is sold in foreign markets and whether the share of sales derived from exports is larger for foreign firms. These data, of course, are more likely to cover product innovations because sales data resulting from process innovations are not readily available.

About 40% of all respondents who provided information[12] on the value of domestic and foreign sales arising from their innovations are exporters. Core-sector firms export more often than do those in the other two sectors.

---

[12] About half of larger innovating firms volunteered figures for domestic and export sales resulting from their most profitable innovation. Almost half of these firms exported.

The proportion of foreign-owned firms that export is larger (57%) than that of Canadian-owned firms (38%) (see Table 10.23).

More importantly, foreign affiliates also export a higher proportion of their sales from an innovation than do Canadian-owned firms. When this ratio is computed for all firms that declared reported sales from an innovation (but did not necessarily export), 33% of foreign affiliates sales but only 27% of sales from an innovation in domestically owned firms are derived from exports. When this is calculated for just firms that do export, foreign-owned firms export a higher share of sales (76%) than do domestically owned firms (53%). Thus, foreign-owned firms are more likely to export than domestically owned firms, and the former export more intensively.

These differences vary to some extent across industry sectors. Canadian-owned firms operating in the core sector are almost as likely to export as are their foreign counterparts. This is not true in the secondary sector, where foreign firms are more likely to export than domestically owned firms. However, the large majority of foreign-owned firms operating in the 'other' sector are not export oriented; only 25% declare exports sales. And this is less than domestically owned firms in this sector. This suggests that most of the foreign-owned firms belonging to the other sector innovate in Canada in order to serve the local market.

Despite differences across sectors in the probability of exporting an innovation, the export intensity of foreign-owned firms is always greater than for domestically owned firms. When an innovation results in export sales, foreign-owned firms are more reliant on these sales than are domestic firms.

10.8.2 The Impact of Innovation on Employment and Skill Requirements

One of the most frequently noted effects of innovation is on labour requirements. Although both foreign and domestic firms are about as likely to report that innovation affected unit labour requirements, these changes may have occurred in different ways. Moreover, innovation may have different effects on total employment in the two populations.

Innovation affects employment in two ways. On the one hand, introducing new or improved production processes increases productivity. The technical change it introduces is likely to reduce unit labour requirements. This tends to reduce the demand for labour. On the other hand,

Table 10.23. *Export Incidence and Propensity, by Industrial Sector and Nationality*

Ownership		All		Core		Secondary		Other	
		Canadian	Foreign	Canadian	Foreign	Canadian	Foreign	Canadian	Foreign
% of firms that exported	Company weighted	41 (4)	57	51 (6)	54	43 (8)	74	32 (6)	25
Exports/sales	Employment weighted	29 (5)		39 (7)		27 (9)		22 (6)	
% of firms that exported	Company weighted	38	57	51	54	32	74	33	25
Exports/sales by ownership									
All firms that declared sales	Employment weighted	27	33	34	43	20	34	20	12
Only firms that exported	Employment weighted	53	76	52	74	57	77	52	77

Note: Standard errors are in parentheses.

a successful launching of a new or improved product may increase sales and lead to increased demand for employment.

Foreign and domestic firms may experience a different effect of innovation either on unit labour costs or on overall sales and thus experience a different effect of innovation on total employment.

Owing to the relative abundance of capital in Canada and in the United States, the home country of the largest proportion of foreign-owned firms operating in Canada, technical change is likely to be labour rather than capital saving. However, relative factor cost differences between the two countries are rather small, and it is unlikely that they would lead to systematic differences between Canadian- and U.S.-owned firms in the extent to which innovation was labour saving.

What is more important in determining differences in the demand for labour in the two groups is their different emphasis on product and on process innovation. Foreign-owned firms introduced pure process innovations more often than did domestically owned firms. Process innovations are more likely to be labour saving than are product innovations. We would, therefore, expect that the effect of innovation in foreign firms is more likely to be labour saving than the product and combined product-and-process innovations typically introduced by domestically owned firms.

This is the case. Innovations introduced by foreign affiliates appear to have a less beneficial effect on job creation than those introduced by locally owned firms. Almost 20% of foreign-owned firms report that their innovation decreased employment of production workers (see Table 10.24). This compares with 11% of Canadian-owned companies. Firms that created jobs are more numerous than firms that reduced them. But again, the positive effect is twice as important in domestically owned firms than in foreign ones (40% of domestic and 22% of foreign).

This conclusion is based on the percentage of firms that indicate either a decrease or increase in employment. When employment-weighted data are used, the differences are not as clear. Now, a greater percentage of employment is in domestic firms that decrease employment compared to foreign firms, and there is little difference in terms of employment increases. Thus, even though fewer Canadian than foreign-owned firms reduced employment, those that did were on average much larger than the foreign ones. The overall employment reduction impact could therefore be larger in the domestically owned sector. Ultimately, we would need data on the actual number of jobs eliminated to decide this issue, and the survey did not try to collect these data.

Table 10.24. *The Effect of Innovation on the Number and Skill Requirements of Workers, by Nationality (% of Innovators)*

Effects on	Ownership	Decrease C.W.*	Decrease E.W.†	Increase C.W.	Increase E.W.	No Change C.W.	No Change E.W.
No. of production workers	Canadian	11	20	40	33	47	44
	Foreign	19	10	22	29	57	59
No. of nonproduction workers	Canadian	3	4	25	23	57	60
	Foreign	7	3	19	22	65	70
Skill requirements of workers	Canadian	1	0.1	38	29	62	71
	Foreign	2	4	40	35	58	61

Notes: Larger firms (IPs) only. Standard errors for the company-weighted estimates for the column 'Increase' range from 3 to 6, while for the column 'Decrease' they range from 1 to 5.
* C.W = company weighted.
† E.W. = employment weighted.

Technological change has been shifting the structure of employment by increasing the share of nonproduction workers in total employment (Baldwin and Rafiquzzaman, 1999). Results of the survey confirm that innovation is one of the causes of this change. The percentage of firms where innovation led to declines in nonproduction worker employment is marginal and similar in both ownership groups (3% and 7% for domestic and foreign firms, respectively) and virtually identical when weighted by employment. The creation of nonproduction worker jobs is also very similar in both Canadian and foreign-owned firms. It increased in about one-quarter of firms, which represents one-quarter of total employment in each group. Thus, the employment impact of innovation on nonproduction or white-collar workers was broadly similar and probably positive in both the Canadian and foreign-owned segments of the industry.

Elsewhere, we have noted that innovation is accompanied by an increase in the skill requirements of firms (Baldwin, 1999). Despite the differences in the innovation profile of domestic and foreign-controlled firms, innovation led to increased skill requirements in about the same proportion of domestically and foreign-owned firms (38% and 40%, respectively). There was a general need to upgrade these requirements as a result of innovation in both sectors.

10.9 OWNERSHIP VERSUS TRADE ORIENTATION

Foreign-owned firms have been shown generally to be more likely to conduct R&D and to introduce innovations than domestic firms that have neither foreign operations nor foreign sales. This does not mean that all domestic firms lack innovative capabilities. The difference between foreign-controlled firms and what we have called internationally oriented domestic firms is less than with domestic firms in general and, in some cases, not significant.

In the previous sections of this chapter, we have defined an internationally oriented domestic firm as one that has either foreign production/R&D facilities or foreign sales. The latter is a rather broad definition of international orientation because it includes all domestic firms who make any foreign sales, no matter how small. To test whether this definition affects the size of the difference between foreign- and internationally oriented domestic firms, we change the definition to include only those firms with a substantial amount of foreign sales – where substantial is defined as 10%, 20%, and 30%, respectively. The R&D and innovation intensities of the pure domestic (no foreign sales or production/R&D facilities), those with less than 10% of sales as exports, and those with foreign production facilities and more than 10%, 20%, and 30% of sales are reported in Table 10.25.

It is apparent that as a domestically owned firm increases its export intensity, it moves closer to the profile of the foreign firms operating in Canada, although this effect generally diminishes beyond the 10% export sales threshold. Thus, a firm that is more intensely involved in exports or one that we can refer to as more globalized resembles a foreign-owned firm operating in Canada quite closely in terms of its innovation profile.

These results indicate that it is globalization, rather than nationality of ownership per se, that affects the degree of innovativeness.

10.10 CONCLUSIONS

Foreign ownership in the manufacturing sector is highest in industries where multinationals can exploit their proprietary technological, marketing, and R&D assets. Previous research has sought empirical evidence of the advantages of multinational investment with measures like labour productivity (Raynauld, 1972; Globerman, 1979) or measures of operational structure (Safarian, 1973). This paper examines the innovation capacity of multinationals operating in Canada and their success at innovation in a more direct fashion.

We find that even though they have a privileged access to their parents' R&D and technology, foreign subsidiaries perform R&D in Canada more

Table 10.25. *Gradations of Innovation Activity, by Degree of Foreign Operations (% of Firms)*

		International Canadian				Pure Domestic: No Foreign Production or Foreign Sales
	Foreign Owned	Foreign Production or Foreign Sales >30%	Foreign Production or Foreign Sales >20%	Foreign Production or Foreign Sales >10%	No Foreign Production and Foreign Sales <10%	
% conducting R&D	89	84	85	86	86	77
% conducting ongoing R&D	53	53	49	48	39	39
% collaborating	35	25	25	24	20	19
% separate R&D	44	30	32	33	26	25
% with innovation	52	47	48	49	16	29
% innovators with world-first	25	23	22	21	13	5

Notes: Larger firms (IPs) only. The standard errors for the first row are lowest (3.0); those for the last row are largest (5).

318

often than do domestically owned firms. A breakdown by sector of activity shows, however, that the technological opportunities and imperatives existing in research-intensive industries force the majority of firms (both domestic and foreign) in the core sector to perform R&D. It is in the secondary and above all in the tertiary other sector where domestically owned firms as a group are less likely to perform R&D than are foreign affiliates. Foreign affiliates are also more likely to complement their own research and development activity by participating in joint ventures and R&D partnerships.

When such determinants of R&D activity as the size of the firm, the sector of activity, and various indicators of the competitive environment are taken into consideration, the probability that a firm performs R&D is higher for firms under foreign ownership. Multinational firms not only exploit their proprietary advantages in Canada but also increasingly develop their own innovation initiatives and tap local sources of technology and scientific research.

One of the ways in which they do so is by collaborating with others on R&D. Because of their involvement in R&D partnerships, there is no evidence to support the once-popular argument that foreign firms do not develop links in Canada and thus create a truncated corporate structure in Canadian manufacturing.

Access to innovative ideas and technological expertise of their parent and sister companies confers a significant advantage to foreign affiliates over the Canadian-owned firms. This privileged source of innovative ideas and technology was, however, used by only about 40% of all foreign-owned firms.

There are two types of foreign affiliates – independent firms that do not rely on affiliates for innovation and those that do so. The technologically more independent foreign affiliates rely more on their own R&D and on sales and marketing in Canada. They are also closer to their customers. The more dependent foreign affiliates are closer to the traditional image of a branch plant, implementing ideas received from parent companies and suppliers. But even here, local facilities exist.

To put their innovative ideas in practice, firms rely on a combination of internal and external sources for technological competence. Independent Canadian-owned companies do not have the luxury of using internal resources that are available within a multinational network, and they generate their own technological expertise through research and experimental development. Foreign-owned firms make greater use of production engineering for this purpose; the difference is particularly important in the R&D-intensive core sector. Another particularity of foreign-owned firms is that they rely more on technology from unrelated firms. The difference is again most notable in the core sector. Even though collaboration with universities and government laboratories is reported relatively less frequently as a source of technological expertise in the innovation process, foreign-owned firms use it more often than do domestically owned firms, especially in the core sector.

Foreign-owned firms innovate in all sectors more frequently than their Canadian counterparts. This is true for most size categories. They especially surpass domestically owned firms in industries that are the recipients of innovations developed elsewhere. These are the industries that are most likely to be the recipients of new products or technologies developed in upstream industries. Here foreign-owned firms do research more frequently than domestically owned firms. Either they are better equipped to absorb innovations developed elsewhere or they are more capable of recognizing the advantages of R&D, even in those industries where technological opportunity is in general less abundant. The most significant laggards are the largest Canadian-owned firms belonging to the other sector. The situation is reversed in the medium-size category of the core sector. In these industries, which also feed technology to the rest of the economy, Canadian-owned medium-size firms surpass their foreign-owned counterparts.

As far as the novelty of innovation is concerned, in keeping with their more frequent involvement in R&D, foreign-owned firms surpass Canadian firms in introducing the more original world-first and Canada-first innovations. They therefore have more use for intellectual property protection than do Canadian-owned companies. The foreign affiliates also have the advantage of arranging for the more efficient continuous transfer of technology, rather than occasional discontinuous transfers.

As far as differences with respect to the effects of innovations are concerned, innovations introduced by foreign affiliates helped them to export and increase their share of foreign markets more often than those introduced by domestically owned firms, which led more often to increased share of the local market. Foreign-owned innovating firms are also more export oriented than are domestically owned firms, and they export innovations more intensively than do domestically owned firms.

All of these comparisons to domestically owned firms as a whole provide evidence that multinational firms do not operate subsidiaries in Canada that are truncated relative to Canadian firms in general. But these comparisons do not deal explicitly with the issue of whether foreign subsidiaries are truncated relative to a standard that is required of competitive global corporations. For this purpose, we have also compared foreign subsidiaries to Canadian corporations that have an international orientation. These additional comparisons to Canadian multinationals show that the two groups of multinationals are quite similar, both with regards to the likelihood that they conduct R&D and that they introduce innovations. It is the international orientation of both foreign and domestically controlled firms that is related to their degree of innovativeness.

ELEVEN

Financing and the Cost of Innovation

11.1 INTRODUCTION

Innovative activity is often characterized as involving classical spillovers that lead to market imperfections. Knowledge that is created by the innovative activity of a firm spills over to other firms who benefit from the activities of the innovator. New ideas and knowledge diffuse across firms not just because new products are copied by imitators. Diffusion also occurs as the new ideas of one firm are improved by others and as general knowledge is incorporated into new production processes. The importance of the diffusion of new ideas has been used to justify governmental support for the innovation process.

A second argument used to justify support is that input markets are imperfect for innovative firms. In particular, innovative firms are characterized as having problems in obtaining financing, partly because their assets are 'soft' and thus do not offer the same collateral as machinery and equipment, partly because their activities are more risky. As a result, financing is characterized as being unduly costly.

In previous chapters, we have provided information on the general importance of innovation, on patterns of diffusion across sectors, on the nature of interactions across firms, on the importance of R&D, and on problems related to appropriability. This chapter focuses on issues relating to the financing of R&D. It examines three issues.

First, we examine the composition of innovation costs. Government policy is often directed at subsidizing expenditures on R&D, since they are seen as the focal point of the innovation process. The effectiveness of this policy is dependent, inter alia, on the relative importance of R&D

in innovation costs. If R&D makes up a relatively small percentage of the total expenditure required to bring innovations to market, then subsidies to this component will have a relatively small impact on total costs. Therefore, the first section examines the distribution of innovation costs across industrial sectors and innovation types.

Second, we examine the evidence that R&D is financed in different ways than other investments that are made by the firm. Earlier studies have argued that R&D is much more dependent on internally generated funds than on assets in general and, therefore, is subject to very different market forces and access problems. This section asks whether internal financing is important and whether this emphasis varies across firms and sectors.

Third, we investigate the extent to which different types of innovators make use of the tax-credit subsidy programs for R&D. In the third section of the paper, we describe which sectors and which types of innovation make use of this program. We also ask whether there is any indication that subsidized R&D is any more successful than the R&D programs that are not subsidized.

Throughout, we compare small and large firms, as well as core to tertiary 'other' industries. Our purpose is to ask whether there are differences in the extent to which financing difficulties emerge in some areas more than in others and whether government programs respond by focusing more intensely on these particular areas.

11.2 THE COSTS OF INNOVATION

Even though R&D expenditures play a key role in debates on innovation, technological change, and industrial policy, they are neither the only nor the most costly component of the innovation process. Expenditures are required in a number of areas in addition to research and development expenditures. Innovations also require expenditures on the acquisition of technological knowledge (patents and trademarks, licenses, consulting services, and disclosure of know-how), on development (engineering, design, prototype and/or pilot plant construction and testing), on manufacturing start-up (engineering, tooling, plant arrangement, construction, and acquisition of equipment), and on marketing start-up activities.

The breakdown of the total cost involved in bringing an innovation to market is provided in Table 11.1 for innovators' most important (i.e., most profitable) innovation. Preproduction tests of the survey discovered that data on innovation expenditures, when they were available, existed for

Table 11.1. *Distribution of Total Innovation Cost, by Stages of the Innovation Process (%)*

Cost Category	All Innovations by Employment Size Class					
	All	World-First	0–20	21–100	101–500	500+
Basic research	8 (1)	5 (1)	11 (5)	8 (1)	9 (2)	5 (1)
Applied research	9 (1)	18 (3)	3 (2)	10 (1)	10 (2)	11 (2)
Acquisition of technology	10 (1)	8 (2)	13 (6)	12 (2)	6 (1)	10 (2)
Development	30 (2)	33 (4)	36 (8)	27 (3)	34 (3)	29 (3)
Manufacturing start-up	34 (2)	27 (3)	33 (9)	35 (3)	32 (3)	34 (3)
Marketing start-up	9 (1)	9 (1)	4 (2)	9 (2)	9 (1)	12 (2)

Notes: Larger firms (IPs) only. These generally have more than 20 employees. Standard errors are in parentheses. The components of innovation costs included herein were defined in the questionnaire as follows: Acquisition of technology (acquisition of patents, trademarks, licenses, specialist consulting services, disclosure of know-how, etc.); development (engineering, layout, design, prototype construction, pilot plant, acquisition of equipment, etc.); manufacturing start-up (engineering, tooling, plant arrangement, construction, pilot plant, acquisition of equipment, etc.); and marketing start-up (marketing costs associated with products).

major projects but were extremely difficult to obtain for the entire range of all innovative activities in the firm. Since innovations vary in scope, originality, and difficulty, meaningful data could only be provided by respondents on the cost distribution of each stage of the innovation process for major innovations of the firm. The major innovation, on average, accounts for between 50% and 65% of the total costs of all innovations carried out by innovating firms.

Even so, the response rate to this question was lower than for most other questions in the section of the survey that examined the characteristics of the major innovation of the firm. Because of difficulties in imputing answers to this question, we report only the unimputed response, contrary to our practice elsewhere in this study.

When all innovators are considered, basic and applied research each account for less than 10% of the total cost of introducing an innovation and together account for about 17% of total cost (Table 11.1). Acquisition of technology accounts for about 10%, a little more than each of the two research components. Development (e.g., engineering, layout, design, prototype construction, pilot plant, acquisition of equipment, etc.) accounts for about 30% of the total and is about twice as important as the combined research expenditures. Manufacturing start-up is even more costly, absorbing about 34% of total innovation costs. Finally, the marketing effort that is required to bring innovations to consumers, at 9% of the total innovation cost, is about as important as each of the research components.

The relatively small proportion of total costs that is accounted for by basic and applied research expenditures indicates that even a very generous subsidy program for those expenditures will have a minor effect on the total cost of introducing an innovation. On average, reducing the basic and applied research portion of innovation costs by 50% would reduce total innovation costs by less than 10%, on average. If subsidy programs support only the basic research portion of R&D, a 50% reduction in these expenses would reduce total innovation expenses by less than 5%. On the other hand, development costs account for over 30% of total costs, and their subsidization will have greater impact on reducing total costs.[1]

The differences in the innovation cost distribution across firm-size categories are not large. The smallest firms (0 to 20 employees) spend a little less on applied research and more on development than the average firm. These differences could simply reflect the fact that the smallest firms have less specialization of function within the firm (the separation of research from development) and that they therefore have greater difficulties in separating expenditures on development from research. Firms in the largest size category spend significantly more on marketing than do smaller firms. This may result from the fact that they are more likely to be introducing multiple product innovations.[2]

World-first innovations spend more on research and somewhat less on manufacturing start-up. This confirms our previous finding that the more original innovations require more R&D effort than do imitative innovations. Nevertheless, it is important to note that the differences here are not large. R&D is a basic part of all innovation processes.

11.2.1 Product-Process Innovation Cost Differences

The cost structure behind innovation is likely to be a function not only of the novelty of innovation but also of the type of innovation. Research is likely to be used more frequently to produce product than process

---

[1] Not all items included under development in Table 11.1 are included in the definition of development that is used by Statistics Canada for its official R&D numbers. In particular, engineering, design, some aspects of prototype, and/or pilot plant construction and testing are omitted. Pilot plants may be included in development only if the main purpose is to acquire experience and compile data. As soon as they begin operating as normal production units, their costs can no longer be attributed to R&D. Prototypes are included only as long as the primary objective is further improvement; otherwise they are excluded. Engineering, design, and drawing are excluded unless in direct support of R&D (Statistic Canada, 1991).

[2] See Chapter 3.

Table 11.2. *Distribution of Innovation Cost, by Stages of the Innovation Process and by Type of Innovation (%)*

	% of Total Costs		
Cost Category	Only Product Innovation with No Change in Man. Technology	Only Process Innovation Without Product Change	Combination of Product/Process Innovation
Basic research	9 (2)	6 (2)	8 (1)
Applied research	16 (3)	8 (2)	8 (1)
Acquisition of technology	7 (2)	17 (4)	8 (1)
Development	39 (5)	24 (3)	29 (2)
Manufacturing start-up	16 (3)	42 (5)	36 (2)
Marketing start-up	14 (3)	3 (1)	10 (1)

Notes: Larger firms (IPs) only. Standard errors are in parentheses.

innovations.[3] In keeping with this hypothesis, firms that only introduced 'pure' product innovations (i.e., product innovation without change in manufacturing technology) devote about twice the percentage of total expenditures to basic and applied research than do firms that are only process innovators (see Table 11.2). Introducing a product innovation also typically requires that a higher percentage of total costs be devoted to marketing start-up costs. On the other hand, the introduction of pure process innovations requires that a relatively higher percentage of innovation expenditures be devoted to technology acquisition and to manufacturing start-up.

The share of the manufacturing start-up cost of process innovations (42%) is lower than reported earlier by De Melto et al. (1980) (50%) and also notably lower than the more recent finding (59%) reported by Kumar and Kumar (1991). One plausible explanation for the lower proportion derived from the evidence presented here is that it covered a broader spectrum of firms. De Melto et al. (1980) cover a small number of industries that are primarily in the core and secondary industries. Kumar and Kumar (1991) examined a smaller group of highly research intensive firms that are less representative than our sample.

Kumar's survey of 105 innovations introduced by large R&D-intensive firms found that Canadian firms conduct less risky, short-term R&D

---

[3] See Chapter 4.

Table 11.3. *Distribution of Innovation Cost, by Stages of the Innovation Process and by Industrial Sector (%)*

Cost Category	Sectors		
	Core	Secondary	Other
Basic research	11 (2)	8 (2)	5 (1)
Applied research	18 (2)	7 (2)	5 (1)
Acquisition of technology	7 (1)	7 (2)	14 (2)
Development	30 (3)	35 (3)	26 (2)
Manufacturing start-up	23 (2)	37 (3)	40 (3)
Marketing start-up	12 (2)	6 (1)	10 (1)

Notes: Larger firms (IPs) only. Standard errors are in parentheses.

projects and give lower priority than their U.S. and Japanese counterparts to updating manufacturing equipment and facilities with state-of-the art process technologies. They reported that the share of manufacturing start-up cost of Canadian product innovations was notably lower (33%) than that reported by Mansfield (1988) in the mid-1980s, and by U.S. and Japanese firms (40% and 54%, respectively). Since the present survey reports an even lower share of manufacturing start-up costs for product innovations in the R&D-intensive core sector (23%), this suggests that manufacturing start-up activities in Canada account for a substantially lower proportion of total innovation expenditure.

Since the proportion of product versus process innovations and the importance of R&D varies from one industrial sector to another, the innovation cost structure should also vary with the industry sector. Table 11.3 indicates that the core sector, and to a lesser extent the secondary sector, rely more on research for innovation, and the tertiary 'other' sector relies more than the first two sectors on technology acquisition.

We have also seen that process innovations are more common in the secondary and the tertiary other sector. Process innovations are associated with higher manufacturing start-up costs than are pure product innovations. As a result, manufacturing start-up represents more than a third of the total innovation cost in the secondary and other sectors, but less than a quarter in the core sector.

The successful introduction of new and improved products requires a successful marketing strategy. In keeping with this requirement, marketing start-up activities account for a larger share of total innovation cost for product innovations than for new or improved manufacturing processes (Table 11.2). But marketing costs are equally high in the core as in the tertiary other sector (Table 11.3).

In conclusion, the structure of an economy will determine just how important R&D is in terms of total innovation expenditures.

11.3 THE SOURCES OF FUNDS FOR INNOVATION

External financing is seen to be more difficult for innovative firms to access and to be more costly for several reasons. First, innovation is risky. Second, the risk is difficult to evaluate because information is asymmetric. Entrepreneurs who are familiar with new technology are in a better position to evaluate its success than are managers in banks and other suppliers of capital. Third, the intellectual property of a knowledge-based firm is less suited to be pledged as collateral for a loan than is standard machinery and equipment.

The difficulties in obtaining external financing are associated with several different theories. Limits to borrowing associated with tax structure and bankruptcy costs have been stressed by Stiglitz (1972); limits resulting from agency costs are the focus of work by Jensen and Meckling (1976); and limits to borrowing caused by the fact that the capital structure acts as a signalling device in markets characterized by asymmetric information are the focus of Myers (1984). In the latter model, managers are expected to consider the interests of existing shareholders and not to dilute their equity until all internal funds are exhausted. The latter has generated the 'Pecking Order' theory of financial structure, which suggests that internal financing will be chosen first, then debt, then equity (Hughes and Storey, 1994).[4]

As a result, it is claimed that innovation will be financed mainly from a firm's internal sources of capital – a source of funds that, at least for small firms, is both expensive and highly variable. Obtaining internal funds is expensive for owners because they must often be raised at high personal-loan rates. It is an uncertain source since internal funds are generated from past success, and these are subject to large cyclical fluctuations arising from business cycles. Moreover, they are sorely lacking at the early stage of a product's life cycle. New start-ups with good ideas therefore face a dearth of financing. Once innovators establish a track record, they can finance innovations from cash flow. But if successful financing is a function of past success, growth at the start of a company's life is restricted.

Previous research has tried to verify the existence of this bias in two different ways. Himmelberg and Petersen (1994) have examined the

---

[4] See also Evans and Jovanovic (1989).

relationship between changes in R&D investment and changes in cash flow over time. Hall (1992) also uses firm data to examine the extent to which R&D investment is influenced by a firm's source of financing. Direct evidence of the connection between sources and uses of funds is provided from a survey of managers of new firms (Baldwin and Johnson, 1999b). Managers indicate that knowledge-based investments are financed internally. Over 69% of firms report using internal funds (either retained earnings or share capital) to finance R&D, over 61% use internal funds to fund technology acquisition, and over 63% use these funds for training. Moreover, firms in R&D-intensive industries are even more likely to finance R&D this way than are firms in non-R&D-intensive industries.

The 1993 Canadian Survey of Innovation confirms the importance of internal funds to innovation. More than two-thirds of all innovating firms (68%) financed their most profitable innovation entirely from internal sources. This is an increase of 10 percentage points in comparison with the 1980 innovation survey (De Melto et al., 1980) that focused on only a subset of the more innovative industries, but it is close to the percentage reported by Baldwin and Johnson (1999b). The smallest firms tend to finance their innovation internally somewhat more often than do the larger ones (see Table 11.4).

There are distinct differences in the extent to which different sectors rely on internal financing. The proportion of firms relying on their own funds is highest in the tertiary other sector and lowest in the core sector. Similarly, less original, other, and Canada-first innovations are financed internally more often than the world-first, more original innovations. These two patterns indicate that firms introducing innovations relying on R&D and exploiting technological opportunities with original innovations in high-tech industries have more success than other firms in tapping external sources of financing. Firms in high-tech sectors may be able to rely more on specialist sources of financing that better comprehend the risks of innovation.

Both the unique product and the unique process innovations have to rely more on outside funds. The more complex type of innovation that combines both product and process innovation is slightly less likely to use outside funds. Thus, both the more complex and the more novel innovations are better served by external sources of funds.

The proportion of Canadian-owned firms that finance their innovations entirely from internal funds is higher than the proportion of foreign-owned firms doing so. This confirms the importance of the advantage that

Table 11.4. *Firms Financing Innovation Entirely from Internal Sources, by Size Class, Nationality, Novelty, and Sector (% of Innovators)*

By employment size	
0–20	75 (3)
21–100	69 (3)
101–500	61 (4)
>500	65 (4)
By Sector	
Core	64 (3)
Secondary	71 (3)
Other	77 (3)
By originality of innovation	
World-first	56 (5)
Canada-first	71 (4)
Other	66 (4)
By Type of Innovation	
Product	69 (6)
Process	69 (5)
Product & process	64 (3)
By ownership	
Canadian owned	74 (2)
Foreign owned	58 (4)

Note: Standard errors are in parentheses.

intrafirm connections provide for foreign-owned firms. These linkages transmit ideas for innovation and provide a needed source of outside funds.

To further examine the financing issue, we focus just on those firms that combine both inside and outside sources of funds. These are the firms that manage to break free from the constraints that face a firm when it is relying entirely on internal financing. In this subsample of innovating firms, internal financing accounts on average for less than half (42%) of total funding requirements (see Table 11.5). In this group, some 8% of funds come from foreign affiliates, another 8% from domestic affiliates, 11% from financial institutions, and 3% from venture capital firms. Governments play an important role – with the federal government providing 12% and provincial governments 9%.

The difference between the company- and employment-weighted figures indicates that large firms rely more on affiliate funding. Funding from research consortia was found only in the largest firms employing

Table 11.5. *The Source of Funding of Major Innovations Not Wholly Funded Internally, by Industrial Sector and Novelty Type (% Distribution)*

Sources of Funds	All		Sectors			Type of Innovation		
	C.W.*	E.W.†	Core	Secondary	Other	World-First	Canada-First	Other
Internal	42 (2)	43 (2)	43 (4)	46 (4)	39 (4)	56 (6)	46 (5)	56 (4)
Other related firms								
Foreign parent or affiliate	8 (2)	18 (2)	12 (3)	10 (4)	4 (2)	18 (6)	11 (4)	2 (2)
Domestic parent or affiliate	8 (2)	17 (3)	5 (2)	4 (2)	13 (4)	3 (2)	8 (3)	8 (3)
Venture capital firms	3 (1)	1 (1)	6 (2)	0	3 (2)	0	3 (1)	0
Other financial institutions	11 (2)	5 (1)	5 (2)	17 (4)	10 (3)	5 (3)	8 (3)	13 (4)
Research consortia	1 (1)	3 (1)	1 (1)	0	1 (1)	2 (1)	2 (2)	0 (2)
Governments								
Federal	12 (1)	8 (1)	17 (2)	13 (3)	7 (1)	12 (2)	13 (2)	15 (2)
Provincial	9 (1)	4 (1)	10 (2)	3 (2)	12 (3)	5 (2)	6 (2)	4 (2)
Other	7 (2)	2 (2)	2 (1)	7 (3)	10 (3)	0	4 (2)	2 (1)

Notes: Large and small firms included in All and in Core, Secondary, and Other Sectors. Large firms (IPs) only included in columns for Type of Innovation. Standard errors are in parentheses.
* C.W. = company weighted.
† E.W. = employment weighted.

over 500 persons, where it accounted for 3% of total innovation costs. On the other hand, small firms are more likely to make use of venture capital funding. Government financing is also used more frequently by smaller firms. This indicates that government programs support innovation in small firms that, owing to imperfections in capital markets, are more likely to encounter difficulties in finding external financing from private sources.

We have previously seen that firms in the secondary and tertiary other sectors are more likely to rely entirely on internal funds. Even in the group of firms that finds external funding, firms in the secondary and other sectors still rely more on internal funds than do firms in the core sector. They also receive a higher proportion of funds from financial institutions and less from related firms and government sources.

In contrast, foreign affiliate funding is more prevalent in the core sector, in keeping with our previous finding that there are closer links to affiliates in this sector than elsewhere.[5] It is also the case that the federal government financed a larger proportion of innovations in the core sector (17%) than in the less R&D-intensive secondary and tertiary other sectors (13% and 7%, respectively). In addition, funds from venture capital sources are directed mostly to firms in the core sector. The largest proportion of firms reporting access to provincial funding is in the tertiary other sector. Funding from financial institutions is also more important in the secondary and other sectors.

Sources of financing also vary by the originality of innovation. Financial institutions and, to a lesser degree, the federal government are more important for the less risky imitative other innovations than the more original ones. By contrast, world-first innovations rely more heavily on the support of parent or affiliated foreign firms.[6]

Venture capital is a non-negligible source of funds for Canada-first innovations only. No support from venture capital is forthcoming to the riskiest world-first innovations. In this respect, venture capital behaves as other risk-averse financial institutions: Their funding is directed more to Canada-first innovations than to the inherently more risky, world-first ones.

Even though the percentage of Canadian- and foreign-owned firms that fund innovation fully from internal sources is identical, innovations introduced by foreign affiliates that are not fully financed internally depend

---

[5] See Chapter 4.
[6] Some 73% of the largest firms (over 500 employees) were able to obtain outside sources for world-first innovations.

much less on internal financing (23%) than comparable innovations created by Canadian-owned firms (45%). While foreign affiliates rely for more than half (56%) of funds on their parent and sister companies, Canadian-owned companies, in contrast, lean more on government funding (23%) than do their foreign counterparts (7%).

In summary, financial constraints can be equated to situations where innovators must rely more on internal than external funds. But surprisingly, firms so constrained are not those that are in the more R&D-intensive sectors; nor are large firms, nor are those producing world-first innovations. Rather, firms that are in the sectors that are consumers of innovations produced elsewhere, that are smaller, and that are domestically owned are more likely to be relying purely on their own sources of internally generated funds. This suggests that markets emerge to solve problems in financing, but that the solutions focus primarily on the highest profile innovators. Those firms that do not capture the spotlight of specialists who provide financing for risky investments in innovation will more likely be constrained in their financing sources.

With this observation, we now turn to how government funding partially fills this void.

11.4 GOVERNMENT FUNDING OF INNOVATION

Public support for innovation-related activities has been justified in several ways. First, governments are responsible for providing new or improved technology for public sector functions (security, health, and communications), and R&D for these ends may be performed in public research laboratories or contracted out to private firms and funded by public revenues. The second justification for public subsidies is to correct for market failures resulting from underinvestment in innovation activities (Arrow, 1962). Owing to the difficulty that firms have in appropriating all the benefits associated with an innovation, it is argued that private firms invest less in innovation than is 'socially desirable'. Other often-adduced reasons for public intervention include high, uninsurable risk and a large minimum-efficient scale required to introduce major innovations. The theory of public policy based on these factors stresses the need for government to provide incentives to private firms to compensate for the gap between the private and social returns to innovation expenditure (in particular to R&D) in order to ensure the socially optimal supply of research and development effort by the private sector.

Others have argued for subsidies that do not directly support the R&D process itself. For Mowery (1983b) and McDonald (1986), the central issue for public policy revolves around the need to subsidize the diffusion of R&D results. In their view, a supply-oriented policy that focuses on subsidizing the performance of R&D neglects that part of the innovative process that renders useful the new knowledge from R&D. High-technology firms are often unable to generate all the information they require from their own R&D and have to ingest information from the work of others. They often decide to release information to rivals in order to receive information in return that will fill in the pieces of what is normally a complex knowledge puzzle. In this case, information that is shared freely may prove more valuable to the firm than information that it chooses to retain entirely to itself (McDonald, 1986).

In the end, most programs of public support for innovation are aimed at supporting research and development directly or indirectly, the premise being that these activities are a necessary input to innovation. Support for R&D rather than for innovations remains the most popular policy support mechanism. The once-popular linear model running from R&D to innovation provides a theoretical underpinning for the policy, and the practicality of tying subsidies and tax credits to formal R&D activity explains the reason that public support is devoted to R&D, rather than directly to innovation.

In this section, we examine the distribution and use of tax credits for R&D, which has become the most important instrument in the arsenal of Canadian government programs supporting the creation and use of new technology.

11.4.1 Tax Credits

Canada has one of the most generous R&D tax-credit programs among major industrial countries. The after-tax cost of $1 Cdn. of R&D expenditure during 1989–91 was less than 50¢ Cdn. in Canada (Warda, 1990). Costs differ across regions because of variations in provincial tax regimes. In Ontario and Quebec, the two provinces with the highest spending for R&D, the after-tax cost of $1 Cdn. of expenditure on research and development was 46¢ Cdn. and 42¢ Cdn., respectively, in 1990 (Palda, 1993).

In 1990, the Scientific Research and Experimental Development Investment Tax Credit (SRED) program delivered nearly $1 billion Cdn. annually in tax credits, more than twice the amount of direct grants to R&D and innovation-related activities. A comparison of federal government funding of R&D made through the SRED in contrast to direct grant programs indicates that the size of tax credits surpassed grants by 1983 and had reached about 18% of business enterprise intramural R&D expenditures (BERD) by 1989. The share of R&D and innovation-related grants remained about constant at 7% of BERD (Hanel and Palda, 1992). Tax credits were, therefore, already the most important and expensive instrument of government policy in support of R&D and innovation in Canada at the time of the survey. According to a recent government report (Finance Canada, 1998), the federal SRED tax credit program was rated as the most important component in the system of Canadian government support to R&D, followed by the refundable federal credit;[7] government grants and contracts received the lowest rating.

Owing to the nature of the program and the confidentiality that surrounds tax-related matters, there is little public information on the distribution of beneficiaries of tax credits and still less on the relationship between tax credits received, R&D, and innovation. The report by Finance Canada (1998) breaks down recipients of tax credits by sector of economic activity only, and does not provide details on the use of tax credits by manufacturing industry subsectors or groups on a two-digit Standard Industrial Classification (SIC) level.

The 1993 innovation survey provides this information and allows us to investigates three issues. The first is whether there are substantial differences in the types of firm that claim tax credits. We ask whether it is the large or small firms, those in core or tertiary sectors, those who produce only world-firsts or Canada-firsts who make use of government programs. In particular, we ask whether tax claims are distributed among sectors and industries in the same proportion as R&D. The second issue that we investigate is the extent to which tax credits appear on the basis of use to be relevant to innovating firms. Here we ask what proportion of innovating firms actually claimed them. Finally, we investigate the efficacy of tax credits by asking whether firms that claimed tax credits innovated more or less than those not claiming tax credits.

11.4.2 Use of Tax Credits

To receive a tax credit for R&D expenditure in Canada, firms must undergo an audit to confirm that their expenditures meet the criteria of the

---

[7] Small Canadian-controlled private corporations can deduct SRED tax credits from federal taxes otherwise payable, and unused credits are refundable.

Table 11.6. *Percentage of Firms Who Conduct R&D and Claim Tax Credits for R&D*

All R&D initiators	16 (1)
Doing R&D as ongoing activity	39 (2)
With a separate R&D department	62 (3)

Note: Standard errors are in parentheses.

tax code. Some firms that report R&D activity may not pass the fiscal criteria nor want the burden of an audit. The percentage of R&D performers that were tax claimants[8] provides a measure that we can use to assess the usefulness of the tax credit program to R&D performers.

Some 65% of firms report that they conduct some R&D. About 16% of the firms reporting R&D in this survey sought a tax credit for the R&D work done during the years 1989 to 1991 (see Table 11.6). Either not all firms reporting R&D met the stringent definitions used under the Income Tax Act or some firms do not find it useful to claim their R&D expenditures. The latter could occur if R&D expenditures are so intertwined with operating expenditures in engineering and production departments that they cannot be separated or because some firms (perhaps the smaller ones) find the costs of complying with the tax program greater than the benefits. Grégoire (1992), in a study of the perception of tax credit programs by industrial firms in Quebec, reports that 45% of firms did not claim tax credits because of red tape; 35% lacked knowledge about the program; and 21% felt the program's definitions were too restrictive.

There are two reasons for expecting that firms with a separate R&D department or with an ongoing program are more likely to claim a tax credit. First, these programs are more likely to be associated with the types of expenditures that are eligible for a tax claim. Second, it is more likely that these types of operations facilitate the separation of research expenditures from other expenditures and, therefore, provide for the type of accounting that is required for the tax claims.

Differences in the incidence of tax-credit claims conform to these predictions. A larger percentage of firms that exhibit a greater commitment to the R&D process claim the R&D tax credit. Over 39% of those firms

---

[8] In addition to the federal SRED program, several provinces have their own tax credit programs. According to the recent overview of R&D tax incentives (Doern, 1995), there were no changes in the program during the period in which the innovation survey took place.

with ongoing R&D claim the tax credit. This increases to 62% for those firms performing ongoing R&D with a separate R&D facility.

11.4.2.1 Sector and Industry Distribution of Tax Credit Claims
In contrast to grants, which are sector- and firm-specific, tax credits are neutral in theory. Any firm that incurred eligible expenses for scientific research and/or experimental development may claim tax credits. However, the incentive to apply for tax credits may differ across industries.

The data show that the applicability of tax credits varies considerable across sectors (see Table 11.7). Some one-third of all R&D performers

Table 11.7. *Firms Conducting R&D and Claiming Tax Credits, by Firm Category and by Industrial Sector (% of Firms)*

	% Conducting R&D		% Conducting R&D Who Claimed R&D Tax Credits	
	C.W.* (1)	E.W.† (2)	C.W. (3)	E.W. (4)
All	65 (1)	85 (1)	16 (1)	48 (1)
Sector				
Core	86 (2)	94 (1)	36 (2)	65 (2)
Secondary	68 (2)	87 (1)	15 (2)	52 (2)
Other	58 (1)	80 (1)	9 (1)	33 (2)
Size				
Small (0–100)	63 (1)	73 (1)	14 (1)	21 (1)
Medium (101–500)	87 (2)	88 (2)	27 (2)	26 (2)
Large (500+)	90 (2)	91 (2)	57 (3)	68 (3)
Core (size)				
Small (0–100)	84 (2)	88 (2)	34 (3)	50 (3)
Medium (101–500)	94 (2)	96 (2)	37 (5)	25 (5)
Large (500+)	97 (2)	96 (2)	77 (5)	85 (4)
Secondary (size)				
Small (0–100)	66 (2)	74 (2)	13 (2)	17 (2)
Medium (101–500)	88 (3)	90 (3)	30 (5)	29 (5)
Large (500+)	92 (3)	92 (2)	60 (6)	76 (5)
Other (size)				
Small (0–100)	57 (2)	66 (2)	7 (1)	11 (1)
Medium (101–500)	83 (3)	82 (3)	20 (3)	23 (4)
Large (500+)	84 (3)	87 (3)	40 (5)	48 (5)

Note: Standard errors are in parentheses.
* C.W. = company weighted.
† E.W. = employment weighted.

in the core sector make use of tax credits. This is twice the proportion of firms who do so in the secondary sector. And these firms claim them again nearly twice as often as do firms in the tertiary other sector. Less than 10% of R&D-performing firms in the tertiary other sector claim tax credits.

These sector differences suggest either that the costs of the program are less or the benefits are much greater in the R&D-intensive industries. It is in the core sector where firms spend more on R&D activities; therefore, they have a higher stake in claiming tax credits. Their cost of learning about the existence and implementation of tax credits is less than their expected benefits, which are higher here since their R&D expenditures are higher. Firms with relatively limited R&D expenditures are more likely to feel that their compliance costs exceed the expected benefits from tax credits, and they make less use of tax claims.[9]

The same argument suggests that a large firm with a proportionally larger R&D budget is more likely than a small one to draw larger benefits from the tax-credit system than the costs it will incur. The proportion of firms with tax claims increases with the size of firm (Table 11.7). Moreover, the percentage of firms that claim tax credits (column 3, company weighted) is generally significantly smaller than their share of total employment (column 4, employment weighted), indicating that large firms claim tax credits in each sector more frequently than do small ones.

The proportion of firms that conduct R&D in one form or another and claim tax credits is particularly low for the smallest firms across all sectors. Only 34% of the smallest firms (0 to 100 employees) in the core sector that reported R&D activity claimed R&D tax credits, while it was 77% for large firms (500+ employees). It was 13% and 60%, respectively, in the secondary sector, and 7% and 40% in the other sector.

In summary, the use of tax credits by sector is positively correlated with the R&D intensity of the sector, and in all sectors, larger firms take advantage of tax credits more frequently than do smaller firms. Firms that are smaller and are located in the secondary and tertiary other sectors are less likely to receive the benefits of government tax programs that provide a subsidy to R&D.

---

[9] According to the Finance Canada (1998) report, the compliance cost for the smallest tax credit claims (less than $100,000 Cdn.) is estimated to be 15% of the value of SRED tax credits claimed.

Table 11.8. *R&D Tax Credit Claims, by Innovation Type and by Industrial Sector (% of Innovators)*

Innovation Type	% of Sector Claiming R&D Credit			
	All	Core	Secondary	Other
World-first	69 (5)	88 (5)	56 (9)	47 (10)
Canada-first	40 (3)	63 (7)	36 (7)	24 (6)
Neither	32 (3)	54 (7)	30 (6)	23 (3)

Notes: Larger firms (IPs) only. Standard errors are in parentheses.

11.4.2.2 Originality of Innovation and Tax Credits

Tax credits might also be expected to have a different applicability by type of innovation. We have previously shown that there is a close association between the originality of an innovation and the importance of R&D as a primary source of information and technology for that innovation.[10] It is more likely, therefore, that world-first innovators would claim these credits than would imitators simply because the former are more likely to rely on R&D. Indeed, the world-first innovators claim tax credits more often than do firms that introduced a Canada-first innovation, and the latter do so more often than their counterparts in the tertiary other sector (see Table 11.8). This ordering is preserved within each sector. Almost 9 out of 10 world-first innovators in the core sector (88%) report claiming tax credits, while only 23% of firms that introduced an 'other' innovation in the tertiary other sector do so. Moreover, all types of innovations in the core sector are more likely to take advantage of the tax credit than are similar types of innovations in other sectors. Thus, even though the criteria of the Scientific Research and Experimental Development Investment Tax Credit program are neutral with respect to industry, the type, and novelty of innovation, its principal beneficiaries in the manufacturing sector are core sector firms, especially those introducing original innovations.

11.4.3 R&D Activity, Tax Credits, and Innovation Performance

11.4.3.1 R&D and Innovation

Our results show that innovation is flourishing, even among firms that do not perform research and development.[11] More importantly, many firms

---

[10] See Chapter 4.
[11] See Chapter 7.

Table 11.9. *Use of R&D and Innovation Success (% of Firms)*

R&D Activity	Innovations		TOTAL
	No	Yes	
No			
% all	17	4	21
% row	80	20	100
% column	33	8	
Yes			
% all	33	46	79
% row	42	58	100
% column	67	92	
TOTAL	50	50	100
	100	100	

Note: Larger firms (IPs) only that answer tax-credit questions.

that perform R&D did not innovate in the three-year period covered by the survey. Both facts are demonstrated in Table 11.9, which provides a contingency table outlining the relationship between the pattern of R&D use and innovation for uses of R&D tax credits.[12]

Almost four out of five firms performed R&D either continuously or occasionally. However, 42% of these firms did not innovate during 1989–91. Only slightly more than half (58%) of R&D performers reported having introduced an innovation in the three years prior to the survey. Examining this from the innovation angle, most innovators performed R&D (92%); only a small fraction of innovators (8%) created an innovation without relying on their own R&D in some form or other. These proportions confirm that R&D is a necessary, though not a sufficient condition for innovation. A non-negligible proportion of R&D activity failed to yield a commercial innovation over the three years covered by the survey.

11.4.3.2 R&D Tax Credits and Innovation

Government support of R&D through tax credits is meant to help Canadian firms accelerate technological change through innovation. The effectiveness of tax credits can be evaluated by examining whether the

---

[12] This information is available only for firms in the integrated portion of the Business Register (IPs).

Table 11.10. *Use of Tax Credits and Innovation Success (% of Firms)*

R&D Tax Claims	Innovations					TOTAL
	No		Yes			
No						
% all	45		33			78
% row		57		43	100	
% column		90		66		
Yes						
% all	5		17			22
% row		23		77	100	
% column		10		34		
TOTAL	50		50			100
		100		100		

Note: Larger firms (IPs) only that answer tax-credit questions.

R&D performers that claimed tax credits were more likely to innovate than were those doing R&D and not claiming tax credits. The relationship between the use of tax credits and innovation performance is provided in Table 11.10, in which R&D-performing firms are cross-classified on the basis of their use of tax credits and their innovation performance.

Less than a quarter (23%) of firms that performed R&D actually claimed them. But over three-quarters (77%) of R&D performers that claim tax credits innovate. R&D performers that do not claim tax credits are much less likely to innovate (43%). Thus, the tax credit is used more by innovative than non-innovative firms within the group that are performing R&D.

Alternately, we can ask what percentage of innovative firms take advantage of tax credits. Here the results are less favourable for the tax credit program. While 92% of innovators performed R&D (Table 11.9), only 34% of innovating firms claim tax credits. The other side of the picture is that nearly one out of four firms that claim tax credits did not innovate during the 1989–91 period. On balance, the tax credit program is selective.

Selectivity is not necessarily bad. Innovations range from the very important to the less important – from world-firsts to Canada-firsts to neither of the above. Tax policy may well have the objective of supporting the most important, rather than the least important, types of innovation and it may accomplish this through the support of R&D.

To investigate this possibility, we examine how R&D, innovation, and the use of tax credits varies across sectors (see Table 11.11). The proportion of innovators claiming tax credits is highest in the core sector and lowest in the tertiary other sector. Firms in industries where R&D is performed more frequently are more likely to innovate and to take advantage of tax credits. In the core sector, 40% of innovating firms claim tax credits, followed by 16% of firms operating in the secondary sector, and only 9% in the consumer goods–oriented other sector.

A breakdown by major industry group further emphasizes the difference between the more and less R&D–intensive industries. Over 35% of innovators claim tax credits in the most research-intensive industries – in pharmaceuticals and electric equipment (which includes electronics and communication equipment) and in machinery. In contrast, only around 6% of firms that innovated claimed tax credits in the wood and furniture industry and even less in printing and publishing (3%), and least of all in leather and clothing (2%).

Thus, the tax credit program has less relevance to the majority of innovating firms in sectors that are not R&D intensive. Since over half of these firms report that they perform R&D, their low rate of take-up on the tax credit program suggests that it is geared to the R&D organization typical of R&D-intensive industries.

Finally, it should be noted that the tax credit system also tends to select large firms over small firms. In firms employing more than 500 persons, close to 90% of those claiming tax credits innovate, irrespective of the industry sector (see Table 11.12). On the other hand, smaller firms (0 to 20 employees) that claim tax credits are less likely to innovate. This size-class difference is most evident in the secondary and 'other' sectors. Differences are even more pronounced between the largest and the smallest size classes in the proportion of innovators that make use of tax claims. This indicates that the 1985 amendment of the Scientific Research and Experimental Development Investment Tax Credit (SRED), which introduced a full refundability to Canadian-controlled private corporations with taxable income of less than $200,000 on the credits earned on the first $2 million of qualifying current R&D expenses, did not fully redress the significant bias in the use of the SRED.

In summary, the tax credit program closely supports successful innovation in large firms. Those firms that perform R&D and take advantage of the tax credit have very high rates of innovation. The same conclusion can be drawn for firms in the core sector. Essentially, Canadian government support has been directed primarily to the large R&D-intensive

Table 11.11. *Industry Differences for R&D, Tax Credit Claims, and Innovation (% of firms)*

SIC		Conducting R&D %	Claiming R&D Tax Credits %	Claimants That Innovate %	Innovate %	% of Innovators Claiming Tax Credits %
All		79 (1)	22 (1)	77 (2)	50 (1)	17 (1)
Sector/industry group						
Core						
31	Machinery	92 (2)	47 (3)	86 (3)	67 (3)	40 (3)
33	Electrical and electronic products	92	45	87	69	40
35	Petroleum refining and coal	95	55	88	68	48
36	Chemicals and pharmaceuticals	81	35	74	48	26
37		89	46	78	61	35
Secondary		83 (2)	21 (2)	77 (4)	49 (3)	16 (2)
15&16	Rubber and plastic	89	29	79	68	23
29	Primary metal	94	25	60	40	15
30	Fabricated metal	78	14	82	42	11
32	Transportation equipment	86	27	89	64	24
35	Non-metallic mineral products	81	23	54	36	13
Other		72 (2)	14 (1)	67 (4)	44 (2)	9 (1)
10–12	Food, beverage, and tobacco	78	18	75	46	13
17&24	Leather and clothing	56	7	35	44	2
18&19	Textiles	71	16	84	52	14
25&26	Wood and furniture	79	12	47	31	6
27	Pulp and Paper	77	23	70	56	16
28	Printing and publishing	58	4	69	48	3
39	Other	76	20	100	63	20

Notes: Larger firms (IPs) only. Sample is those firms answering tax-credit questions. Standard errors are in parentheses.

Table 11.12. Percentage of Innovators Claiming Tax Credits, by Size Class and Industrial Sector

Size Class – Employees	All Sectors		Core		Secondary		Other	
	% Claiming Tax Credits That Innovate	% Innovators Claiming Tax Credits	% Claiming Tax Credits That Innovate	% Innovators Claiming Tax Credits	% Claiming Tax Credits That Innovate	% Innovators Claiming Tax Credits	% Claiming Tax Credits that Innovate	% Innovators Claiming Tax Credits
All	77 (2)	17 (1)	86 (3)	40 (3)	77 (4)	16 (2)	67 (4)	9 (1)
21–100	77 (4)	15 (2)	88 (5)	44 (5)	77 (8)	12 (2)	63 (8)	8 (2)
101–500	71 (4)	18 (2)	81 (6)	28 (5)	73 (8)	22 (4)	58 (9)	11 (3)
500+	88 (4)	45 (3)	90 (4)	67 (6)	87 (6)	48 (6)	85 (5)	29 (4)

Notes: Larger firms (IPs) only. Standard errors are in parentheses.

firms. Other types of innovation are taking place that do not focus on R&D. Small, domestic firms outside of the core sector have a greater problem accessing external funds for their innovations and have little access to R&D tax grants because their innovation system is not geared to this type of input.

11.5 CONCLUSION

Government policy in most Western countries is aimed at encouraging R&D. Canada is no exception. It employs several instruments to encourage firms to pursue research and development. Setting R&D subsidies in the context of total innovation expenditures allows us to evaluate the extent to which subsidies for this activity facilitate innovation. If R&D makes up a small portion of total innovation costs, subsidies for R&D will have a relatively small impact on total innovation costs.

Although R&D is an important part of the innovation system, this chapter has shown that it makes up only a small part of the resources required to bring an innovation to market. Fundamental and applied research account for about 16% of total innovation expenditures. The expenditures that are required for the development of the innovation, for the construction of pilot plants, and for the acquisition of new technologies to revamp production processes are many times larger than those devoted to pure research activities. In addition, of course, there are other expenditures that we have not considered here. Often new forms of organizational structure have to be developed. In addition, new skills have to be imparted to the workforce.

Research expenditures make up the largest proportion of total innovation costs in the core sector. Therefore, general research subsidy programs will favour innovation in the core sector. In contrast, the costs associated with manufacturing start-up account for a greater percentage of total innovation costs in the secondary and other sectors, and in more imitative innovations. These are sectors that therefore stand to benefit less from subsidy programs aimed at research.

Financing innovation offers particular challenges since the success of an innovation program is difficult to evaluate and the assets associated with knowledge-based investments do not provide the type of collateral that lenders generally require. As a result, innovative firms are frequently forced to rely on internal sources of funds for investments that create innovations. More than two-thirds of all firms finance innovation entirely from internal sources.

But some external funding is provided by lenders who develop specialized capabilities in evaluating the likely success of innovative activities. It is interesting to note that the proportion of firms relying exclusively on internal financing is highest in the tertiary other sector and lowest in the core sector. Less original innovations are internally financed more often than the more radical ones. This suggests that financial markets are less capable of evaluating the least innovative sectors. Innovative opportunities are most difficult to spot in these sectors, and the development of specialist lenders in these sectors may therefore have lagged behind other sectors.

The proportion of Canadian firms that finance their innovations internally is higher than for foreign firms. There are other differences between foreign and domestic firms that obtain outside funds. Owing to financial support from their parent and sister companies, foreign affiliates in the group that obtain some form of external financing depend much less on internal financing than do comparable Canadian-owned firms. Even though the latter obtain some outside funds, they still rely more on internal funds than do those foreign-owned firms receiving some outside funds. Moreover, this group of Canadian-owned firms rely more on government funding.

In general, firms that find outside funding (especially those that are Canadian controlled) rely more on the federal government than on other external sources. Financial institutions, provincial governments, and related firms follow by order of importance. Venture capital is generally unimportant as a source of funds, except in the core sector.

Federal support in Canada was used more often to help with innovations in the R&D-intensive core than in the secondary and tertiary other sectors. To the extent that there are spillovers from the core sector to the rest of economy, these public expenditures benefit more than just the core sector – but only indirectly. On the other hand, the federal expenditures do not focus particularly on only one type of innovation. They are equally important for world-first, Canada-first, and other types of innovations.

Canada has one of the most generous programs of tax credits for R&D of all industrialized countries. Nevertheless, while almost two-thirds of firms performed R&D, only about one out of seven claimed tax credits. Firms in the more R&D-intensive sectors make greater use of this program. This is not simply because R&D intensity differs across sectors. For example, while 95% of electronics firms conduct R&D, 55% of them claim tax credits; in contrast, while more than half of firms in leather and clothing and in printing and publishing industries perform R&D, 7% or less claimed tax credits. Thus, the interindustry distribution of tax credit claims is not proportional to the use of R&D. Either the firms in less R&D-intensive sectors are less aware of the existence of government programs or their lower R&D expenditures do not outweigh the costs of implementing the tax credit program.

The relationship between the use of tax credits and innovation is even more asymmetrical. Of those firms that perform R&D and claim tax credits, 23% do not innovate; of those innovating, 66% do not claim tax credits. The proportion of claimants is highest among innovators belonging to the core sector (40%), less important in the secondary sector (16%), and even lower in the tertiary other sector (9%). Overall, larger firms and those that introduced original world-first innovations are more likely to benefit from tax credits than other firms are. Again, the most frequent users of tax claims are innovators in the high-tech electronics sector. At the other extreme, less than 3% of innovators in leather and clothing and in printing and publishing claim tax credits. Since over half of the latter group of firms conduct R&D, their low rate of utilization of tax credits suggests that the fiscal criteria used to determine the type of activity allowed for the tax credit leads to selectivity differences. The program is geared to the R&D organization typical of R&D-intensive industries.

To summarize, firms in the core sector focus more of their innovation expenditures on R&D, which means that they naturally benefit more from subsidy programs that support R&D. But they tend to make greater use of these programs than would be expected from their greater orientation to R&D. R&D performers in the core sector make more use of tax credits than do R&D performers in other sectors. In addition or because of these differences, firms in the tertiary other sector rely more on government for funding of innovations. Innovators in the core sector are also more likely to receive outside funding.

In contrast, firms in downstream industries have greater difficulties in accessing outside funding. They are more likely to have to rely on their own internal funding. And if they receive outside funding, it is less likely to come from government sources, and they make less use of tax credits.

There are, therefore, substantial differences in the workings of the financial markets between the upstream and downstream industries in the innovation process. Government programs help the upstream sectors more than the downstream sectors. And there is evidence that downstream innovators face greater difficulties in accessing outside funds in general. This may be desirable if there are large externalities flowing from large to small firms and from the core to the other sectors. But equally, it may indicate that innovation opportunities in other areas are more constrained by financing problems. The economic benefits from innovation in both the secondary and other sectors are also important, but their financing profiles are very different.

Finally, it should be noted that the tax-credit subsidy program tends to favour larger firms. This is partially because the percentage of total costs that is devoted to R&D is higher for larger firms and in the core sector. Therefore, general R&D subsidy programs will favour innovation in the core sector and in larger firms. But of those firms performing R&D, once again larger firms tend to make greater use of tax credits. And the tax credits appear to be more effective in the larger group. Of those R&D performers receiving tax credits, large firms are more likely to be reporting that they have successfully innovated.

Small firms were substantially more likely to have to fund their innovations internally. And among those firms who managed to obtain outside funding, they were less likely to obtain funds from affiliates or from financial institutions. They were, however, more likely to obtain some federal funding. This pattern indicates that government grant programs are helping to alleviate capital market imperfections by easing financial constraints experienced by small innovators.

TWELVE

The Diffusion of Innovation

12.1 INTRODUCTION

Economic activities can be organized through markets via arm's-length market transactions or outside of markets within the confines of firms. The boundary that separates the two is determined by the relative efficiency of each (Williamson, 1975). Externalities that are unpriced affect these boundaries because they provide an incentive for internalization – by shifting the boundaries of a firm.

Innovation is commonly seen to involve substantial interactions between firms that arise during the course of knowledge diffusion. Previous chapters have described the nature of the diffusion process. The innovation system involves a considerable flow of ideas among firms. Market transactions serve to diffuse many innovations. Industries in the core sector tend to diffuse the innovations that they have produced to downstream industries, which buy either innovations in the form of machinery and equipment or intermediate products from the core sector. As previous chapters have stressed, the most important sources of ideas for innovations involve customers and suppliers – a diffusion that is associated with commercial transactions. But knowledge that is not imbedded directly in products is also transferred via commercial contracts. And contractual problems can be severe when it comes to the transfer of knowledge.

The transfer of knowledge involves particularly difficult assessment and enforcement problems that are sometimes best handled within a firm, rather than through commercial arm's-length transactions. When arm's-length contractual problems arise, the boundaries of a firm are often extended so as to allow intrafirm organizational efficiencies to facilitate the resolution of these problems. We might, therefore, expect to find that the transactions associated with knowledge transfer would require the adaptation of new forms of contracts or of organizational structures.

In this chapter, we examine several different types of transactions that are used to transfer ideas from one firm to another. In the first section, we investigate the importance of the transfer of technology from one firm to another, the transactions that primarily exchange knowledge about production processes. We ask whether there are particular problems that have to be resolved and the nature of the response. In particular, we ask whether the extension of innovation networks via internal linkages between related firms resolves the difficult problem of writing long-term contingent contracts in the area of technological uncertainty.

An alternative or complementary strategy to technology transfer is offered by the joint development of the technology needed for innovation. One such linkage involves collaborative projects with other firms, public and private research institutions, and universities. In the second section, we explore the importance of joint ventures – a commercial transaction that allows a firm to exploit scale economies in the innovation process. This avenue allows firms temporarily to extend their boundaries for the purpose of specific innovation projects. We investigate whether this particular institution is more suited to some industries, some types of firms, or some types of innovations. Differences in use reveal where the problems of forging joint partnerships prevent their widespread application.

In the third section, we investigate the extent to which innovations are transferred from one firm to another via direct sale – not through the sale of a product or intermediate good but through the sale of a right to produce an innovation. By examining the importance of this practice, we may infer the extent to which markets for innovations are well or poorly developed.

12.2 TECHNOLOGY TRANSFER AGREEMENTS

The introduction of innovations often requires access to external sources of technical expertise. Technical expertise is bound up in the specific processes that can be applied to production – either in terms of blueprints, organization, or other forms of production-specific knowledge that is proprietary and not easily transferred. Technology can be purchased outright through the acquisition of investment equipment, but it frequently also requires this general form of technological knowledge.

This transfer of knowledge is generally accomplished through a form of license agreement that protects the intellectual property rights of the developer. Like any market transaction, technology transfer via a license agreement presents a set of contractual challenges. A licensee firm will benefit if it gains access to new technology at a lower cost or more quickly than if the firm were to develop the technology itself. Acquiring technology that has successfully been in operation elsewhere greatly reduces the risk for the licensee, especially when the agreement provides for adequate technical assistance.

A licensee will try to maximize the risk-adjusted discounted present value of the future net revenues generated by the new technology. In this respect, acquiring new technology is similar to acquiring any other revenue-generating asset. The value of the technology will change as new technologies are introduced. New vintages often reduce and sometimes eliminate the revenue-generating capacity of the established technology.

The speed with which technology is likely to improve in the future is one of the important determinants of its present value for the licensee and for the licensor. The transfer of technology can be part of a continuing relationship between the licensor and the licensee or a one-time transaction. Differences in the environment will affect the desirability of different types of contracts.

At one end of the spectrum is a one-time transfer of technology for a specific process, product, or combination thereof; on the other end is a continuous transfer, including access to future technology developed by the licensor. Interests of licensees and licensors are very different with respect to these two cases. When a party acquires new technology to complement its own expertise in an ad hoc fashion, it may be satisfied with a one-time transfer of technology. When the technology is mature, stationary, or slowly evolving and the licensee has adequate technological capabilities, a one-time transfer may be all that is needed. On the other hand, when technology is evolving more rapidly, and the licensee firm uses technology transfer as a substitute for research and development, it is likely to desire a more continuous arrangement.

The interests of an arm's-length licensor often differ with those of the licensee. A licensor will be interested in transferring technology when the discounted net revenues from the transfer exceed the discounted stream of risk-adjusted net revenues that it could generate by exploiting the new technology itself. It is often difficult or impossible to establish the market value of new knowledge or technology without revealing it to the prospective client. Entering into licensing negotiations is therefore risky. In the face of contractual uncertainty associated with imperfect information, licensors will prefer to transfer older technology whose value is already established or from which they have already extracted as much profits as possible, or technology that they do not wish to exploit by themselves. If the owning firm can exploit the new technology commercially by itself, it has little incentive to license it. When it does, it is likely to prefer a one-time transfer that gives it as much freedom for its further use and development as possible.

The cost of arranging the technology transfer is an important determinant of the net benefits that can be expected from the technology transfer. This cost will depend on the experience the licensor has accumulated in transferring the particular technology, which usually increases with its age (Teece, 1977). It also will depend crucially on the technical capabilities of the licensee, as well as on its learning and research competence. Transfer costs for more advanced technologies are higher because negotiations in these instances must consider more contingencies due to the rapid changes that take place in the environment. Alternately, in order to reduce the up-front costs of specifying responses to a large number of contingencies, a relationship can be forged that requires other types of negotiations of an ongoing nature. While a one-time transfer may be an advantageous way to extract additional profits from an old technology, a continuous transfer may be the only feasible way to keep the transfer costs manageable in the case of rapidly evolving modern technologies.

Since continuous technology transfer reduces the technological lead of the licensor over the licensee, thereby potentially reducing the advantages of the transfer to the licensor, it is more likely to be used between related parties than between potential competitors. Forging ownership links between firms is one of the prime methods used to extend the boundaries of the firm when contractual problems need to be solved. When the required technical expertise for full exploitation of the new technology is tacit and difficult or impossible to codify, it is most easily transferred within related organizations, where mechanisms are already in place to handle the problem of firm-specific knowledge that is difficult to codify. Companies with access to technical resources of related firms have in these cases an advantage over independent firms that have to negotiate technology transfer at arm's length with unrelated firms.

In addition to determining whether a contract will be one time or continuous, the licensee and licensor need to specify a number of other rights. These rights determine the territory where the licensee can manufacture and/or sell the product. They may be exclusive or not. If a licensor chooses to limit potential competition, it will grant the licensee only those rights that do not interfere with its competitive strategy. Typically, the rights granted to unrelated parties are likely to be more restrictive than those granted to affiliates.

Technological transfer occurs not only between technologically unequal partners; it may also involve a cross-licensing of different technologies between two partners that have both developed a lead in their own, often narrowly defined, fields, especially in cases of complex products.[1] Cross-licensing enables each of the partners to acquire complementary capabilities needed for the pursuit of their technological strategy. It also allows firms to use patents for blocking purposes that in complex products create incentives for profit sharing (Cohen et al., 2000).

In this section, we examine how innovating firms use technology transfer to help them produce an innovation. Access to sources of technical expertise outside the firm depends on the type of innovation, the nature of the firm, its links with other firms, the strategy it is pursuing, and a host of conditions specific to the industry, geography, and sociopolitical environment in which it operates. Therefore, we discuss the incidence and the type of technology transfer, relating both to the size of firms, the sector where they operate, and the type of innovation involved. Next we examine restrictions and conditions that are attached to these technology transfers and the regions to which they apply.

12.2.1 Characteristics of Technology Transfer Agreements

The innovation process depends heavily on the transfer of technology. More than one-third of all larger innovating firms[2] acquired new technology required for their major innovation through a licensing agreement or other form of technology transfer agreement. Of course, not all innovators used outside sources of technology during their innovation process. But if we restrict the population to just those firms reporting that they relied on external sources of technology for their innovation, a license agreement is reported in 40% of the cases.

The pattern of technology transfer revealed by the survey shows important intersectoral differences. It is the core sector where the largest

---

[1] A complex product technology designates products that are comprised of a large number of patentable elements (e.g., electronic products). In contrast, a 'discrete' product technology is one that includes few patentable elements (e.g., chemical compounds, drugs) (Cohen et al., 2000).
[2] The information is available only for IPs.

percentage of firms using external sources for technology report a technology transfer via license (46%). Similarly, world-firsts (47%) are more likely to report a formal transfer license than are non-world-firsts (37%). Thus, the sectors and the types of innovations that are more research intensive are more likely to make use of the type of technology transfers that involves licensing.

A transfer can specify the continuity of the agreement, various rights regarding the geographic regions where manufacture or sale are permitted, and the intellectual property rights that can be used. Continuous agreements imply a longer-term relationship. Permissions or lack thereof to sell abroad protect the licensor from competition. Permissions to use property rights like patents allow the licensee to further develop technology.

For firms using external sources, slightly more involved those with continuous (54%) than with one-time transfers (46%) (see Table 12.1). But firms that relied upon related entities for external information on technologies were much more likely to utilize a technology transfer agreement specifying that the transfer of knowledge was to be continuous (74%), thereby confirming that internal transfers are the method used to handle the contingent contract problem. Firms that used unrelated third parties for technological information are relatively more likely to use one-time transfers (50%) than continuous transfers (45%). There is no meaningful difference in the split between continuous and one-time transfers across firms of different sizes. Cross-licensing, an option open only to original innovators that have developed new technologies, appears to be an option rarely used, and if so, then mostly by the largest firms.

Firms in the core sector are more likely to make use of continuous (52%) rather than one-time transfers (39%) (see Table 12.2). And the reason is that firms that make use of technology transfers from related firms are much more likely to use continuous technology transfer clauses in the core sector than in others. However, it is in the area of Canada-first innovations that one-time transfers are relatively more likely to occur.

12.2.2 Rights and Restrictions Associated with Technology Transfer

Transfers of technology specify the rights and responsibilities of both licensor and licensee. Rights conferred make the transfer more valuable to the purchaser, whereas restrictions reduce the chances that the licensor will lose sales in its own markets.

Table 12.1. *Characteristics of Technology Transfer Agreements for Those Firms Using External Technology Sources, by Size Class (% of Firms with Technology Transfer)*

Type of Transfer	All firms Using External Sources	Firms Using Related Firms as Sources	Firms Using Other External Sources	Firms 0–100 Employees Using External Sources	Firms 101–500 Employees Using External Sources	Firms 500+ Employees Using External Sources
Continuous transfer	54 (8)	74 (15)	45 (9)	54 (13)	50 (13)	44 (11)
One-time transfer	46 (8)	21 (9)	50 (9)	42 (13)	42 (13)	48 (11)
Cross-licensing	4 (4)	6 (6)	5 (5)	4 (4)	8 (4)	7 (4)

Notes: Larger firms (IPs) only. Most IPs are over 20 employees. Standard errors are in parentheses.

Table 12.2. *Type of Technology Transfer Agreement, by Sector and Novelty of Innovation (% of Firms with Technology Transfer)*

Type of Transfer	Sector			Novelty of Innovation		
	Core	Secondary	Other	World-First	Canada-First	Other
Continuous transfer	52 (13)	83 (17)	38 (10)	61 (17)	48 (4)	58 (11)
One-time transfer	39 (9)	11 (6)	57 (14)	33 (17)	48 (12)	37 (6)
Cross-licensing	9 (4)	6 (9)	5 (5)	6 (6)	4 (4)	5 (5)

Notes: Larger firms (IPs) only. Standard errors are in parentheses.

Table 12.3. *Rights Associated with Technology Transfer Agreements, by Industrial Sector (% of Firms with Technology Transfer)*

Right	All	Sector		
		Core	Secondary	Other
Any production or selling right	17 (2)	23 (3)	14 (3)	14 (3)
Right to manufacture	11 (2)	17 (3)	6 (2)	10 (2)
Right to sell	10 (1)	16 (3)	6 (2)	8 (2)
Right to use in manufacturing	10 (1)	14 (3)	7 (3)	8 (2)
Intellectual property rights	16 (1)	21 (3)	13 (3)	15 (3)
Right to use patents	7 (1)	15 (3)	3 (2)	3 (1)
Right to use industrial designs	5 (1)	8 (2)	2 (2)	5 (2)
Right to use trademarks	4 (1)	8 (2)	3 (2)	2 (1)
Right to use trade secrets	6 (1)	8 (2)	4 (2)	7 (2)
Other rights	3 (1)	2 (1)	3 (2)	4 (2)

Notes: Larger firms (IPs) only. Standard errors are in parentheses.

About 17% of firms that turn to outside sources for technology answered that they have one or other of the rights to manufacture or sell embedded in a licensing agreement (see Table 12.3). Technology transfer agreements specified in about similar proportion the right to manufacture (11%) and the right to use in manufacturing (10%). The right to sell was granted with about the same frequency (10%).

Core sector firms acquired the rights to manufacture more often than those in the other two sectors. There are few significant differences in the frequency with which these rights are specified across world-firsts, Canada-firsts, and other innovations.

The acquisition of technology may also involve the right to use a patent or some other form of intellectual property right. These property rights protect the investment in the intellectual capital embodied in the technology. The effectiveness of various property rights varies significantly from industry to industry, as demonstrated in Chapter 9. Patents tend to be used more frequently in product than in process innovations. Patents are more effective than other means of intellectual property (IP) protection in the chemical and pharmaceutical industry and less effective elsewhere. Trade secrets are generally employed relatively more than patents in smaller than larger firms. Appropriability conditions vary from industry to industry and they affect incentives for technology transfer.

In about 16% of firms that make use of outside sources of technology, the transfers involved confer intellectual property rights. About 7% of the licensees acquired the right to use patents, 6% trade secrets, 5% industrial designs, and 4% trademarks (Table 12.3). Acquisition of rights varies with the size of the firm. Most notably, the right to use patents increases and the right to use trade secrets decreases with the size of the firm. These differences are in keeping with the overall differences in usage that were reported in Chapter 9. The preference of large firms for using patents rather than secrecy was also observed in the European Community Innovation Survey (Arundel, 2001).

The difference between the core sector and the other two sectors outlined in Table 12.3 is striking. The core sector obtains intellectual property rights more frequently than the other sectors. Since technology transferred to firms operating in the core sector is likely to be more R&D intensive and therefore more original and patentable, associated intellectual property rights are, for the licensor in this sector, a more important means of appropriation of innovation benefits than in the less R&D-intensive secondary and tertiary other sectors. These differences basically arise only in firms that rely on related firms for technology. In this group, technology transfer is accompanied by the transfer of intellectual property rights in the core sector – but not in other sectors.

There is also a higher frequency of the transfer of intellectual property rights associated with Canada-first innovations (Figure 12.1).

12.2.3 Restrictions Attached to Technology Transfer Agreements

Technology licenses may include two additional types of restrictions. First, restrictions on geography or sales territory may be applied. These specifications allow the licensor to control territorial impact and to tailor contracts to the sales potential of different markets. Second, guarantees of exclusive jurisdiction can be granted, allowing the licensee to exploit the

Table 12.4. *Territorial Restrictions and Exclusivity Clauses in Licensing Agreements (% of Firms with Licensing Agreement)*

Clause Designating	Size Class (Employees)			
	All Firms	0–100	101–500	500+
The territory of manufacturing	19 (3)	23 (6)	19 (6)	10 (4)
The territory of sales	27 (4)	34 (7)	24 (7)	10 (4)
The exclusive right to manufacture	38 (4)	33 (7)	51 (8)	33 (7)
The exclusive right to sell	31 (4)	38 (7)	22 (6)	25 (6)
The source of inputs	13 (3)	12 (5)	17 (6)	12 (4)

Notes: Larger firms (IPs) only. Standard errors are in parentheses.

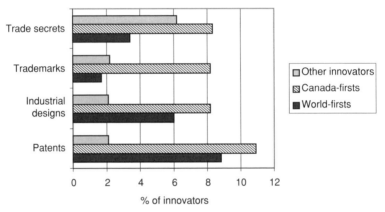

Figure 12.1. Transfers of Intellectual Property Rights, by Type of Innovator

value of the technology more fully. Finally, requirements may be placed on the inputs that must be purchased. These requirements may be methods of monitoring the value of the technology and extracting rent for its use via tied selling.

More than half of all licensing agreements included one or other of these clauses (see Table 12.4). The agreements conferred most frequently the exclusive right to manufacture (38%) and the exclusive right to sell (31%). The territorial restrictions for manufacture and selling were specified in 19% and 27% of cases, respectively. In 13% of technology transfer agreements, the licensee was obliged to purchase inputs from a specified source.

Agreements concluded by the largest firms contain territorial limitations for selling and manufacturing less often than do those signed by the smaller ones. These size-class differences are even more apparent in those firms that receive their technology from related firms. The smaller firms that receive technology from related firms are much more likely to face constraints than are the larger ones; and it is probably these differences that affect the ability of these firms to grow. This pattern suggests that the largest firms that are part of a family of firms are more able to obtain licensing conditions that allow them to operate with a world mandate. Smaller firms that rely on unrelated parties also are more likely to face specified sales or manufacturing territories, but they are no less likely to negotiate exclusive rights to sell or manufacture than are larger firms.

The licensing and other technology transfer agreements of core-sector firms are subject to various restrictions (specified manufacturing or sales territory and restricted sources of input procurement) and exclusive rights (to manufacture and/or to sell) more frequently than are those contracted by firms in the tertiary other and the secondary sectors. Because of the inherent novelty of new technology, it is in the core sector that the licensee is both restricted more frequently and at the same time given more security in the markets where it operates. By way of contrast, territorial restrictions for manufacturing and sales are more likely to accompany technology transfers associated with Canada-first innovations, while the world-first innovations more often are given exclusive rights to manufacture and to sell.

There are no very significant differences between Canadian- and foreign-owned firms as regards most types of restrictions, except for the exclusive right to manufacture, which is granted more often to foreign-owned firms. Since exclusivity is a source of market power, it is not surprising that it tends to be associated with other sources of market power, such as that bestowed by technological exclusivity of world-first innovations and by technological leadership of many firms in the R&D-intensive core sector.

Understandably, territorial restrictions specify Canada most often (see Table 12.5). The United States, followed by Europe and Pacific Rim countries, are mentioned less frequently. But it should be stressed that prohibitions are not frequent.

Technology transfer agreements to firms in the core sector more frequently designate specific restrictions, as well as specific authorizations. Firms in the core sector tend to receive authorization to manufacture and sell in Europe and in the Pacific Rim markets more often than firms in other sectors. This may reflect the fact that core firms sign technology

Table 12.5. *Regions to Which Territorial Restrictions Apply (% of Firms with Licensing Agreement)*

	May Manufacture	May Not Manufacture	May Sell	May Not Sell
Canada	58 (6)	4 (2)	59 (6)	5 (2)
United States	39 (5)	6 (3)	44 (6)	8 (3)
Europe	15 (4)	13 (4)	26 (5)	9 (3)
Pacific Rim	12 (4)	11 (3)	18 (4)	11 (4)
Other	4 (2)	7 (7)	7 (3)	4 (2)

Notes: Larger firms (IPs) only. Standard errors are in parentheses.

transfer agreements most often with U.S. firms, which are more likely to restrict access to their domestic market than to Europe and Pacific Rim countries.

12.2.4 A Probabilistic Model of Technology Transfer via Licensing

The complex interactions among all the characteristics that affect whether and how technology is transferred can be analyzed by using a multivariate framework. The decision or the opportunity to acquire technology in the innovation process will depend on a number of factors that determine the net benefits of doing so. Among the factors that would be expected to affect these decisions are the type of innovation, its originality, and the technical competence of the acquiring firm.

Type of innovation will be important if the licensing of new technologies is particularly crucial to the development of product, rather than process technologies. Cohen and Klepper (1996a, 1996b), for example, have claimed that process technologies are particularly difficult to transfer and therefore are more likely to be developed internally by larger firms. Of interest is whether the technology transfer that is difficult in the best of circumstances tends to accompany product innovation more than process innovation.

Technology transfer may also be unevenly distributed across innovations that differ in terms of complexity or originality. On the one hand, technology transfer might at first glance be most closely associated with more imitative innovations, since they by definition require the ingestion of ideas that are developed elsewhere. On the other hand, technology transfer is difficult and therefore costly. This may mean that it is only where there are very high payoffs that transfer will occur. If so, we would expect to find more transfer via licensing in those innovations that are world-first, are in the core sector, or involve more complex combined product and process innovations.

Technological competencies differ substantially across small and large firms. Small and large firms develop different competencies in acquiring information on the most successful technologies. As we have demonstrated in Chapter 7, small firms feel particularly disadvantaged when it comes to information on technologies. We might therefore expect them to engage less in licensing agreements that bring new technologies into the firm.

There are other, more direct measures of technological competency of firms. Firms differ in terms of their use of a research group, or of a development department, or of engineering capabilities. In some cases, this may hinder the internal creation of technologies and force a firm to go outside for new technologies. In other cases, this may provide the necessary complementary capabilities that allow for the assessment of the usefulness of external technologies and thus lead to more licensing agreements.

Competition may also have an influence on the use that is made of technology licensing agreements. If a competitive environment is more likely to force firms to experiment with new techniques, then we may find that technology licenses are taken out more frequently where there is more competition.

Finally, we might expect that impediments have an effect on technological transfer via licensing. When firms have internal problems in developing new technologies because of a lack of skilled workers, they may turn to outside technologies. If they cannot collaborate with others in the development of new technologies, they may once more turn to licensing agreements.

To uncover the statistical association between industry and firm characteristics and the likelihood that a firm resorts to technology transfer and the type of transfer, we estimate two separate probit models.

The first model identifies variables that are related to the probability that an innovative firm will resort to a technology transfer that involves a licence. The second model estimates the probability that a firm has a continuous technology transfer as part of a licensing agreement, rather than a one-time transfer or a cross-licensing arrangement.

The probability that an innovative firm reported that it employed a technology transfer was estimated by using the regressors: employment size; number of competitors; whether the main competition came from the United States; the sector (core, secondary, tertiary); whether R&D

Table 12.6. *Probit Regression Coefficients for Licensing and Joint Venture Models*

Variables	Model 1 – Any Licence	Model 2 – Licence Involving Continuous Transfer	Model 3 – Joint Venture or Alliance
Intercept	1.8287	−0.4333	−2.2191
Employment	0.0001		
6–20 competitors	0.0024		
Main competition from USA		0.3048	
Core sector			−0.0026
Secondary sector	−0.0063	0.0132	
Other sector	−0.0043		
R&D occasional	0.0035		
R&D collaboration		−0.0065	0.0064
Process innovation	0.0039	−0.0078	0.0037
Product & process innovation	0.0053		0.0064
World-first innovation	0.0048		0.0039
Canada-first innovation	0.0064		0.0055
Principal source of technology			
Research	−0.0047		
Experimental development			−0.0002
Engineering	0.0034	−0.0049	
Impediments to innovation			
Lack of skilled personnel	0.0029		
Lack of market information	−0.0038		
Barriers to collaboration			0.0035
Location of collaborative partners			
Customers in USA			0.0029
Affiliates in USA	0.0025*		
Log likelihood	−1135.2	−245.4	−989.5

Notes: All regression coefficients are significant at least at the 1% level (except *, which is significant at the 2% level). The regression is weighted by the sample company weights. Larger firms (IPs) only.

is being done; whether the innovation is product or process; whether the innovation is world-first, Canada-first, or other; whether the principal source of technology is R&D, engineering, or experimental development; and whether the impediments to innovation come from lack of skilled personnel, lack of market information, or barriers to collaboration.

The variables with significant coefficients are presented in Table 12.6. The estimated coefficients of model 1 are expressed relative to the default reference case of a firm operating in the core sector, which introduced a pure product innovation that was neither a world-first nor a Canada-first.

A negative sign attached to a particular coefficient indicates that the category would be less likely to use technology transfer than the default case.

The results show that firms in sectors with lower technological opportunity (firms operating in the tertiary other or in the secondary sector) are less likely to adopt technology via a license transfer than firms operating in the research-intensive core sector industries. The probability of transferring technology increases linearly with the employment size of the firms, and it is higher in markets where the innovating firm faces more than 6 and fewer than 20 competitors. Process innovations and even more complex innovations combining both new product and process features are more likely to make use of technology transfer than pure product innovations. A firm performing R&D on a regular basis is less likely to use transfer of technology than firms that do not regularly perform R&D, and is more likely to do so if its principal internal source of technology is production engineering, rather than its own R&D. Finally, firms that experience difficulties in finding skilled personnel are more likely to resort to the transfer of technology; those complaining about difficulties in finding market information are less likely to use technology transfers. The latter are probably those firms whose information systems are inadequate, both for purposes of technology acquisition and for product development.

In summary, the analysis indicates that technological transfer is undertaken in the more complex situations (core sector and combined product/process innovations) and by firms that have developed more technological competencies (large firms with engineering capabilities). But it also occurs where there are specific problems associated with the lack of skilled labour. It requires the development of complementary engineering capacities and is not heavily dependent upon internal R&D capabilities. Finally, like innovation in general,[3] intermediate levels of competition stimulate the transfer of technology.

12.2.5 Probability of Continuous Versus One-Time Transfers of Technology

The second model is estimated to determine when a technology transfer was continuous, rather than a one-time or a cross-licensing agreement. The same set of explanatory variables was used.

There are considerable differences between the correlates of the likelihood of a technology transfer and the likelihood of conducting it

---

[3] See Chapter 14.

continuously. First, firm size and the competitive environment as measured by the number of competitors are not significant in the second case, though they were in the first. Second, the likelihood of a license in general was higher in the core sector, but the likelihood that a firm uses continuous types of technology transfer is higher in industries belonging to the secondary sector, particularly for firms facing strong competition from the United States. In contrast to the case for a license agreement, ongoing transfers of technology are less likely in the case of process innovations than in the case of product or a combination of product and process innovation. This suggests that process innovations are more likely to involve one-time transfers, whereas product transfers require the ongoing improvement of process technology. Firms that have R&D collaboration agreements and those that rely on internal expertise in production engineering are less likely to utilize ongoing technology transfers. When technology transfer is ongoing, firms tend not to develop internal engineering capabilities.

It is evident, then, from both sets of multivariate regression results that transfers of technology, especially those involving an ongoing relationship, are substitutes for the performance of R&D. But continuous transfers appear to be closely related to the lack of internal engineering facilities.

12.3 PARTICIPATION IN JOINT VENTURES AND STRATEGIC ALLIANCES

The types of networks that are used in the innovation process vary considerably. The previous section delineated how they were used to transfer technology. A different type of network – a joint venture – is also used to create innovation. Joint ventures are collaborative efforts where all parties contribute to the creation of new ideas.

The increasing cost of research and development, the complexity of new technology, and increasing global competition lead firms to develop new external sources of research and development expertise. These efforts have resulted in the expansion of R&D collaboration with other firms, research centres, and universities in Canada and abroad. This phenomenon is recent and fairly important in the United States, Europe, and Japan (Mowery and Rosenberg, 1989; Freeman, 1991). Even though Canadian firms actively participate in various forms of collaborative ventures, there are conflicting indications about how widespread these new forms of collaborative research are in Canadian industry (Kumar, 1995;

Niosi, 1995a). Firms engage in collaborative ventures to share and thus reduce their own cost of basic and precompetitive research, to learn from their allies and competitors about unfamiliar technologies, to avoid costly duplication, and to reduce the risk of the failure of research programs. By internalizing knowledge externalities, cooperative ventures may reduce the market failure–related disincentives to invest in R&D. They also may permit firms to realize scale economies.

To become a member of a research consortium, firms have to contribute complementary expertise. And in order to benefit from these cooperative projects, participating firms have to complement their participation by a considerable research effort aimed at the efficient absorption of results. The scope and potential for efficient and beneficial cooperative research projects thus depends not only on the capacities of firms but also on the nature of technology, industry, and the research system of the country.

We previously visited the extent of R&D collaboration in Chapter 6. But that information only reveals whether firms are jointly exploring the frontiers of research. Here we examine the extent to which the major innovation of the firm resulted from a joint venture.

12.3.1 The Incidence of Joint Ventures and Strategic Alliances

A joint venture or strategic alliance to produce an innovation may not be strictly limited to collaboration in R&D, although the definition of technological alliances suggests that they share very similar objectives with research and development partnerships.[4] In fact, the terms 'technological alliances', 'partnerships', and 'collaboration' are sometimes used interchangeably. Nevertheless, an innovation may be developed outside a collaborative research framework.

While about 16% of firms reported that they performed collaborative R&D, some 10% of innovating firms[5] participated in a joint venture or strategic alliance in the process of creating their most profitable innovation (see Table 12.7). The most active participants in cooperative research are medium-size firms.

Contrary to the case of collaborative research, the frequency of research joint ventures and strategic alliances is not very different across

---

[4] Niosi (1995a) defines technological alliances as long-term contractual arrangements between two or more firms with the purpose of developing new or improved products and/or new or improved processes or a combination thereof.
[5] Both large and small firms.

Table 12.7. *Proportion of Innovating Firms That Formed a Joint Venture or Strategic Alliance to Produce Their Innovation, by Size Class*

	Size Class (Employees)			
	All	0–100	101–500	500+
Joint venture or strategic alliance	10 (1)	9 (1)	19 (3)	13 (3)

Note: Standard errors are in parentheses.

Table 12.8. *Proportion of Firms That Formed a Joint Venture or Strategic Alliance to Produce Their Innovation, by Industrial Sector and Novelty of Innovation*

		Sector		
Originality of Innovation	All	Core	Secondary	Other
All	10 (1)	13 (2)	9 (2)	9 (2)
World-first	16 (4)	19 (6)	14 (6)	14 (7)
Canada-first	18 (3)	12 (4)	22 (6)	21 (5)
Other	12 (2)	5 (3)	7 (4)	18 (4)

Notes: Larger firms (IPs) only. Standard errors are in parentheses.

industry sectors (see Table 12.8). Even though firms in the core sector report using joint ventures marginally more than those operating in the other two sectors, the difference among sectors is not statistically significant. The same is generally true of the frequency of strategic alliances for innovations that differ in terms of originality. The Canada-first and world-first innovators reported participating in joint ventures and strategic alliances more often than did firms that introduced 'other' innovations, but again the difference is not statistically significant. Except for firms operating in the core sector, the smallest firms rarely use joint ventures. It is, however, noteworthy that Canada-firsts in the secondary and tertiary other sectors make the greatest use of joint ventures. The process that brings innovations into Canada depends most heavily on joint ventures.

12.3.2 The Relationship Between R&D Collaboration, Joint Ventures, and Innovation

The increasing speed of technical change has caused accelerating technical obsolescence through ever-shorter life cycles of products and processes. The increasing complexity of most new technologies increases the cost of R&D, uncertainty, and risk for companies of all sizes. Technical alliances

Table 12.9. *The Cross-Classification of Joint Ventures and Collaborative Research (% of Innovators)*

Collaborative Research	Joint Venture		Total
	No	Yes	
No			
% all	62	6	68
% row	92	8	100
% column	74	39	
Yes			
% all	23	9	32
% row	72	28	100
% column	26	61	
TOTAL	85	15	100
	100	100	

Note: Large firms (IPs) only.

are a way of coping with these problems and of reducing at least some of the uncertainties by acquiring or accessing knowledge from complementary fields (Niosi, 1995b).

Innovative firms tend both to have a joint venture and to do collaborative research (see Table 12.9). Some 61% of firms that formed a joint venture or strategic alliance to produce their innovation also participated in R&D collaboration, presumably with the same partners. The association between R&D collaboration and joint venture (or strategic alliance) is positive, though only weakly significant. The actual percentage of firms that collaborate in R&D and formed a joint venture or alliance to produce their innovation is 9%, more than three times as large as the expected proportion (3%).[6]

Of course, innovations can be produced without R&D, and strategic alliances that have been successful in producing innovations may or may not be accompanied by collaborative research agreements. There are 39% of large innovators that engage in a joint venture but do not perform research on a collaborative basis.

It is, therefore, useful to associate information on the use of a joint venture or a strategic alliance in the innovation process with information on the incidence of R&D collaboration by sector (Figure 12.2). R&D collaborative agreements are slightly more important in the core sector than

---

[6] A chi-squared test rejects the hypothesis that the incidence of joint ventures and collaborative research are independent of one another.

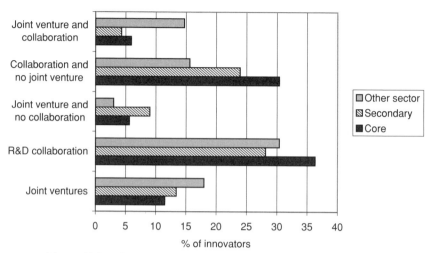

Figure 12.2. Joint Ventures and Collaboration, by Industrial Sector

elsewhere. In contrast, joint ventures are more important in the tertiary other sector. As a result, there are more joint ventures with collaborative agreements in the tertiary other sector and more collaborative agreements without joint ventures in the core sector. Joint ventures are not synonymous with R&D collaborative agreements in those sectors where R&D intensity is lower. But they are in the R&D-intensive sector. The complementarity between joint ventures and R&D collaboration therefore differs across sectors.

Finally, we break down the percentage of firms engaged in joint ventures and collaborative research by industry in Table 12.10. Joint ventures are particularly popular in the highly concentrated oligopolistic industries (paper, primary metals, refined petroleum, transportation equipment) and practically nonexistent in the more competitive industries (leather and clothing, wood and furniture). Some industries like paper, primary metals, and transportation have both joint ventures without R&D collaboration and R&D collaboration without joint ventures. In these industries, both forms of firm interaction are important. In machinery, chemicals and pharmaceuticals, and rubber and plastics, R&D collaboration is important but joint ventures are not.

12.3.3 Modelling the Probability of a Firm Using a Joint Venture

Like many other features of the innovation process, the decision to use a joint venture or a strategic alliance is a complex one. To examine the

Table 12.10. Individual Industry Differences in the Importance of Joint Ventures and Collaborative Research

Industry Sector	% of Innovators				
	Joint Venture	Collaborating on R&D	Joint Venture & No Collaboration	Joint Venture & Collaboration	Collaboration & No Joint Venture
Core					
Electrical and electronic products*	11 (2)	36 (4)	6 (2)	6 (2)	30 (40)
Chemicals & pharmaceuticals (37 & 374)	16	43	4	12	31
	9	40	7	2	38
Machinery	10	30	6	4	26
Refined petroleum & coal	20	41	9	11	30
Secondary	13 (3)	28 (4)	9 (2)	4 (3)	24 (40)
Fabricated metal	9	16	9	0	16
Rubber & plastics (13 & 8)	8	35	2	7	28
Nonmetallic mineral products	12	36	10	1	36
Primary metal	30	53	20	10	44
Transportation equipment	18	24	12	7	17
Other	18 (3)	30 (3)	3 (1)	15 (3)	16 (3)
Food & beverage (10–12)	20	39	4	17	23
Leather & clothing (17 & 24)	6	7	0	6	2
Paper	39	45	16	22	22
Printing & publishing	16	23	0	16	7
Textiles (18–19)	16	25	0	16	10
Wood & furniture (25, 26)	4	9	4	0	9
Other	27	46	2	26	21

Notes: Larger firms (IPs) only. Standard errors are in parentheses.
* Includes instruments.

factors that are related to the use of joint ventures, we estimated a probit model for the sample of all (i.e., small and larger firms) and for the sample of larger firms only. The explanatory variables are basically the same as used previously to model the use of technological transfer (firm size, degree of competition, industry sector, the importance of R&D, type of innovation, and source of technology). But we also investigated whether the geographic source of collaborative partners impacted on the probability that a joint venture would be used.

The determinants of the probability of a firm forming a joint venture or a strategic alliance to produce its innovation are generally the same for both the all-firm and the larger firm samples, except for the effect of the industry sector. Therefore, only the results for the large-firm sample are reported in Table 12.6. When the probit equation is estimated for the total sample, firms operating in core-sector industries have a higher probability of forming a joint venture or alliance; in the sample of larger firms, belonging to the core sector reduces the probability of a joint venture. Most of the small firms that form joint ventures and strategic alliances are in the core sector, while larger firms in this sector are less likely to adopt the same strategy than are larger firms elsewhere.

As was the case for technological transfer, producing more complex innovations increases the probability that innovators form joint ventures or alliances with competent partners. Firms that introduced a process innovation, particularly the most complex ones associated with new or improved product features, are more likely to team up with other firms. Confirming this picture is the finding that a joint venture or strategic alliance is most likely in the case of a Canada-first and least likely when a firm introduces an imitative 'other' innovation. This, too, is similar to the case of licensing. Thus, both instruments are used more in the case of complex innovations.

Collaborative R&D increases the likelihood of a joint venture. And those who rely upon experimental development rather than R&D for their sources of technology are less likely to engage in joint ventures. Joint ventures are therefore more closely tied to the R&D process than were technology transfers that were accomplished via licensing agreements. Moreover, we find that joint ventures are more frequent when R&D collaborative partners are customers and affiliates in the United States. This strongly suggests that when the joint venture is used along with R&D, it is facilitating cross-border diffusion of the benefits of R&D.

It should also be noted that joint ventures were more likely than technology transfers to be complements to R&D, especially in the sectors that are not R&D intensive. It is in these sectors that innovation involves more than R&D, and it is here that the joint venture takes on added importance to facilitate innovation.

12.4 DIFFUSION OF INNOVATION

The economic impact of innovations depends on their diffusion throughout the economy. Creation of new or improved consumer products satisfies existing consumer needs and creates new ones. Other innovations reduce the cost or increase the quality of industrial materials, inputs, and machinery. They increase productivity and profitability in other firms and institutions in the rest of the economy. The combined effect of these outcomes of technological change is reflected in increased productivity, consumption, and economic growth.

In this section, we examine two aspects of the diffusion process – the sectors in which innovations are used and to whom they are transferred when the right to produce or use is sold outright.

12.4.1 The Users of Innovations

An innovator may capture the benefits of innovating in one of three ways – by either selling or licensing the right to exploit the innovation, or by producing and selling products embodying the innovation. It has long been argued that difficulties in evaluating the value of an innovation make the former two routes less likely than the latter. This section provides quantitative information on how little the former route is used.

Innovations primarily diffuse throughout the economy because the products in which they are imbedded are sold to others – as final products for consumers, as machinery and equipment, or as intermediate materials that other firms use to produce other products. The distribution of the sales arising from an innovation shows that most are destined for other businesses (see Table 12.11). About 22% are sold directly to households (i.e., consumers). Only 6% goes to government.

The interindustry and intersectoral distribution of innovation sales follows the input/output structure of the economy. Firms in the core and secondary sectors sell primarily to other firms, that is, to the business sector (73% and 81%, respectively). Their direct sales to households represent only about 13% of their innovation-related sales. Even though direct sales of innovations to consumers are much more important (35%) in the more consumer goods–oriented 'other' sector, the main customers

Table 12.11. *Distribution of Innovation Sales to Types of Purchasers, by Industrial Sector*

	Sector of Purchaser			
Sector of Origin	Households	Business/Industry	Government	All
All	22 (2)	71 (2)	6 (1)	100
Core	14 (3)	73 (3)	13 (2)	100
Secondary	12 (3)	81 (3)	6 (2)	100
Other	35 (4)	63 (4)	2 (1)	100

Notes: Larger firms (IPs) only. Standard errors are in parentheses.

for innovations produced in the tertiary other sector still are firms in the business sector, accounting for 63% of innovation-related sales. The public sector (government) purchases innovation-related sales mainly from firms operating in the core sector (see Table 12.11). This is related both to defence expenditures and general spending on computer-related equipment.

Only firms in the food and beverage industry sell more than one-third of their new or improved products to households (52.2%). The government consumes more than 20% of sales of new or improved products from firms in the petroleum refining (27%), rubber (25%), and electronic industries.

12.4.2 Interindustry Flows of Innovations

While most innovations are diffused through embodied products, diffusion also occurs when the right to produce a new product or use a new process is sold to other firms. This occurs in only a minority of situations. Less than 2% of the larger innovators reported that they transferred their major innovation in this way.

Innovations can be sold as machinery and equipment or in the form of new materials that others can use. In addition, intellectual property rights may also be transferred. The importance of these three routes for those firms who transferred their innovations is depicted in Table 12.12, where innovations are classified according to whether they were sold to other firms as intermediate products, processes, or capital goods or whether they were transferred as intellectual property to other users.

When firms in the core and secondary sectors transfer their innovations, they do it mainly through sales of intermediate products and to a lesser degree through sales of capital goods or processes. The transfer of intellectual property rights was highest (18%) for firms in the R&D-intensive core sector.

Table 12.12. *Method of Transfer of Innovations to Other Firms, by Industrial Sector (% of Innovators)*

Sector	Sale of Intermediary Products	Sale of Processes and/or Capital Goods	Transfer of Intellectual Property
All	46 (6)	48 (6)	10 (3)
Core	51 (9)	36 (8)	18 (7)
Secondary	59 (11)	43 (11)	1 (3)
Other	26 (9)	72 (9)	7 (5)

Notes: Larger firms (IPs) only. Standard errors are in parentheses. The rows sum to more than 100 since a transfer can be accomplished in more than one way.

Innovators also provided information on the industry or sector to which the major innovation was sold or otherwise transferred. The diffusion of an innovation can be quite extensive since commodity flows are quite complex – as input/output tables illustrate.

The flow of innovations created by a new microprocessor produced by an electronics firm is illustrative. Consider a microprocessor designed for the needs of a machinery producer that is incorporated in a new pulp and paper machine that will produce a new variety of plastic-coated paper for printing photographs. In this example, the chain of innovations flows from the electronic branch to nonelectrical machinery, from there to the paper industry and eventually to a photo-finishing service. Or consider a new chemical substance developed and produced by a chemical firm for a drug manufacturer who incorporates it in a new veterinary drug used in agriculture.

Throughout the present study, we have used the sectoral classification based on a survey of sectoral patterns of production and use of innovations in the U.K. (Robson et al., 1988), which shows increasing intensity of innovative activity from the tertiary other, to the secondary, to the core sector. The direction of the diffusion process runs mainly from the core to the secondary to the other sector. Similar results were obtained by earlier studies of patents, both in the United States (Scherer, 1982b) and in Canada (Séguin-Dulude, 1982; and more recently Hanel, 1994).

The Canadian intersectoral pattern of the diffusion of those innovations that are transferred to others is depicted in Table 12.13. The present survey covers a shorter period than that of Robson et al. (1988), and the number of observations is less than in their study. It is also lower than Hanel's (1994) work that is based on all Canadian patents granted to

Table 12.13. *Production and Use of Innovations, by Industrial Sector (Employment Weighted)*

Indicator	Sectors		
	Core	Secondary	Other
Percentage of all innovations produced	47	25	28
Percentage of all innovations used	31	29	55
Ratio of innovations: produced/used	1.5	.9	.5
Ratio of innovations: produced/used – Robson	3.3	1.3	.5

Notes: Larger firms (IPs) only. Source for line 4 is Chapter 3, Table 3.3.

Canadian and foreign firms (56,007 patents granted from 1986 to 1989). Moreover, the survey includes many imitative innovations that are neither a world-first nor a Canada-first, whereas Robson's study covered only major innovations and Hanel's was based on patented inventions (world-first innovations by definition, but not necessarily innovations that were produced and used). Nevertheless, the matrix of intersectoral innovation that flows across sectors (Table 12.13) confirms the broad patterns that have been found previously.[7]

As in the case of the Robson et al. (1988) study of major innovations introduced in the U.K. over the 1945–1983 period, there is a clear hierarchy of sectors. The core sector introduced the majority of product innovations, and it was also the main source of innovations used in other sectors of the economy. The ratio of the number of innovations created to the number of innovations used in the core sector is 1.5 (Table 12.13), significantly higher than in the two other sectors (0.9 and 0.5, respectively, for the secondary and tertiary other sectors). For the sake of comparison, the results from Robson's study are presented in the last line of Table 12.13.

12.5 CONCLUSION

Previous chapters have depicted the extensive network that is used to spread innovations from one firm to another. This transfer of knowledge occurs generally through normal commercial relations – but the creation of networks of related firms is an important adjunct to arm's-length transactions because of difficulties associated with handling the type of contingent, asymmetric contract problems that accompany knowledge diffusion. This chapter has investigated these themes.

---

[7] See also Table 9.16 in Chapter 9.

It has shown that most firms exploit their innovations themselves. The market for the transfer of innovations is sufficiently imperfect that few innovations are sold outright to other firms. And this occurs despite the fact that most transactions are interfirm, rather than to the final consumer. Almost three-quarters of sales resulting from innovations are directed to the business sector, about one-fifth to consumers, and the rest to government.

Innovations are rarely transferred to others outright. Rather, the transfer takes the form of sales of intermediate products or machinery and equipment that contain the innovation. Less than 2% of firms reported an arrangement where the fruits of the innovation, that is, the technical information including the right to use the innovation, were sold to other firms. Where this is done, the intersectoral diffusion pattern follows the diffusion pattern that others have found – from the core to the secondary to the tertiary other sector.

This does not mean that transfers of knowledge are not important – rather that the difficulties in evaluating the effectiveness of the knowledge embedded in an innovation leads to other forms of contracts or other forms of institutions to overcome the contingent contracting problem.

The first such example studied in this chapter is the licensing of technical knowledge. Transfers of technical knowledge are an important part of the innovation process. Over 60% of firms make use of some form of outside knowledge for technology acquisition during their innovation process, and a good 40% of these firms obtain their technology from outsiders via a licensing agreement. A licensing agreement can be used either between arm's-length parties or between related members of a large multinational organization. Problems in negotiating the terms of these agreements in the face of rapidly changing technologies make ongoing transfers more appropriate for transactions between related partners who have a greater ability to adapt to changing circumstances.

The evidence shows that these technical transfers differ between related and unrelated partners. Related partners are more likely to transfer technical knowledge through contracts that allow for continuous exchange, whereas unrelated partners are more likely to rely on one-time transfers.

In our investigation of the sources of information, we isolated several different models of the knowledge-creation process. Some firms are R&D oriented, with or without an external focus on providing input from customers and the sales department for innovation ideas. Another set of firms focuses on the production and engineering departments for their technological development.

Technical transfers through licensing arrangements are used more frequently in situations where R&D is not the chief source of the innovation process. These transfers are also concentrated in the core sector and on Canada-first innovations. Thus, these transfers help related firms bring technology into Canada for the first time in order to support the adaptation of innovations that have already been developed elsewhere in the world.

It is interesting to note that the tendency to use these arrangements is about the same across all firm-size classes. This is an area where the largest firms do not particularly benefit from informational advantages, and in fact, these technical transfers allow small firms to compete more favourably with large firms. Since technical transfers allow related sets of firms the benefit of continuous transfer of technology, they facilitate the extension of a firm's boundaries. Firms that coalesce or merge are more likely to be able to utilize this form of knowledge diffusion for innovation.

An alternate instrument is available for the creation of knowledge that also extends the boundaries of the firm – though in a different way. Joint ventures and strategic alliances allow firms to enjoy the benefits of scale and scope economies in the production of knowledge without formal mergers. Joint ventures bring together the resources of several companies to create the technology that each needs for innovation.

Joint ventures are utilized by about 10% of firms that successfully innovated. Contrary to the technology transfer process that is accomplished with licenses, joint ventures are more closely related to R&D intensity. They complement the collaborative R&D process. They are less frequent where the engineering department is the primary source of ideas for innovation. Thus, joint ventures and technology licensing are substitutes, supporting competing models of innovation. Moreover, although licensing supports Canada-first innovations in the core sector, joint ventures are more common in Canada-first innovations in all sectors. Although joint ventures tend to be supported by or are related to collaborative research, they are not more concentrated in the most R&D-intensive core sector. Indeed, they are quite evenly distributed across sectors. The reason is that only the smaller firms in the core sector make very intensive use of joint ventures; large firms in the core sector do not. Once more this demonstrates a substitutability across instruments – this time across firm-size classes.

In conclusion, this chapter has demonstrated that the problems of knowledge diffusion are aided by several devices that allow firms to internalize the problems associated with the transfer of knowledge. While there are similarities in the areas where licensing and joint ventures are used, there are also differences. They both tend to be used in situations that involve more complex innovations. But they are not used in exactly the same way. And while they both solve the need to extend the boundaries of the firm to handle contractual problems, they do so in quite different ways. Technology licensing is a substitute for R&D; joint ventures tend to be used more in conjunction with an R&D facility – but in the secondary and tertiary other industry sectors where R&D is quite different from the core in that it tends to fall outside the definition required to qualify for a tax subsidy. This suggests that both are really used as substitutes for the narrowly defined R&D process and as such, provide an insight into how the innovation system makes use of non-R&D forms of instruments to facilitate innovation.

THIRTEEN

Strategic Capabilities in Innovative Businesses

13.1 INTRODUCTION

Innovation is traditionally equated with research and development. While it is true that in this study, we have shown a close relationship between R&D and innovation, we have also emphasized that there are alternate routes, including a focus on engineering, the adoption of advanced technology, and networks with customers and suppliers. Nevertheless, even here, our attention on these routes has implicitly focused on the scientific capabilities of firms.

Although the degree of inventiveness is commonly regarded as being determined by the underlying capabilities of a firm with regards to technical and scientific progress, the successful introduction of inventions requires a firm not only to develop core skills in the scientific area, but also to acquire a set of complementary skills in order to facilitate the process of commercialization.

Few firms can successfully innovate if they only discover new products and processes. They must develop successful marketing divisions; they must acquire the capacity to finance the risky activity of innovation; and they must attract and develop the knowledge base of their workforce.

Studies of the innovation system attempt to understand the forces that generate or that are associated with innovation. This is often done by delineating differences between innovators and non-innovators. Previous work has attempted to do so in one of two ways. On the one hand, Acs and Audretsch (1988) have asked what industry characteristics are associated with the existence of large-firm innovators and what industry characteristics are associated with small-firm innovators. An industry-based

Introduction

study does not focus directly on underlying firms and the heterogeneity of firm capabilities. Others use microeconomic multivariate exercises (Kleinknecht, 1996a) to ask whether certain firm-related characteristics, like the performance of R&D, along with industry characteristics are related to a firm's being an innovator. However, while the latter type of study focuses more on the firm, it tends not to delve very deeply into the characteristics of the firm – except to look at certain science-related activities. It rarely asks whether other competencies of firms differ between science-based and non-science-based firms and how these competencies are related to successful innovation.

Alternately, an entirely separate management literature focuses on the types of strategies that successful innovators emphasize. For example, Teece (1986) argues that firms must develop a set of assets that are complementary to their technological expertise in order to protect their intellectual assets. These assets serve to reduce the probability that competitors will readily bring to market imitations of their innovations. The complementary skills that must be developed lie in marketing, competitive manufacturing techniques, and after-sales support. The evidence adduced in Chapter 9 for Canada and in related work reported by Levin et al. (1987) for the United States confirms that innovators report that strategies, such as providing sales and service, are often more important than patent protection in reaping the benefits of innovation.

Other research focuses on complementary skills in other areas. Studies by McGuiness and Little (1981), Utterback et al. (1988), Napolitano (1991), and Rosenbloom and Abernathy (1982) have also suggested that complementarities exist between firms' technical capabilities and skills in such functional areas as human resources, management, marketing, production, and financing. In particular, Rosenbloom and Abernathy (1982) argue that much of the success of Japanese electronic firms is the result of their having combined ongoing innovation with concurrent investment in manufacturing systems, attention to employee relations, full utilization of employee skills, and a strong commitment to improving quality and productivity.

Two recent studies using data on small and medium-sized firms in Canada confirm that innovative firms develop a range of competencies to complement innovation. Baldwin and Johnson (1996b) use a survey of growing small and medium-sized firms to examine the difference in emphasis given by innovative and non-innovative firms over a wide range of functional areas. Innovative firms give greater emphasis to human resources, marketing, financing, and management skills and practices. Gellatly (1999) also demonstrates this complementarity for small firms in the business services sector. Innovative firms take a more balanced approach to their business operations by striving for excellence in a number of different areas.

In a second study that examines just entrants, Baldwin and Johnson (1999a) also find that innovative firms tend to develop a wide range of business competencies. Not only do they place greater emphasis on technical resources; they are also more likely to stress marketing, production, management, and human-resource competencies. This difference in emphasis is confirmed by differences in the intensity of activities. Innovators are twice as likely to have a business plan. They are more than twice as likely to have a financial plan. They are much more likely to have a formal training plan. They are significantly more likely to invest in R&D, technology, market development, and training. With respect to their product strategies, innovators in new firms emphasize customizing products, offering a wide range of products, and introducing new/improved products more frequently than do non-innovators.

In this chapter, we examine whether the evidence from the 1993 Canadian Survey of Innovation confirms these differences between innovators and non-innovators.

13.2 STRATEGIC CAPABILITIES AND COMPETENCIES

The 1993 Canadian Survey of Innovation was careful to avoid focusing exclusively on the scientific capabilities of firms. Three sets of questions were included to allow a comparison of the differences between innovators and non-innovators in the emphasis given to different product-based strategies and in the importance given to strategies in specific functional areas like marketing and human resources.

The first set of questions examined the importance given to different product-based competitive strategies. Management of respondent companies rated their position *relative to their competitors* with respect to quality of products, customer service, flexibility in responding to customer needs, range of products offered, and the frequency of the introduction of new products. For each of these variants of product strategy, firms ranked their competitive position on a Likert scale of 1 to 5 (with 1 corresponding to 'behind', 2 to 'somewhat behind', 3 to 'about the same', 4 to 'somewhat ahead' and 5 to 'ahead'). The emphasis given by competitors to each of these strategies was chosen as the point of comparison so as to provide a common frame of reference since most firms constantly compare their performance to this standard.

Second, respondent firms indicated whether their prices were higher, lower, or about the same as their competitors. Relative price serves as a proxy for quality of product. It also offers evidence on whether competition focuses more on innovation or on costs. Firms that are in the latter stages of a product life cycle are more likely to focus on price competition. Firms in the earlier stages of a product life cycle, when product innovation is intense and products turn over rapidly, compete more on the basis of innovative new products and less on prices. Firms in the latter stages of the life cycle focus more on costs because products have become more homogeneous and competition with respect to price is more intense.

In the third set of questions, respondent firms highlighted the importance that they gave to various specific strategies, covering five different functional areas – marketing, technology, production, management, and human-resource practices. In each case, firms assessed the importance of strategies that were closely connected to innovation activities – for example, whether marketing strategies were aimed at the current market or at new markets, at current products or at new products. Once again, firms provided their responses on a Likert scale of 1 to 5, this time indicating whether they considered a particular strategy to be 'not important' (1), 'slightly important' (2), 'important' (3), 'very important' (4), or 'crucial' (5).

Answers to the emphasis that firms placed on strategies all use scores on a Likert scale of 1 to 5. As such, these questions provide information that is different from the questions that ask whether the firm possesses an R&D unit. The questions that probe the emphasis that a firm gives to a specific strategy are perceived to involve more subjectivity than a question that asks whether a firm is performing R&D – although it should be noted that social scientists outside economics have long found these types of questions to be a useful part of a research tool kit.

The usefulness of the subjective scores that are assigned to capabilities can be addressed by comparing the answers to the questions using the relative ranking of firms on the Likert 1–5 scales to their ranking using data on the intensity of certain activities. The inclusion of parallel questions on activity in the innovation survey provides an independent check on the validity of the answers to the subjective questions on competitiveness and capabilities. For example, answers to the question about the emphasis given to R&D can be compared to the answers regarding the existence and type of R&D unit. Answers to the emphasis given to skill development can be compared to answers about the existence of training. When these comparisons have been made previously, firms that place a higher valuation on the importance of a given strategy have been found to be more likely to perform the activity or perform it more intensively (Baldwin and Johnson, 1996b; Johnson, Baldwin, and Hinchley, 1997). This is one reason that we regard the scores on the importance of specific strategies as good proxies for either the intensity of resources devoted to these areas or the level of competencies developed therein.

The second reason we regard the scores as good proxies is that the questions have been carefully tested and used successfully in a series of surveys that have investigated the innovation process (Baldwin, Chandler et al, 1994; Johnson, Baldwin, and Hinchley, 1997; Baldwin, Gellatly, Johnson, and Peters, 1998).[1] While the answers to these questions are subjective, the issues are ones that business managers must constantly evaluate. Questions that deal with the magnitude of competitive forces and the importance devoted to strategies in various areas of the firm are those that business managers ask themselves on an ongoing basis. Competitive forces constantly require firms to compare themselves to their competitors. The practice of benchmarking, for instance, has led many firms to continuously assess themselves against industry leaders. Corporate planning requires firms to decide on the emphasis that will be placed on different areas. The questions posed in the survey are thus not unfamiliar to most managers. And extensive preproduction tests confirmed this.

In what follows, we compare the emphasis that firms of different sizes and nationalities give to the various product strategies in different functional areas by using extreme scores on particular questions – that is, the percentage of firms that indicate they gave a particular competency a score of 4 or 5 on a 5-point scale. This allows for a parsimonious summary of the percentage of the population that felt strongly enough about this competency to rank it as being of greater than median importance (a score of 3).[2] We equate differences in strategic emphasis to differences in competencies or capabilities in specific areas. We compare the competencies and capabilities of the innovative group to the non-innovative

---

[1] These include the 1992 Survey of Growing Small and Medium Sized Enterprises, the 1996 Survey of Operating and Financial Practices of New Firms, the 1996 Survey of Innovation in Service Industries, and the 1998 Survey of Advanced Technology in the Food Processing Industry.

[2] We use this raw score, rather than the top n percentile of a standardized score, because the alternative makes little difference to our conclusions.

group by asking whether the percentage choosing a score of 4 or 5 differs significantly between the two.

In the following sections, innovators are defined broadly. They are firms that indicated they had introduced a product or process innovation, had filled out a set of questions about their major innovation, or had sales from a major innovation. Non-innovators are firms that do none of these.

13.3 INNOVATION AND TECHNOLOGY STRATEGIES

In this section, we compare the stress that is placed on innovation strategies between innovative and non-innovative firms.

Firms that have produced an innovation and those that have not done so do not necessarily have to differ in terms of the stress that is placed on innovation. In a world where most firms are trying to innovate but only a few lucky ones are successful, there will be little difference in the emphasis placed on innovation by innovators and non-innovators, since most firms will be pursuing an innovative strategy. On the other hand, if firms divide into a group who are attempting to innovate and a group who place little emphasis on innovation and who do not innovate, we should expect to find considerable differences between innovators and non-innovators in terms of their emphasis on innovation.

In an associated study of the service sector (Baldwin, Gellatly, Johnson, and Peters, 1998), the latter model was found to be more appropriate. In communications, financial services, and business services, the population divides into two groups – firms that had innovated successfully and, at the same time, exhibited an ongoing commitment to innovation programs and those that did neither. The group of firms that had introduced an innovation was basically the same group reporting that they had innovation programs that were ongoing.

In order to investigate how innovative and non-innovative firms differ across a wide range of innovative activities, we compare the two groups with respect to various innovative and technological strategies and confirm the differences in strategic emphases by also examining differences in the intensity of several innovative activities – those reporting an R&D unit and the percentage of sales derived from innovations (see Table 13.1).

Innovative firms are more likely to emphasize spending on innovation, spending on R&D, R&D management, and the management of intellectual property. The largest difference between innovative and

Table 13.1. *Emphasis Given to Innovation and Technology Strategies (% of Firms Scoring 4 or 5)*

	Innovative	Non-innovative	Difference
Leading competitors in:			
Spending on innovations	34 (1)	11 (1)	23
Spending on R&D	31 (1)	8 (1)	23
R&D management	25 (1)	7 (1)	18
Management of intellectual property	20 (1)	8 (1)	12
Strategic emphasis on:			
1) Technology			
Developing new technology	45 (1)	17 (1)	28
Improving others' technology	34 (1)	15 (1)	19
Using others' technology	38 (1)	21 (1)	17
Improving own technology	65 (1)	32 (1)	33
2) Intellectual property			
Improving intellectual property management	28 (1)	14 (1)	14
Innovative activities			
% of firms conducting R&D regularly	47 (2)	17 (1)	30
% R&D performers with separate R&D unit	29 (2)	22 (1)	7
Sales from Innovations			
% sales coming from major innovations introduced 1989–91	24 (1)	0	24
% sales coming from minor improvements made during 1989–91	18 (1)	9 (1)	9
TOTAL	42 (1)	9 (1)	33

Notes: All differences between innovative and non-innovative firms are significantly different from zero at the 1% level. Standard errors are in parentheses.

non-innovative firms lies in the areas of spending on innovation. Innovation requires a purposive investment, and those firms who are innovating recognize this fact.

Innovative firms are also more likely to emphasize developing new technology, improving the technology of others, using others' technology, and improving their own technology. The differences between innovative and non-innovative firms are greatest for the category – 'improving own technology'. This confirms Hollander's (1965) finding that making a string of incremental innovations is an important part of the innovation process. But it should also be noted that the second largest difference between innovators and non-innovators is in the category 'developing new technology', a much more radical strategy. Both radical and incremental technological change are an important part of innovation. Innovative firms also place greater stress on intellectual property management – a key capability that is used to protect the investment that is required for innovation.

In summary, the population does not consist of firms all of whom are pursuing some form of innovative activity, with some succeeding and others failing. Firms that introduced an innovation during the period 1989–91 do not differ from those who did not innovate simply as a matter of luck. Innovators clearly have a different strategy with regard to innovation. They are more likely to have R&D units and to stress this aspect of firm strategy. They have developed superior technological capabilities. They give greater emphasis to the management of intellectual property.[3]

Non-innovative firms do place a positive value on innovation-related activities – but it is much lower than for innovative firms. They tend not to place the same emphasis as do innovative firms on innovation spending, on R&D spending, or on the strategy of developing new technologies. They benefit from minor, not major, innovations – but even here, minor innovations are less important to this group than to the group of more innovative firms.

13.4 PRODUCTION STRATEGIES

Production strategies are important for all firms, but particularly so for innovative firms. The importance of maintaining a competitive position in the production arena is emphasized by Teece (1986), who noted that 'innovating firms without the requisite manufacturing and related capabilities may die, even though they are the best at innovation'. Firms cannot survive by simply offering technologically advanced products. They must offer them at acceptable prices, and competition will force them to continually improve the efficiency of their production process in order to do so.

A firm's production strategy relates to how it makes its products. Firms can improve their production process by making it more efficient or by using better inputs. The goal of efficiency can be achieved by developing new processes or by improving existing processes. Firms can also emphasize the use of new materials or they can emphasize general cost reduction.

---

[3] The latter is important because it indicates that non-innovators during the period 1989–81 have little in the way of intellectual property based on innovative activity prior to 1989.

Table 13.2. *Emphasis Given to Innovation and Production Strategies (% of Firms Scoring 4 or 5)*

	Innovative	Non-innovative	Difference
Leading competitors in:			
Frequently introduced new processes	32 (1)	12 (1)	20
Use of advanced manufacturing processes	35 (1)	15 (1)	20
Costs of production	29 (1)	23 (1)	6
Production management	34 (1)	22 (1)	12
Strategic emphasis on:			
Using new materials	40 (1)	25 (1)	15
Using existing materials more efficiently	60 (1)	42 (1)	18
Cutting labour costs	65 (1)	55 (1)	10
Reducing energy costs	50 (1)	46 (1)	4

Notes: All differences between innovative and non-innovative firms are significantly different from zero at the 1%. Standard errors are in parentheses.

These differences in emphasis are examined here by comparing the score that innovators and non-innovators give to innovative new processes and new materials, on the one hand, and the emphasis on cost cutting in general, and labour or energy savings in particular, on the other (see Table 13.2).

Innovative firms were more likely to stress the introduction of new processes and the use of new materials. Both are at the heart of the innovation process. Moreover, the differences between innovative and non-innovative firms were largest here.

Innovators also place greater emphasis on production costs and on production management than do non-innovators. In this area, innovators give the greatest emphasis to the need to use existing materials more efficiently and to cutting labour costs. Both are important inputs into the cost of production for innovative firms – labour costs because of the higher wage rates that are paid for skilled labour in technologically advanced plants (Baldwin, Gray, and Johnson, 1997; Baldwin and Rafiquzzaman, 1999). But the differences between innovators and non-innovators are larger on the materials than on the labour cost side. All firms focus on reducing labour costs, but innovators go beyond this strategy and focus more intently on materials costs as well. This result was also found in an earlier survey (Baldwin, Chandler et al., 1994).

Innovators are thus more likely to differentiate themselves from non-innovators on the technology and the production end of the innovation process. Once again, this demonstrates the importance of the production/technological process to innovation. R&D is important, but so too is technology, and the development of the latter, as we have seen, is often done outside of R&D scientific establishments.

13.5 MARKETING AND PRODUCT-BASED STRATEGIES

Innovations rarely sell themselves. Marketing skills are important prerequisites if firms are to persuade consumers to exploit new products and technologies.

Examples of the importance that has been given to the marketing of new products are provided by business historians. Main's (1955) description of the evolution of the nickel industry and of the steps that the International Nickel Company took in order to persuade steel makers to adopt nickel as a hardener illustrates the difficulties of penetrating new markets. Chandler (1977) and Stuckey (1983) describe how the Aluminum Company of America integrated downstream into the production of consumer goods and developed a large sales force in order to market aluminum when it was still a comparatively new product.

Others have also recognized the key importance of marketing skills for the commercialization of innovations. McGuiness and Little (1981) argue that marketing skills are key competencies that complement innovation. Utterback et al. (1988, p. 24) maintain that 'marketing activities play a pivotal role in the success of small firms'. Additionally, marketing efforts that result in the penetration of foreign markets are important. Edmunds and Khoury (1986) note that exporting is critical to success. Developing a significant export market allows firms to reduce risks by diversifying across dissimilar markets and to prolong the marketability of their products.

Marketing strategies are inextricably tied to the product offering of the firm. Firms may focus on a core or existing market. They can compete primarily with respect to price, focus on quality, strive to provide superior customer service, or offer greater flexibility in meeting customers' needs. Alternately, firms can try to alter their product line and seek new customers in existing markets, or extend their operation into new markets. In doing so, firms may choose to customize their products, develop a product line that carries a wide range of related products, or continually expand and update their product line by frequently introducing new or improved products.

Table 13.3. *Emphasis Given to Marketing Strategies (% of Firms Scoring 4 or 5)*

	Innovative	Non-innovative	Difference
Leading competitors in:			
Quality of products	70 (1)	55 (1)	15
Customer services	69 (1)	58 (1)	11
Range of products	47 (1)	34 (1)	13
Flexibility in responding to customer needs	73 (1)	61 (1)	12
Frequency of introduction of new products	37 (1)	17 (1)	20
Marketing	26 (1)	15 (1)	11
Price level relative to competitors			
Higher	23 (1)	12 (1)	11
About the same	59 (1)	61 (1)	−2
Lower	12 (1)	16 (1)	−4
Strategic emphasis on:			
Maintaining current production in present markets	71 (1)	60 (1)	11
Introducing new products in present markets	59 (1)	30 (1)	29
Introducing current products in new markets	58 (1)	36 (1)	22
Introducing new products in new markets	46 (1)	25 (1)	21

Notes: All differences between innovative and non-innovative firms are significantly different from zero at the 1% level. Standard errors are in parentheses.

In order to examine differences in the product emphasis of innovative and non-innovative firms, we focus first on a firm's competitive position with regard to price, quality of products, customer service, flexibility in responding to customer needs, range of products, frequency of introduction of new products, and marketing. Innovative firms perceive themselves to be significantly more competitive with respect to all these specific elements (see Table 13.3). But the largest differences between innovators and non-innovators occur in the 'frequency of new product introduction' and the emphasis that is placed on quality. Quality, of course, often involves improvements in product offerings.

Innovative firms also feel superior to their competitors with regard to marketing capabilities – perhaps because this is so essential to the general strategies implemented to protect intellectual property rights. Brand identification is one of those strategies that firms use to protect their innovations from those of their competitors who are tempted to copy them.

Innovative firms are almost twice as likely to have a higher price relative to their competitors than are non-innovative firms. Innovation permits the differentiation of products and fetches higher prices from the consumer.

Both innovators and non-innovators alike place the most emphasis on their core markets–'maintaining current production in current markets'. And differences between the two groups of firms are not large here. Rather, innovative firms are primarily distinguished from non-innovative firms by the importance that they attribute to more aggressive policies involving either 'developing new products' or 'seeking new markets'.

In summary, innovative firms are more competitive in terms of the core tools (customer service and flexibility of responding to customer needs), but have an even greater advantage over non-innovative firms in innovation-related areas like the frequency of introduction of new products. Innovators have somewhat higher prices, likely due to the strong product differentiation afforded to them by their innovativeness and the greater attention that they devote to marketing capabilities. Innovators attribute greater importance to market and product strategies, particularly the aggressive policies involving the development of new products and the penetration of new markets.

13.6 HUMAN RESOURCE STRATEGIES

The successful integration of new technologies into innovative firms is critically dependent on increasing workers' skill levels, motivating workers, and involving employees in decisions related to the adoption of new technologies.

Many studies have emphasized that innovation requires a human resource strategy that stresses training. Matzner, Schettjat, and Wagner (1990) argue that the introduction of more flexible forms of production can be best achieved through a more highly skilled workforce. Reshef (1993) contends that successful innovation requires a human resource strategy that involves workers implementing change.

Employee relations and participation have been cited as a central factor in the success of business in Canada (Kochan, 1988) and in the United States (Mowery and Rosenberg, 1989). Rosenbloom and Abernathy (1982) cite attention to employee relations and the full utilization of employee skills, in conjunction with innovation, as one of the factors behind Japanese success in the consumer electronics industry.

There is extensive Canadian evidence of a special need for skilled workers during the innovation process. Using the 1989 Survey of Technology, Baldwin, Gray, and Johnson (1996) find that managers of manufacturing establishments that use advanced manufacturing technologies, such as flexible manufacturing systems, robots, computer-based design and engineering systems, generally report that the skill requirements in their plants increased after the implementation of these technologies. In turn, technology-using plants were more likely to have a formal training program.

In Chapter 6, we reported that over 50% of innovators indicated that their skill requirements increased as a result of innovation, while virtually no firms reported a decrease in skill requirements. Firms that developed more novel innovations (those that introduced world-firsts) were more likely to report an increase in skill requirements than those that introduced innovations that had already been introduced elsewhere.

Perhaps more indicative of the importance of skill requirements is the lack of skilled personnel, the impediment to innovation that is cited by the largest percentage of innovators in the manufacturing sector. Over 46% of firms report that a lack of skilled personnel acted as an impediment to innovation – significantly more than reported that they lacked information on technologies, markets, technical services, or interfirm cooperation (Chapter 7).

Similar results emerge from the 1998 Canadian Innovation Survey of service industries. Over 30% of firms in financial services, business services, and communications report that their most important innovation increased skill requirements, and virtually none reported decreases (Baldwin, Gellatly, Johnson, and Peters, 1998).

The manufacturing plants in the 1993 Survey of Innovation also indicated that labour problems were an important impediment to technology adoption (Baldwin, Sabourin, and Rafiquzzaman, 1996). Some 39% of technology users indicated that one or another of shortage of skills, training difficulties, or labour contracts provided a significant impediment to the introduction of advanced technologies. Some 25% focused on skill shortages as being the most important impediment. Labour-related problems were second in importance after the cost of capital.

It is noteworthy that differences in skill shortages exist not just between non-innovators and innovators – but also between firms that differ in terms of the intensity or novelty of the innovation. Skill shortages are always higher in firms that are 'more-innovative'. As previously noted, skill shortages are reported more frequently by innovators that had introduced world-first innovations than by those that had introduced an imitative innovation. These differences indicate that during the innovation process, innovators unearth skill-related problems that have to be solved. The more advanced the innovation, the more frequent are the problems that have to be solved. Baldwin, Sabourin, and Rafiquzzaman (1996) find that almost twice as many users of advanced technologies report a labour-related problem (40%) as do nonusers (23%).

In summary, more innovative firms are hypothesized to place greater emphasis on the type of human resource strategies that are needed to obtain, develop, and retain skilled workers. We should, therefore, expect that innovative firms are more likely to emphasize the hiring of workers with advanced skills or to develop them through training programs. They also must devise remuneration schemes that retain highly skilled, mobile workers.

In order to examine the extent to which innovators place greater stress on all three areas, we examined how firms rank their capabilities relative to their competitors with regards to their spending on training, their labour skills, and their labour climate (see Table 13.4). We also compared the emphasis given to three human resource strategies – continuous staff

Table 13.4. *Emphasis Given to Innovation and Human Resource Strategies (% of Firms Scoring 4 or 5)*

	Innovative	Non-innovative	Difference
Leading competitors in:			
Spending on training	23 (1)	10 (1)	13
Labour climate	47 (1)	35 (1)	12
Skill levels of employees	46 (1)	40 (1)	6
Strategic emphasis on:			
Continuous staff training	44 (1)	26 (1)	18
Innovative compensation packages	22 (1)	14 (1)	8
Staff motivation in other ways	48 (1)	32 (1)	16
Activities			
% of staff covered by collective agreement	16 (1)	13 (1)	3

Notes: All differences between innovative and non-innovative firms are significantly different from zero at the 1% level. Standard errors are in parentheses.

training, innovative compensation packages, and other methods of staff motivation.

Innovative firms place more emphasis than non-innovative firms do on all of these areas. Innovative firms differ most from non-innovative firms with regards to their emphasis on training. They are more likely to emphasize training as a strategy and are more likely to stress that they are more competitive in terms of spending on training. Innovators are also more likely to emphasize the development of innovative compensation packages.

The least differences between innovators and non-innovators are found in the emphasis that skills receive. Both innovative and non-innovative firms appreciate the need for skilled workers; however, innovative firms implement internal systems to ensure that they obtain skills that they need. This represents either differences in capabilities to solve problems or differences in needs. Innovative firms may need very different firm-specific skill sets and, therefore, have to develop training programs to fill these needs.[4]

Finally, innovative firms are more likely to feel that they have a competitive labour climate. Accompanying this finding is that slightly more innovative than non-innovative firms are covered by collective agreements.

In conclusion, this evidence confirms the previous findings of Baldwin and Johnson (1996a, 1996b, and 1999a) that innovators place more emphasis on human resource strategies.

13.7 SPECIALIZED MANAGEMENT STRATEGIES

Management capabilities are required to develop a superior set of competencies with regards to innovation strategies, marketing capabilities, human resource activities, and production programs.

This observation has considerable support from special studies. McGuiness and Little (1981) observe that management skills, generally defined, are an important factor behind success. Others point to the need for a close interaction between management and employees for the successful adoption and integration of new technologies (Reshef, 1993). Rosenbloom and Abernathy (1982) also suggest that the close contact between top-level executives and persons developing new technologies is part of the reason for the success of Japanese firms in the consumer

---

[4] Evidence that the latter appears to be the case is found in Baldwin, Gray, and Johnson (1996) and Baldwin and Peters (2001).

Table 13.5. *Emphasis Given to Innovation and Management Strategies (% of Firms Scoring 4 or 5)*

	Innovative	Non-innovative	Difference
Strategic emphasis on:			
Improved management compensation schemes	26 (1)	15 (1)	11
Innovative organizational structure	35 (1)	17 (1)	18
Improved inventory control	48 (1)	33 (1)	15
Improved process control	53 (1)	36 (1)	17

Notes: All differences between innovative and non-innovative firms are significantly different from zero at the 1% level. Standard errors are in parentheses.

electronics industry. In contrast, management failure is the chief cause of firm bankruptcy (Baldwin, Gray, Johnson, et al. 1997).

Our previous results that show a greater emphasis on R&D, technological, production, marketing, and human resource competencies demonstrate that innovators must have developed a wide range of management competencies to excel and to coordinate functions in all these areas. In order to extend our understanding of management differences, we examined four specific areas that are relevant to the innovation process and asked whether innovative firms place greater emphasis on each (see Table 13.5).

The first area pertains to management incentive compensation programs. These programs have been found to be particularly important in innovative companies, where profits often are inadequate to fund incentive programs early in a firm's life cycle, but equity participation that promises future payoffs from the successful commercialization of inventions provides the same incentive. Innovative companies in the manufacturing sector place more emphasis on compensation programs than do non-innovative companies (Table 13.5).

The second management strategy that we examined is the use of innovative organizational structures. Innovation requires not only new technology and skills but also new forms of organizations. The implementation of new processes requires specific business practices to exploit the advantages of new technologies. In turn, the implementation of these practices often involves new organizational forms.

For example, Cad/Cam design and engineering systems are best exploited by concurrent engineering practices or rapid prototyping. In the agrifood processing sector, the implementation of advanced manufacturing technologies is often accompanied by the adoption of advanced practices (Baldwin, Sabourin, and West, 1999). In turn, these practices increase the need for coordination between divisions and thus the need for new organizational structures.

The importance of the need for the adoption of innovative organizational structures is confirmed by the difference between the proportion of innovative firms and non-innovative firms stressing this strategy.

The third and fourth strategies that we examined are specific areas that innovative firms might be expected to emphasize – inventory control and process control. Inventory control is often particularly important in the early stages of the innovation life cycle (such as machinery manufacturing for electronic circuit board.) when the unique nature of the products being produced essentially means that the manufacturing operations are craft-type production exercises with unique and multiple parts. Standardizing and reducing parts requirements can yield substantial gains to innovators – above and beyond the advantages that all firms enjoy with just-in-time inventory programs. Similarly, managing process control is important to all firms, but particularly to innovators who possess more complex processes.

Here again, the evidence confirms that innovative firms place considerably greater emphasis on both inventory control and process control in the Canadian manufacturing sector. This parallels earlier findings for Canadian firms (Baldwin and Johnson, 1996b, 1999a).

13.8 CONCLUSION

The importance of innovative activity has led to a search for a better understanding of the strategies and activities emphasized by innovative firms. Previous chapters focus primarily on the importance of R&D and the technological capabilities of innovative firms. But these chapters also stress the importance of other key areas like marketing. Innovative firms are more likely to be active in global markets. A larger portion of their sales is exported, and they are more likely to operate in more oligopolistic markets.

The competitive forces facing innovative firms result in a product-market strategy that involves continually offering higher quality products and developing increasingly superior technology. Innovative firms consider themselves to be more competitive in all the functional areas examined – markets and products, production processes, innovation, and human resources.

The most important finding of this chapter is that strategic emphases differ considerably between innovative firms and non-innovative firms.

Innovators place greater weight on innovation-related strategies – R&D, technology, production processes, and aggressive marketing of new products in new markets. In addition to actively developing new products and penetrating new markets, they are more active in developing and refining technology. Innovative firms are not those that serendipitously stumble across inventions. Innovators differ from non-innovators in that they adopt a purposive stance to find new products and to adopt new processes. The Canadian manufacturing sector is not a world where most firms are engaged in innovative activity, where some are rewarded by chance and others are not. It is a world that divides into firms that heavily stress an innovation strategy and those that do not.

In Chapter 9, we found that innovative firms were more likely to make use of intellectual property rights. While patents, trade secrets, and trademarks can play a role in protecting innovation from being copied, most firms feel that complementary strategies are more important. 'Being first in the market' and other strategies that bundle services with innovative products are more important in allowing firms to reap the benefits of their investment in innovative ideas. One of the most important ways of protecting an innovation is to develop a marketing strategy that provides special value to consumers.

For this and other reasons, innovative firms place greater stress on marketing strategies and activities than do non-innovative firms. Innovators are more competitive than non-innovators in traditional areas, such as customer service, flexibility in responding to customer needs, and product quality. But innovators attribute greater importance to aggressive marketing strategies that emphasize new products and new markets. Innovativeness fetches higher prices in the market.

Innovative firms are not simply concerned with being at the leading edge of product development. As demonstrated in Chapter 3, most innovations involve combinations of both product and process change. Thus, innovative firms place greater stress on the development of new production technologies. They introduce new processes more frequently, are more likely to use advanced manufacturing processes, and possess superior capabilities in production management. They also place greater emphasis on the need to use new materials, to use existing materials more efficiently, and to reduce labour costs. Innovation is multifaceted.

Innovative firms are also more likely to be more concerned about human resources. They value their employees' skills more highly, but they exhibit a much stronger commitment than non-innovative firms to enhancing those skills, to motivating their employees through a variety of means, and to involving their employees through collective agreements. They also boast a superior labour climate.

Given that the implementation of human resource strategies, marketing programs, and production capabilities involves management competencies, innovative firms also place greater emphasis on management and develop superior management skills.

The results of this survey confirm earlier findings (Baldwin and Johnson, 1996a, 1996b, 1999a; Gellatly, 1999) that innovative firms place more emphasis on a broad range of strategies than do non-innovative firms. Our findings here confirm not only that innovative firms develop superior capabilities in the core areas of research and development and technological capabilities, but that their capabilities also extend beyond a narrow scientific bent. Their marketing strategy goes beyond maintaining consumers in core markets. They aggressively seek out new markets for new products. Innovation also requires highly skilled workers. The greatest impediment to innovation that is perceived by firms in the manufacturing sector is the shortage of skilled labour. In response to this problem, innovators are more likely to stress the development of skills within their firm through training programs.

All of this indicates that innovative firms must master a set of complementary skills. Although their competitive advantage lies in innovation-related areas, such as spending on innovation, spending on R&D, and R&D management, in other functional areas they also emphasize special competencies. As such, innovative firms develop a balanced approach that makes them compleat firms.

FOURTEEN

Determinants of Innovation

14.1 INTRODUCTION

The topic of innovation has garnered the interest of a select group of economists from Schumpeter (1942) to Nelson and Winter (1982), who have stressed that it is the key to economic growth. However, until the advent of panel data sets, there was little empirical evidence to link the innovation stance of firms and their performance. Recent work that links dynamic panel data sets on the performance of firms and special surveys on the strategies that are being pursued by firms has demonstrated the importance of innovation to growth. Baldwin, Chandler et al. (1994) demonstrate that in small and medium-sized Canadian firms, a measure of success that is based on growth, profitability, and productivity is strongly related to the emphasis that firms place on innovation. Baldwin and Johnson (1998a) use a sample of entrants to show that growth in new firms is related to whether the firm innovates. Crépon, Duguet, and Mairesse (1998) find that innovation in French firms is associated with increases in productivity.

While we have evidence, therefore, on the connection between success and innovation, there is less evidence on the factors that condition whether a firm adopts an innovation policy. Not all firms innovate despite the benefits of doing so. Research has, therefore, been aimed at understanding the conditions that are associated with being innovative.[1] A number of questions have been posed in this literature – the extent to

---

[1] For related studies covering France, Germany, Italy, Holland, and Switzerland, see Crépon, Duguet, and Kabla (1996), Felder et al. (1995), Leo (1996), Sterlacchini (1994), Brouwer and Kleinknecht (1996), and Arvanitis and Hollenstein (1994, 1996).

which the intellectual property regime stimulates innovation; whether the exclusive emphasis that is given to R&D ignores the importance of other inputs; the effect of firm size and market structure on the intensity of innovation; and the extent to which multinational firms are more innovative. In this chapter, we use data from the 1993 Canadian Survey of Innovation and Advanced Technology to study the differences between firms that innovate and those that do not innovate, and we use multivariate analysis to address the following issues.

The first is the extent to which the intellectual property regime stimulates innovation. Patents are seen to be a key form of protection for innovation, but Mansfield (1986) and Levin et al. (1987), using data from firm-based surveys, have presented empirical work that suggests that patents may not be that important in many sectors. The evidence in Chapter 9 for Canada and a similar study for the United States (Cohen, Nelson, and Walsh, 2000) also indicate that other ways of protecting intellectual property, such as being first in the market, using trade secrets, and developing complex designs, are perceived by managers to be more effective in protecting intellectual property rights than are patents.

The second issue is whether the existence of an R&D unit is essential to innovation. While it is traditional to emphasize the importance of R&D facilities to the innovation process, Mowery and Rosenberg (1989) have stressed that a good deal of innovation comes out of engineering departments and production facilities. In this chapter, we ask whether possessing an R&D unit increases the probability of innovating, or whether it does so more than, say, having an engineering facility increases the probability of innovation.

Third, we investigate whether there are other competencies besides the development of a research and development unit that are closely linked to innovation. Not only do successful innovators have to develop new products; they also have to develop the technology to produce them; they have to put together successful marketing programs; they have to master production logistics. Baldwin and Johnson (1996b) and Chapter 13 demonstrate that innovators and non-innovators differ with regard to the emphasis that they place on a wide range of competencies – from marketing to human resources. Here we ask whether the emphasis given to different competencies is associated with innovation.

The fourth issue is the extent to which a larger average firm size and less competition stimulate innovation. Often described as the Schumpeterian hypothesis, it is sometimes claimed that innovation is fostered by a climate where firms are large or in industries where there is less competition. While there is mixed evidence that either matter (Scherer, 1992), the issue continues to receive attention (Cohen and Klepper, 1996a, 1996b). This chapter investigates whether the competitive environment in Canada is associated with innovation intensity, and in addition, whether competition has the same effect on world-first and other types of innovations.

The fifth question asks whether foreign-owned firms are more innovative than domestically owned firms. Both Dunning (1993) and Caves (1982) have stressed the special role of the multinational firm in transferring innovation skills and advanced technology from one nation-state to another. The role of multinationals in Canada is particularly important since they control over half of the manufacturing sector. McFetridge (1993) stresses the importance of linkages from Canada into the world innovation system that are accomplished through multinationals. In previous chapters, we have shown that multinationals are more likely to be performing R&D. Of key interest here is whether they also are more likely to innovate, after the effect of R&D on innovation is taken into account.

Finally, we examine the extent to which the domestic scientific infrastructure that is stressed by Tassey (1991) is used extensively by innovators. The environment facing each industry is seen to condition a firm's ability to innovate. On the one hand, the availability and quality of education, private and public technical services such as test laboratories, and standardization institutes, as well as research institutes, favour innovation. On the other hand, firms also need educational infrastructure to take account of new knowledge. The state of a country's higher educational facilities will affect a firm's ability to digest new information. Therefore, we investigate whether firms that make greater use of the infrastructure provided by university facilities are more likely to innovate.

Throughout the exercise, we examine these issues for different types of innovations. Innovations vary considerably both in terms of their nature – product versus process – and their novelty. As Chapter 8 demonstrates, the innovation regime – the extent to which R&D is used, the types and sources of information, the use that is made of intellectual property – varies considerably by novelty of innovation. Simply categorizing firms as innovative or non-innovative risks aggregating different types of innovators in a way that may hide important relationships. Therefore, in this chapter, we investigate the extent to which the factors associated with innovation differ across innovation types.

The chapter is organized as follows. Section 14.2 describes the empirical model used for the analysis. Section 14.3 contains the results for the model that estimates the determinants of innovation for the Canadian manufacturing sector. We first examine the result for innovators in general, and then we extend the analysis by estimating the model for product and process innovation separately. Finally, we examine the determinants of innovation activity for innovations that differ in terms of novelty.

14.2 EMPIRICAL MODEL

14.2.1 The Model

Firms innovate in the expectation of receiving an increase in profits. The expected postinnovation return to innovation activity r_i^* for firm i is taken to be a function of a set of firm-specific and industry-specific exogenous variables x_i. This may be expressed formally as:

$$r_i^* = bx_i + u_i \tag{1}$$

Even though r_i^* is not directly observable, we can observe whether firm i innovated or not. We assume that when the expected return from innovation is positive, firms innovate. The observable binary variable I_i takes a value of one when the firm is an innovator and zero otherwise. Thus, we can write:

$$I_i = 1 \quad \text{if } r_i^* > 0$$
$$I_i = 0 \quad \text{otherwise}$$

The expected return from innovation, given the characteristics of the firm and of the industry to which it belongs, is

$$E(r_i^*|x_i)$$

Thus, the probability of observing that a firm is innovative is given by:

$$\text{Prob}(I_i = 1) = \text{Prob}(u_i > -bx_i) = 1 - F(-bx_i)$$

where F is the cumulative density function for the residuals u_i.

The choice of the statistical model depends on assumptions about the form of the residuals u_i. If the cumulative distribution of residuals is normal, the probit model is the appropriate choice; if it conforms to a logistic function, the logit model is appropriate.

Differences in expected profits from innovation and, therefore, differences in profitability are hypothesized to be related to differences in firm size, market structure, appropriability conditions, technological opportunities, technological competency, and R&D activity.[2]

14.2.2 Innovation Variable

Innovation surveys allow us to examine the determinants of the output of the innovation process. In this respect, they differ from previous studies using R&D expenditures (Levin and Reiss, 1984) or patents (Pakes and Griliches, 1984). Innovation surveys ask whether a firm has produced an innovation, and then proceed to explore the various firm and industry characteristics associated with innovation.

Innovations differ in several aspects. They vary in nature – product versus process. They differ in terms of importance – the radical in contrast to the more imitative. Innovations also vary in terms of novelty – world-firsts, Canada-firsts, and other types of innovations.

In order to test the sensitivity of our results to alternative definitions, innovation is measured with three different dependent variables in this chapter. First, the incidence of innovation is captured by a dichotomous variable that measures whether or not firms introduced an innovation of any type within three years prior to the survey date of 1993. In this case, the dependent variable takes a value of one for innovative firms, and zero for non-innovative firms. Firms are classified as being innovative if they introduced or were in the process of introducing a product or process innovation during the period 1989 to 1991 (question 3.1), or if they listed product or process innovations (question 3.2).

A second set of dependent variables capture the product/process distinction. The first variable takes a value of one if a firm produced a product innovation – either a product-only innovation or a product innovation that required a change in production processes – and zero if the firm has not produced any innovations of any type.[3] The second dependent variable contrasts process-only innovators against non-innovators.

A third set of variables capture innovations that differ in terms of novelty – world-first innovators versus non-innovators; Canada-first innovators versus non-innovators; 'other' innovators versus non-innovators.

---

[2] See Cohen (1996) for a review.
[3] We experimented with an alternate version that used product only and obtained results that were broadly similar.

14.2.3 Explanatory Variables

Innovation is generated by capabilities internal to the firm and stimulated by the environment in which a firm finds itself.[4] Therefore, innovation is postulated here to be a function of both firm-specific and industry-specific variables. Firm-specific variables include characteristics variables – such as firm size and ownership – and activity variables – such as R&D and patenting. Industry-specific variables include the number of competitors that a firm faces and technological opportunity. The summary statistics for the variables that are used in this chapter are provided in Table 14.1.

14.2.3.1 Firm Activities and Characteristics

RESEARCH AND DEVELOPMENT. Although neither a necessary nor a sufficient condition for innovation (Äkerblom, Virtaharju, and Leppäahti, 1996), R&D is an important input into the innovation process. Firms that have established an effective R&D program are more likely to innovate for several reasons. First, R&D directly creates new products and processes. Second, firms that perform R&D are also more receptive to the technological advances made by others (Mowery and Rosenberg, 1989). A binary variable was constructed to capture this effect, taking a value of one if the firm engages in R&D, zero otherwise. This variable, in turn, was divided into two components – those firms that performed R&D continuously and those firms that performed R&D only occasionally.

OTHER FIRM COMPETENCIES IN MARKETING, TECHNOLOGY, PRODUCTION, AND IP PROTECTION. While size is often used as a proxy for scale effects, it is also related to differences in the internal competencies of firms. Large firms do not differ from small firms in that they are simply scaled up versions of the latter, a requirement if size captures only scale effects. Scale economies refer to differences that arise from an equal percentage increase in all factors. However, large firms use factors in very different proportions than do small firms. Their capital/labour ratios are generally higher. The production process between large and small firms is different since technology use is not the same (Baldwin and Sabourin, 1995). Not only are large firms more likely to adopt an advanced technology, but they also combine greater numbers of advanced technologies. The observed differences between large and small firms come from a host of activities that change as firms grow.

---

[4] See Bosworth and Westaway (1984).

Empirical Model

Table 14.1. *Summary of Dependent and Explanatory Variables Used in Multivariate Analysis*

Variable	Description	Mean
1. Dependent		
Innovation		
INNOV	Innovator or non-innovator	0.44
PRODINV	Product innovator only	0.28
PROCINV	Process innovator only	0.19
WORLDINV	World-first innovator	0.08
CANFIRST	Canada-first innovator	0.13
OTHERINV	Other innovator	0.17
2. Firm characteristics		
Size	Employment size	
ENTSIZE1	0 to 99 employees	0.54
ENTSIZE2	100 to 499 employees	0.27
ENTSIZE3	500 or more employees	0.19
Ownership		
FOREIGN	Foreign owned	0.23
Strategies		
TECH_STR	Technology strategy importance of developing and improving technology	1.0
MRKT_STR	Marketing strategy importance of new products and new markets	1.0
PROD_STR	Production strategy importance of new materials importance of improving inventory/ process control	1.0
IP_STR	Intellectual property management strategy importance of intellectual property management	1.0
3. Firm activities		
R&D activity		
R&D	Conducts R&D	0.84
R&DCONT	Conducts R&D continuously	0.39
R&DOCC	Conducts R&D occasionally	0.45
Intellectual property rights		
PATENTS	Use of patents	0.22
TRADSECR	Use of trade secrets	0.15
4. Industry characteristics		
Competition	Number of Competitors	
COMP1	5 or fewer competitors	0.28
COMP2	6 to 20 competitors	0.35
COMP3	More than 20 competitors	0.36
Technological opportunity		
TECH_OPP	Technological opportunity	14.1

Few innovation studies consider many firm-specific competencies, outside of R&D, as contributing factors to innovation. Yet, over time, firms build up a range of competencies that are crucial for their overall growth and development. Those firms best able to develop certain key competencies relating to innovation might be expected to be more likely to innovate. In a recent study, Baldwin and Johnson (1996b) used data from a survey of small and medium-size businesses and found that more-innovative firms place a greater emphasis on marketing, finance, production, and human resource competencies than do less-innovative firms. We confirm this finding in Chapter 13. It is, therefore, important to include in our multivariate analysis a measure of the extent to which a firm has developed key capabilities in areas required for the implementation of a successful innovation strategy.

A set of questions on the 1993 innovation survey that were discussed at length in Chapter 13 allow us to examine the extent to which innovation is associated with greater competencies in a number of areas. In these questions, firms are asked to indicate the importance that they give to various marketing, technology, production, and human resource strategies.

Competency variables are constructed from the firms' responses to questions about the importance they give to strategies in each of these areas. Each is scored on a Likert scale that ranges from 1 (not important) to 5 (crucial). For *market strategy,* three questions are used – the extent to which a firm focused its marketing activities on introducing new products in present markets, current products in new markets, or new products in new markets. For *technology strategy,* three questions are used – the importance a firm attributes to developing new technology, improving technology developed by others, and improving on their own existing technology. For *production strategy,* four questions are used – the importance that a firm gives to using new materials, using existing materials more efficiently, improved inventory control and improved process control. Under *human resource strategy,* two questions are used – whether a firm values continuous staff training, and whether it uses innovative remuneration schemes. For *intellectual property rights strategy,* the importance that a firm gives to improving intellectual property management is employed.

The answers to these questions are used to capture the emphasis given to a particular area or the importance of this input to the production function. While these are subjective questions, they have been used successfully in other surveys (Baldwin, Chandler, et al., 1994; Johnson, Baldwin, and Hinchley, 1997). An aggregate score for each of the strategies was constructed by summing the scores of the individual strategies. For example, the sum of the scores of three factors – the importance of developing new technology, improving technology developed by others, and improving on their own existing technology – was used for the aggregate score for technology strategy. Since the number of factors considered varies across strategies, the results were averaged within each category to correct for this variation and standardized to a mean of 1.

An alternative approach is to use principal component or factor analysis to define a set of competency variables from the firm scores in the areas of marketing, technology, and production strategies. This was tried and it was found that the resulting conclusions were similar. But these results are more difficult to interpret and are not reported here.

SIZE. A measure of firm size is included to test whether there are inherent advantages associated with size. Large firms, it is often argued, tend to be more innovative than their smaller counterparts.[5] Suggested reasons include scale advantages of large firms, a greater likelihood of engaging in risky projects, and economies of scope (Cohen, 1996). Larger firms have easier access to financing, can spread the fixed costs of innovation over a larger volume of sales, and may benefit from economies of scope and complementarities between R&D and other manufacturing activities. Counterarguments, however, exist to suggest that as firms grow large, their R&D becomes less efficient. Levin and Reiss (1989) review the empirical evidence and observe that it is inconclusive.[6] Economies of scale and scope may exist, but they appear to be exhausted long before a firm reaches the largest size classes.

Size is measured here by the total number of employees in a firm, including both production and nonproduction workers. Firms are classified as belonging to one of three size categories – less than 100 employees, 100 to 499 employees, and 500 employees or more. Based on this classification, three binary variables have been constructed to capture size effects.

NATIONALITY OF OWNERSHIP. Because of its size and proximity to the United States, Canada has a mixture of both Canadian-owned and foreign-owned firms. Existing studies, relying on R&D intensity, are

---

[5] See Crépon, Duquet, and Kabla (1996); Levin, Cohen, and Mowery (1985).
[6] The recent research as reviewed by Cohen and Levin (1989) tends to regard the failure of the empirical literature to obtain robust results on how innovation is related to size of firm and to market structure as an indication that these relationships are more complex than previously believed.

inconclusive as to whether or not the nationality of ownership of a firm has an impact on its innovative activity. Caves et al. (1980, p. 193) suggest that foreign activity reduces the rate of R&D activity in Canada. However, lower R&D intensity may not signify less innovation if multinational subsidiaries import innovations from their parents. Using a survey for a limited number of firms in five industries, De Melto et al. (1980) reported that foreign firms operating in Canada were less R&D intensive than their domestic counterparts, but that they accounted for a disproportionately large percentage of process innovations.

In order to investigate whether foreign-controlled firms were more likely to be innovative, a binary variable – taking a value of one if the firm is foreign-owned and zero otherwise – is included.

APPROPRIABILITY AND INTELLECTUAL PROPERTY RIGHTS. We also included a variable to capture the extent to which the activities of the firm with regard to protecting its intellectual property rights was related to whether it was more likely to innovate.

Firms commercialize new products and processes expecting, in return, certain rewards – usually an increase in profits. If inventions are easily copied by competitors, there is little incentive to innovate. To protect their innovations from being copied, firms use various forms of intellectual property protection, such as patents, trade secrets, copyrights, and trademarks.

Despite the widespread belief that the existence of intellectual property protection is critical to the innovation process, empirical evidence as to the beneficial effects on innovation activity is sparse (Cohen, 1996). Indeed, there is empirical evidence to suggest the opposite. In a study examining the effectiveness of patents in protecting intellectual property rights, Mansfield (1986) found that only in the pharmaceutical and chemical industries did managers feel that patents played an important role. Levin et al. (1987) also found that product patents were more important for pharmaceuticals and chemicals. Moreover, Levin et al. (1987) found that other forms of intellectual property rights protection were perceived by managers to be more effective than patents. Complementary marketing activities and lead time were found to be the most effective in protecting product innovations. For process innovations, patents were found to be much less effective, while secrecy was found to be the most effective. Cohen (1996) concludes that although there is growing evidence of interindustry differences in appropriability conditions, there is less empirical evidence as to the beneficial effect of these conditions on innovation activity across a wide range of industries.

In this study, we construct two binary variables to test whether appropriability conditions within a firm affect innovation.[7] The variables are based on whether or not a firm uses patents, or whether it uses trade secrets to protect its innovations. This is a direct measure of the extent to which the firm found these to be important, or the degree to which it was able to devise a strategy to protect its intellectual property. Learning how to do this is not straightforward and requires the development of specific competencies – legal skills, design skills, marketing skills, and service skills. Each variable takes a value of one if the particular property right is used and a value of zero if it is not.

Other studies have tended to define appropriability conditions at the industry level. We choose to move to the level of the firm, because there is evidence that firms, even within well-defined industries, are idiosyncratic when it comes to their tendency to develop a capacity to protect their ideas. Appropriability will be partially conditioned by the nature of the industry – whether the product is sufficiently distinguishable from others that it can be patented. But even within those industries where patents are generally not used, there will be some firms that develop a strategy of protection, and there will be other firms who do not. Firms make appropriability work and this requires conscious action. Appropriability may partially be exogenous in that it stems from some inherent product characteristic that varies considerably across industries; but a substantial part of the appropriability environment stems from individual decisions taken at the firm level to develop product characteristics that are patentable or that can be protected by trade secrets, or to develop legal expertise that protects what otherwise might not be protected.

We also experimented with an alternative variable to capture the effect of the amenability of the intellectual property regime of a firm on its innovation success. We used the score (on a scale of 1 – not very effective – to 5 – extremely effective) that a firm gave to the efficacy of patents, trade secrets, and other intellectual property rights as a means of protecting their innovation. This effectiveness variable provides a measure that does not depend directly upon past activity – that is, while the patent-use variable at the firm level reflects past innovative activity, the firm's attitude towards intellectual property right protection and its skill in protecting intellectual property (the patent-effectiveness variable) is

---

[7] We experimented with a third variable that includes any other intellectual property right – trademarks, copyrights and industrial designs – but it was not found to be significant.

a more direct measure of existing attitudes towards the value of patent protection – though existing attitudes are, of course, conditioned by past innovation activity and experience with the effectiveness of intellectual property rights.

Although we experimented with this efficacy variable, we chose not to focus our results on it for two reasons. First, fewer firms answered this question, and those who did so were not representative. Firms that answered this question were much more likely to have taken out a patent. Therefore, average efficacy scores at the industry level really reflected differences in the propensity to patent – the variable that is used here at the firm level.[8] Second, a regression of the average score on the efficacy of patent protection shows a strong relationship to the propensity to patent (Baldwin, 1997a). Since the patent-efficacy score was capturing the propensity to patent and there were more observations on the latter variable, the propensity to patent is used here.

14.2.3.2 Industry Effects

TECHNOLOGICAL OPPORTUNITIES. Technological opportunities differ across industries since the scientific environment provides more fertile ground for advances in some industries than in others. As a result, the technical advance generated per unit of R&D is greater in some industries than others (Cohen, 1996).

Two proxies that were suggested by Levin et al. (1987) have been used in various studies. Sterlacchini (1994) uses the percentage of those firms investing in R&D that have collaborative projects with universities. Arvinitis and Hollenstein (1994) use the extent to which outsiders, like competitors and customers, contributed to the innovation. The first is a measure of the extent to which an industry relies on science-based research, whereas the second really measures the extent to which an industry relies on external sources of knowledge, such as customers and suppliers, for technological advance.

In this study, we use the first approach since it comes closer to the concept of the advanced scientific knowledge base that is available to a firm. The second is more a function of the ease with which knowledge flows from firm to firm, and represents the extent to which knowledge is easily transferable, rather than differences in the underlying scientific environment.

---

[8] See Chapter 9.

To capture the first concept, technological opportunity is measured here by the percentage of R&D performers within an industry that have collaborative R&D agreements with universities, colleges, or external R&D institutions.

COMPETITIVE CONDITIONS. Firms that are active in highly concentrated markets have been hypothesized to be more likely to innovate. Monopoly power, it is claimed, makes it easier for firms to appropriate the returns from innovation and provides the incentive to invest in innovation. However, this view is far from universal. Others (Fellner, 1951; Arrow, 1962) have argued that the gains from innovation at the margin are larger in an industry that is competitive than under monopoly conditions. Moreover, insulation from competitive pressure can breed bureaucratic inefficiency (Scherer, 1980). Finally, if market structure is determined largely by the life cycle of an industry, and if innovation is more intensive in the early stages of the an industry when the industry structure is more atomistic, we should expect innovation to be higher when markets are less concentrated.[9]

Since the intrinsic concept that we want to measure is the degree of competition faced and concentration is a poor proxy for this (Baldwin and Gorecki, 1994), we choose to measure the potential competition that a firm faces by the number of competitors the firm tells us it faces. Firms are grouped according to whether they face 5 or fewer competitors, 6 to 20 competitors, or 20 or more competitors. Three binary variables are used to capture these competitive categories.

14.2.4 Estimation Procedures

While we are primarily interested in the effects on innovation of a number of variables, one of which is patent use, we recognize that patent use itself may be a function of the extent to which innovation occurs.

Our model therefore consists of two equations – one for innovation and one for the use of intellectual property.[10] These equations contain two variables that we treat as endogenous – innovation (INNOV)

---

[9] For a discussion of the relationship between innovation and structure, see Abernathy and Utterback (1978), Rothwell and Zegveld (1982), Gort and Klepper (1982), Klepper and Millar (1995), and Klepper (1996). Empirical evidence on the relationship between concentration and innovation is mixed (Cohen and Levin, 1989). See also Levin and Reiss (1984, 1988) and Cohen and Levinthal (1989).

[10] See Lunn (1986) for recognition of the importance of adopting a simultaneous equations framework in the case of innovation regressions.

and intellectual property use (IPUSE) – and a number of exogenous variables.[11]

$$\text{INNOV} = \alpha_0 + \alpha_1 \text{ SIZE} + \alpha_2 \text{ COMP} + \alpha_3 \text{ R\&D} + \alpha_4 \text{ IPUSE} + \alpha_5 \text{ TECH\_OPP}$$
$$+ \alpha_6 \text{ FIRM STRATEGIES (TECH\_STR, MRK\_STR, PROD\_STR)} \quad (2)$$

$$\text{IPUSE} = \beta_0 + \beta_1 \text{ SIZE} + \beta_2 \text{ TECH\_OPP} + \beta_3 \text{ FOREIGN}$$
$$+ \beta_4 \text{ INNOV} + \beta_5 \text{ IP\_STR} \quad (3)$$

INNOV is a binary dependent variable indicating whether or not a firm is innovative. IPUSE measures the use of patents – though in the variant of the innovation equation that uses process innovation, IPUSE captures trade secret use.[12]

R&D measures whether a firm engages in R&D activity. FIRM STRATEGIES represents other complementary activities that develop competencies of a firm. The firm strategies that are included in the innovation equation are TECH_STR, which measures the importance that a firm attributes to improving its technology; MRKT_STR, which measures the importance given by a firm to the marketing of new products and penetrating new markets; PROD_STR, which measures the importance of a progressive production strategy. The firm strategy variable that is included in the appropriability equation is IP_STR, which measures the importance that a firm attributes to intellectual property management. SIZE is the employment size of a firm, while FOREIGN measures whether a firm is controlled from abroad. Technological opportunity (TECH_OPP) measures whether a firms employs university partnerships, making use of basic science. Competition (COMP) is the number of competitors a firm faces.

Three issues arose in choosing the estimation procedure – issues associated with the use of a dichotomous dependent variable, those associated with the use of survey data, and problems arising from possible simultaneity.

First, since the dependent variable that is used for the innovation equation is a binary variable, we experimented with both a logit and probit regression and obtained similar results. We report the results of the probit formulation here.[13]

---

[11] While we recognize that other variables, such as market structure, may be endogeneous, we do not consider these relationships here because of the limitations of a single cross-sectional database for this purpose.
[12] This division results from the results of Chapter 9 that suggest trade secrets are used more for process innovation.
[13] We also experimented with ordinary least squares (OLS), using a generalized routine to account for the heteroscedastic error term. OLS resulted in qualitatively similar results.

Second, since the data used for the analysis come from a survey that randomly samples a population stratified by region, industry, and size, we take into account the sampling weights attached to each observation in our multivariate analysis. Survey data must be weighted by sampling weights if they are to give an accurate picture of the population. Multivariate analysis of survey data, if it is to represent the behaviour of the population, also needs to take into account the sampling weights attached to each observation – unless the variables that are included in the analysis and the functional form are correctly specified. In the latter case, the unweighted results will be the same as weighted results. Since it is unlikely that we could meet the rigid conditions for relying completely on the unweighted results, we experimented with both routes. The formulation reported here produced quite similar results when we used the logit formulation, and thus we have confidence in the results. We report the unweighted results here.

Third, a method for dealing with endogeneity must be chosen. Innovation is taken to be a function of the extent to which a firm can appropriate the benefits of innovating – as measured by its use of patents – as well as by a set of firm-specific and industry-specific characteristics. Firms that can effectively protect their innovations – through the use of patents, trade secrets, or other forms of intellectual property rights – are expected to have a greater likelihood of being innovative. They are more likely to have established capable legal departments for handling patent applications, or perhaps their organization is better suited for preventing the disclosure of trade secrets.

Intellectual property use, such as patents, on the other hand, is likely to be a function of innovation and a set of firm-specific and industry-specific characteristics. Once an innovation is discovered, a firm may turn to patents to protect its innovation from being imitated.

Innovation and intellectual property use are, therefore, not independent of each other. Because of this, the use of a single equation model can lead to biased and inconsistent estimates. We employ a simultaneous equation model to overcome this difficulty, recognizing that a poorly specified simultaneous model can worsen, rather than improve, bias in the estimates. It is primarily for this reason that we do not extend our model to other variables.[14]

---

[14] For this purpose, we made use of a statistical package (MECOSA) that was developed by Arminger, Wittenberg, and Schepers (1996). This program builds on Amemiya's (1978) article that develops the method to estimate a simultaneous probit model.

Since the coefficients of the probit model by themselves do not reveal fully the magnitude of the effects of these variables, we also report the probability values associated with every variable. Probability values provide a quantitative estimate of the partial effects of an activity or firm characteristic on the likelihood of introducing an innovation. The probabilities are calculated by estimating the value of the probit function using the sample means of other variables and the parameter estimates.

For binary explanatory variables, two probabilities are calculated. The first estimates the probability of innovating when the explanatory variable takes a value of one; the second when the explanatory variable takes a value of zero. The difference in the two probabilities provides the quantitative effect of this variable on the decision to innovate.

For the continuous variable (technological opportunity), we adopt a different approach. The probability of introducing an innovation is first estimated for a given explanatory variable at its mean value. It is then estimated at its mean value plus/minus one standard deviation.

14.3 REGRESSION RESULTS

The regression results for innovation as a whole are reported in Table 14.2 and the probability values in Table 14.3. All regressions are estimated against an excluded firm that is small, is Canadian owned, and faces few competitors. All results are for large firms only – those firms that are found in the profiled set of firms in the Statistics Canada Business Register.[15]

14.3.1 Incidence of Innovation

Firms that perform R&D are more likely to innovate. R&D activity has a positive and significant impact on innovation. Firms not performing R&D have only a 1% probability of innovating. The probability of reporting an innovation is 11% for those performing R&D only occasionally and 33% for those doing so continuously.[16] The importance of R&D that is revealed by this estimate accords with other studies (Cohen and Klepper, 1996a; Baldwin, 1997b).

---

[15] Only large firms were asked all of the questions about the variables used in the analysis. The sample used in the regression consists of 1,253 observations by respondents who answered all of the questions used in the analysis out of a total of 1,593 in the survey. The proportion of the total that was accounted for by this group was about the same across the strata used in the survey – that is, they are broadly representative of the sample.

[16] Brouwer and Kleinknecht (1996) also report for Holland that firms with continuous R&D facilities have a higher probability of innovating.

Table 14.2. *Regression Coefficients for the Determinants of Innovation Activity*

	Innovation	Patents
Intercept	−3.07 (−6.78)***	−1.05 (−10.47)***
Intellectual property use		
Patents	−0.75 (−4.91)***	–
Innovation activity		
Innovation	–	0.82 (7.06)***
Firm characteristics		
Firm Size		
100–499 employees	0.46 (3.29)***	0.30 (2.43)**
500 or more employees	1.24 (6.83)***	0.22 (1.38)
Ownership		
Foreign owned	–	0.12 (0.96)
Firm activities		
R&D activity		
Any R&D	1.10 (4.67)***	–
Continuous R&D	0.78 (6.12)***	–
Strategies		
Technology	0.35 (4.53)***	–
Marketing	0.25 (3.64)***	–
Production	−0.15 (−2.27)**	–
Intellectual property		0.12 (2.19)**
Industry characteristics		
Competition		
6 to 20 competitors	0.31 (2.43)**	–
Over 20 competitors	−0.16 (−1.16)	–
Technological opportunity		
Collaborative agreements	0.01 (1.00)	–
SUMMARY STATISTICS		
N	1290	
$Q_T(\theta): \chi^2$	11.3	
R^2	0.31	0.31

Notes: *t*-statistics are in parentheses. For larger firms (IPs) only. These firms are generally larger than 20 employees.
*** Significant at the 1% level.
** Significant at the 5% level.

Firm size also has a positive impact on the probability of innovating. The largest firms have a significantly higher probability of being innovative (59%) than do medium-sized ones (29%) and small firms (15%).

Nationality of ownership was included in the first round of estimates, but it was found to have no significant effect on the probability that a

Table 14.3. *Estimated Probability of Introducing an Innovation and Using Patents*

	Innovation	Patent Use
Intellectual property use		
Patent user	25	–
Nonuser	31	–
Innovation		
Innovator	–	49
Non-innovator	–	20
Firm activities		
R&D activity		
Occasional R&D	11	–
Continuous R&D	33	–
No R&D	1	–
Complementary strategies		
Technology (mean)	25	–
+ standard deviation	38	–
− standard deviation	16	–
Marketing (mean)	25	–
+ standard deviation	34	–
− standard deviation	18	–
Production (mean)	25	–
+ standard deviation	21	–
− standard deviation	31	–
Intellectual Property (mean)	–	31
+ standard deviation	–	36
− standard deviation	–	27
Firm characteristics		
Firm size		
0 to 99 employees	15	28
100 to 499 employees	29	39
500 or more employees	59	28
Ownership		
Foreign	–	31
Canadian	–	31
Industry characteristics		
Number of competitors		
0 to 5	22	–
6 to 20	32	–
Over 20	22	–
Technological opportunity		
Technological opportunity (mean)	25	–
+ standard deviation	25	–
− standard deviation	25	–

Note: For larger firms (IPs) only.

firm is innovative and was subsequently omitted from the final results. It should be noted that nationality does become significant if size or R&D is omitted. Foreign-owned firms are larger and are more likely to perform R&D, but once these factors are taken into account, there is no additional effect of nationality on innovation.

Technological opportunity has a positive but statistically insignificant impact on innovation. It should be noted, however, that the significance of this variable was sensitive to the formulation chosen. In other formulations (the linear generalized least squares (GLS) and the logit), this variable was weakly significant. Firms in industries relying on science-based research are more likely to be innovative. This finding corroborates other research that reports greater opportunities for innovation in industries for which basic science is important (Arvanitis and Hollenstein, 1994; Crépon et al., 1996).

The most striking result is the negative sign attached to the use of patents. Patent use, a mark of a firm that has learned how to manage its intellectual property, is not positively associated with the reporting of innovations. It should be noted that this result only appears in the simultaneous model. In the single equation model, intellectual property use in the form of patents is positively associated with innovation. The use of a nonsimultaneous framework yields a very different conclusion about the nature of the relationship between the intellectual property regime and innovation, and this difference emphasizes the importance that should be given to adopting a simultaneous framework.

Several of the competency variables that represent other inputs to the innovation process are highly significant.[17] The emphasis that is placed on both technology and marketing strategy is positively related to innovation and highly significant. In each case, the probability of innovating increases by about 20 percentage points going from the midteens to the mid-30s as the competency variable increases from its mean value minus one standard deviation to the mean value plus one standard deviation.

Those firms that place more emphasis on their technology strategy are more likely to innovate. As Mowery and Rosenberg (1989) emphasized, R&D is not the only important input to the innovation process. Similarly, firms that develop the capacity to market new products and target new markets for existing products are more likely to innovate. Conversely, the

---

[17] They also represent fixed effects. Firms that are more likely to perform R&D are more likely to be advanced in many other respects. Without inclusion of these other competencies, the R&D variable will partially capture these fixed effects.

incidence of innovation is found to be negatively related to the emphasis that a firm places on production strategy. Firms that stress the importance of using new materials and improving inventory and process control are less likely to innovate. This suggests that an emphasis on cost cutting is a substitute for innovating for some firms. The emphasis on human resources was not found to be significantly related to the probability of innovation and has been excluded from the regression results.

Innovation is also significantly related to the number of competitors that a firm faces, but the relationship is not monotonic. Firms facing moderate competition – 6 to 20 competitors – are significantly more likely to innovate. Firms facing the stiffest competition – more than 20 competitors – are the least likely to do so. Firms with 6 to 20 competitors have a 32% probability of being innovators. Firms in the most atomistic groups have only a 22% chance of innovating.

Various alternatives were tried in order to test whether our results are affected by other specifications of the innovation equation. We moved to the industry level to define appropriability conditions. That is, we defined the environment in which the firm operated as being determined by the average intensity of patent use at the two-digit industry level. This has two effects. First, it means that the environment is more likely to be exogenous and, therefore, the need for a simultaneous equations framework is less persuasive. Second, it overcomes the criticism that the use of patents at the firm level may reflect how a firm engineers not so much its intellectual property environment as its past innovation success. However, when patent use is included at the industry level, it is found to be insignificant.

We also experimented with an alternative measure of the importance of the intellectual property environment – the score (from 1 – not very effective – to 5 – extremely effective) that the firm gave to the effectiveness of various instruments in 'preventing competitors from bringing to market copies of its new product or process technology'. The instruments considered included seven formal options – copyrights, patents, industrial design, trade secrets, trademarks, integrated circuit designs, and plant breeders' rights. In addition, the scores given to other strategies – complexity of product design, being first in the market – were included. When the average scores for patents, trade secrets, and other strategies are included at the two-digit industry level, the patent score is insignificant, but the score on trade secrets is positive and significant. In industries where trade secrets are seen to be effective, the probability that innovation occurs is higher. When patent and trade secret use, as well as the efficacy variables, are included, the use variables remain insignificant, but both the efficacy scores given to trade secrets and other strategies (product complexity and trade secrets) remain positive and significant.

Industry effects were also included. For this purpose, we broke our sample down into three broad groups that differ in terms of the innovation intensity of the industry. Industries were classified in one of three groups – core, secondary, and other – based on the classification system used in earlier chapters. When we included binary variables for these three classifications, very little changed. Thus, the inclusion of broad industry effects, which we know are associated with innovation tendencies, does not affect our results.

We also estimated the innovation equation by using efficacy scores at the firm level. The sample for which these scores are available is considerably smaller than that for which patent use is available. In this formulation, strategies relating to complexity of product and lead time are significant – much as they are in the equation that summarizes these at the industry level.

We also included the score given to whether a firm gets its innovation ideas from its competitors or from its customers. Arvanitis and Hollenstein (1996) have included a variant of this variable to capture the presence of technological opportunities, but this variable could also approximate conditions under which information flows are easily transmitted and difficult to protect. This hypothesis is confirmed by its negative coefficient in the formulation that includes it.

In summary, the results of these various experiments suggest that appropriability stimulates innovation. But it is not patents that matter; rather, trade secrets and other strategies allow a firm to appropriate the fruits of its investments in intellectual capital. This is in keeping with Schankerman's 1991 finding that at least 75% of the private returns to inventive activity are obtained from sources other than patents. Moreover, in industries where ideas are easily taken from competitors or transferred from others, innovation is less likely.

14.3.2 Patent Use

Patent use is related strongly to whether a firm is an innovator. The probability of using patents increases from 20% for a non-innovator to 49% for an innovator. Patent use also is a function of whether the firm has a well-developed intellectual property strategy. It increases from 27% for those putting little emphasis on intellectual property to 36% for those putting a heavy emphasis on intellectual strategy. Being an innovator has a much greater effect on the probability of patent use than emphasizing competencies in this area (Table 14.3).

Both innovation and patent use are related to size. But the effect of size on the probability of patent use is much less than on innovation (Table 14.3).

14.3.3 Type of Innovation

Innovations are commonly categorized as being one of two types – product or process. To investigate whether the characteristics of innovators differ between those who produce one type or the other, we divide innovators into two groups – those who reported a product innovation (either a product innovation only or a product innovation that required a change in processes) or a process-only innovation. The same models are estimated – though, in this instance, we use patent use in the product innovation equation and trade secret use in the process innovation equation. We also combine R&D conduct into one variable – the performance of any form of R&D. The parameter estimates of the probit regressions for each of these innovation-types are reported in Table 14.4. The probability values associated with the estimates are reported in Table 14.5.

R&D activity has a positive and significant impact on both forms of innovation. If a firm does not perform R&D, it has an insignificant chance of introducing a product innovation, whereas a firm performing R&D has a 17% chance of introducing a product innovation. The same probabilities for process innovation are 0% and 2%. While being neither a necessary nor a sufficient condition for innovation, R&D increases the probability of introducing a product innovation by 17 percentage points and the probability of introducing a process innovation by only 2 percentage points. This confirms the greater importance of R&D for product innovation.

Firm size is also highly statistically significant for both product and process innovation. Small firms have a 6% probability of introducing a product innovation; large firms have a 33% chance of introducing a product innovation – a difference of 27 percentage points. Small firms have a 1% chance of introducing a process innovation, while large firms have a 11% chance of doing likewise – a difference of 10 percentage points.

Cohen and Klepper (1996a, 1996b) suggest that size should be more closely related to process than to product innovation. This, they suggest, is because of the difficulty a firm faces in realizing the return on a process innovation except through own-firm production. Process innovation is too firm-specific to allow for easy transfer of information. In contrast,

Table 14.4. Regression Coefficients for the Determinants of Product and Process Innovations

	Product Innovation		Process Innovation	
	Innovation	Patent Use	Innovation	Trade Secret Use
Intercept	−3.66 (−7.11)***	−0.73 (−5.98)***	−3.45 (−5.51)***	−0.51 (−2.25)**
Innovation				
Product innovation only	—	0.70 (5.61)***	—	—
Process innovation only	—	—	—	0.83 (4.58)***
Intellectual property use				
Patents	−0.51 (−2.80)***	—	—	—
Trade secrets	—	—	−0.74 (−2.72)***	—
R&D Activity				
Conducts R&D	1.79 (6.08)***	—	0.84 (3.21)***	—
Complementary strategies				
Technology	0.41 (4.79)***	—	0.40 (3.57)***	—
Marketing	0.23 (3.29)***	—	0.11 (1.27)	—
Production	−0.10 (−1.57)	—	0.04 (0.46)	—
Intellectual property	—	0.11 (1.98)**	—	0.14 (2.13)**
Firm Size				
100–499 employees	0.37 (2.51)**	0.30 (2.48)**	0.23 (1.40)	−0.19 (−1.26)
500 or more employees	1.12 (5.48)	0.31 (1.93)*	1.09 (5.36)***	−0.22 (−1.02)
Nationality				
Foreign	—	0.18 (1.49)	0.16 (1.08)	−0.07 (−0.48)
Competition				
6 to 20 competitors	0.23 (1.84)*	—	0.38 (2.25)**	—
Over 20 competitors	−0.17 (−1.30)	—	0.09 (0.51)	—
Technological opportunity				
Collaborative agreements	0.01 (1.60)	—	—	—
SUMMARY STATISTICS				
N	1196		1078	
$Q_T(\theta): \chi^2$	7.2		24.2	
R^2	0.40	0.34	0.25	0.19

Notes: t-statistics are in parentheses. For larger firms (IPs) only.
*** Significant at the 1% level.
** Significant at the 5% level.
* Significant at the 10% level.

Table 14.5. *Estimated Probability of Introducing a Product Innovation and Using Patents, and Introducing a Process Innovation and Using Trade Secrets*

	Product Innovation		Process Innovation	
	Innovation	Patent Use	Innovation	Trade Secret Use
Intellectual property use				
Patent user	10	–	–	–
Non–patent user	13	–	–	–
Trade secret user	–	–	2	–
Non–trade secret user	–	–	2	–
Innovation				
Product innovator	–	58	–	–
Process innovator	–	–	–	68
Non-innovator	–	31	–	36
Firm activities				
R&D Activity				
Conducts R&D	17	–	2	–
No R&D	0	–	0	–
Complementary Strategies				
Technology (mean)	10	–	2	–
+ standard deviation	20	–	4	–
– standard deviation	5	–	1	–
Marketing (mean)	10	–	2	–
+ standard deviation	15	–	2	–
– standard deviation	7	–	2	–
Production (mean)	10	–	2	–
+ standard deviation	10	–	2	–
– standard deviation	10	–	2	–
Intellectual property (mean)	–	39	–	43
+ standard deviation	–	43	–	48
– standard deviation	–	35	–	38
Firm characteristics				
Firm Size				
0 to 99 employees	6	34	1	43
100 to 499 employees	12	45	1	43
500 or more employees	33	46	11	43
Ownership				
Foreign	–	39	2	43
Canadian	–	39	2	43
Industry characteristics				
Number of competitors				
0 to 5	9	–	1	–
6 to 20	13	–	3	–
Over 20	9	–	1	–
Technological opportunity				
Technological opportunity (mean)	10	–	–	–
+ standard deviation	10	–	–	–
– standard deviation	10	–	–	–

Note: For larger firms (IPs) only.

the fruits of a product innovation are more easily realized by selling it to others. Our evidence is not consistent with their hypothesis.

The results for product in contrast to process innovation confirm the importance of the complementary strategies for both types of innovations. The emphasis on technology at the firm level is significantly related to innovation in both types of innovations – product and process innovation. Firms that introduce new product innovations without a change in manufacturing technology still stress the general development of technology. And the importance of this emphasis ranks with the emphasis placed on R&D. However, the emphasis that is given to marketing (a proxy for marketing competencies) is significant only for product innovation.

As before, the effect of intellectual property use is negative – although this time it is negative for patent use with regards to product innovation and trade secret use for process innovation. It should be recalled that the intellectual property use variable captures two effects – previous success on the one hand and the appropriability environment on the other. The negative relationship tends to indicate that reliance on past success has a stronger negative effect than the positive effect hypothesized to arise from the appropriability environment. This accords with the finding of Geroski, Van Reenan, and Walters (1997), who report that very few innovative firms are persistently innovative. We have previously reported that only a segment of Canadian firms are pursuing an innovative strategy that produces major innovations. Our findings here suggest that within this group, leadership roles change significantly over time. Those who hold patents as a result of past discoveries are less likely to be reporting innovations in the period of the survey.

Competition is significantly related to both product and process innovation. In both cases, the nonlinearities are the same, with competition being greatest for situations where there are between 6 and 20 employees. Firms in this category have a four percentage point higher probability of introducing a product innovation and a two percentage point higher probability of introducing a process innovation. Process innovation tends to be associated generally with cost cutting. Product innovation is associated with improvements in quality. Therefore, competition tends to have more of an effect on the quality changes associated with innovation in product markets.

14.3.4 Novelty of Innovation

Innovations vary in terms of importance. Some are the first of their kind in the world (world-firsts), some are the first of their kind in Canada

(Canada-firsts), while yet others are new only to the firms introducing them (other innovations).

Since the novelty of world-first innovations is greatest, we might expect the difference between non-innovators and these firms to be greater than for other types of innovators. In order to evaluate this hypothesis, we estimate three different probit regressions. In the first case, the dependent variable takes a value of one for a world-first innovation and zero for non-innovators. The second takes a value of one for a Canada-first innovation and zero for non-innovators. The third takes a value of one for the other (imitative) innovation category and zero for non-innovators. By comparing the results of the three formulations, we can draw inferences about whether the more novel category – world-firsts – requires different capabilities than do the less novel categories – Canada-firsts and all 'other' categories The estimated parameters from the probit regressions are reported in Table 14.6. The probability value associated with the simultaneous probit is presented in Table 14.7.

Size is important in the case of all categories – but it has the greatest impact on the probability of introducing Canada-first innovations.

It is also the case that R&D matters for all types of innovations. The importance of R&D, even in the case of Canada-firsts and other innovators, shows that R&D is essential to the adaptation of ideas from abroad and from other firms. Not only are R&D laboratories used to create and develop absolutely novel innovations, but they also serve to keep the firm informed about related innovation activity done by other firms and help to adapt it to Canadian conditions.

While competition (at least modest amounts) was important for all innovators taken together (Table 14.2), the effect of the number of competitors disappears for world-firsts in Table 14.6 in the simple probit. It is significant for the other two categories. This means that changes in the degree of competition are more effective in stimulating firms to imitate others than they are in affecting the introduction of the more novel forms of innovation.

A stress on marketing is also closely associated with both world-firsts and other innovators but has the greatest impact on the probability of introducing other innovations. In the latter case, marketing is essential if firms are to catch up with leaders. In the former case, the emphasis on marketing simply confirms the point that adapting to conditions both at home and abroad is important for innovators that are producing brand new products.

Table 14.6. *Regression Coefficients for Determinants of World-First, Canada-First, and Other Types of Innovations*

	World-First	Canada-First	Other
Intercept	−4.10 (−5.87)***	−3.32 (−6.87)***	−3.10 (−6.63)***
Patent use	−0.53 (−2.15)**	−0.33 (−1.77)*	−0.56 (−3.16)***
R&D activity			
Conducts R&D	1.44 (3.53)***	1.48 (5.11)***	0.97 (4.23)***
Complementary strategies			
Technology	0.47 (3.75)***	0.42 (4.73)***	0.14 (1.68)*
Marketing	0.17 (1.75)*	0.09 (1.28)	0.22 (2.93)***
Production	−0.22 (−2.45)**	−0.12 (−1.58)	0.10 (1.39)
Firm size			
100–499 employees	0.43 (2.27)**	0.19 (1.12)	0.32 (1.99)**
500 or more employees	1.44 (5.03)***	0.93 (4.46)***	0.64 (3.32)***
Competition			
6 to 20 competitors	0.16 (0.93)	0.16 (1.11)	0.42 (2.96)***
Over 20 competitors	−0.33 (−1.70)*	−0.37 (−2.29)***	0.01 (0.10)
Technological opportunity			
Collaborative agreements	0.01 (1.22)	0.01 (0.45)	0.01 (0.67)
SUMMARY STATISTICS			
N	899	967	1023
$Q_T(\theta): \chi^2$	13.7	8.0	14.9
R^2	0.38	0.39	0.20

Notes: t-statistics are in parentheses. For larger firms (IPs) only.
*** Significant at the 1% level.
** Significant at the 5% level.
* Significant at the 10% level.

14.4 CONCLUSION

Because Canada is a small, developed country with an open economy and a substantial amount of foreign investment, it might be expected to have a different innovation system than other countries. An innovation system is made up, on the one hand, of institutions that govern the way in which knowledge is created and disseminated and, on the other hand, of the firms that make up the economy.

This chapter has focused on various characteristics of the firms that are innovators. Several findings are of note:

First, while there is a close connection between innovation and the appropriability climate or patent use, the causal relationship is much stronger, going from innovation to the decision to use patents than from the use of patents to innovation. This extends the other findings based on

Table 14.7. *Estimated Probability of Introducing a World-First, Canada-First, or Other Innovation*

	World-First Innovation	Canada-First Innovation	Other Innovation
Intellectual Property Use			
Patent user	1	4	4
Nonuser	1	5	5
Firm activities			
R&D Activity			
Conducts R&D	2	7	6
No R&D	0	0	1
Complementary strategies			
Technology (mean)	1	4	4
+ standard deviation	2	9	5
− standard deviation	0	1	3
Marketing (mean)	1	4	4
+ standard deviation	1	4	6
− standard deviation	0	4	3
Production (mean)	1	4	4
+ standard deviation	0	4	4
− standard deviation	1	4	4
Firm characteristics			
Firm size			
0 to 99 employees	0	3	3
100 to 499 employees	1	3	5
500 or more employees	0	17	10
Industry characteristics			
Number of competitors			
0 to 5	1	5	3
6 to 20	1	5	7
Over 20	0	2	3
Technological opportunity			
Technological opportunity (mean)	1	4	4
+ standard deviation	1	4	4
− standard deviation	1	4	4

Note: For larger firms (IPs) only.

survey evidence (Levin et al., 1987; Mansfield, 1990; Cohen, Nelson, and Walsh, 1996), that patents are not seen by firms to be a very efficacious means of protecting innovations, even though they tend to be used once an innovation occurs.

Second, as has been found elsewhere, R&D is an important factor contributing to innovation. Firms that have an R&D capability have a higher probability of introducing an innovation, compared to those that do not. Nevertheless, conducting R&D is not a guarantee of success. It only increases the chance of some form of success to less than 33%.

Third, while developing an R&D emphasis is important, developing capabilities in a number of different areas is also generally a prerequisite for innovation. In particular, firms that give a stronger emphasis to technological capabilities and to marketing competencies are more likely to be innovators. The emphasis on technology is found for all three types of innovations – world-firsts, Canada-firsts, and other. But it is more important for world- and Canada-firsts. Marketing is found to be positively related to world-first and other innovations.

While technological and marketing capabilities are closely related to the probability that a firm will be an innovator, human resources capabilities are not and were not included in the analysis – a finding that is contrary to results found for new firms (Baldwin, 1998). The emphasis on strategies that were classified as involving the production process were found to be negatively related to innovation, particularly for world-first innovations.

Fourth, the two variables that are often used in testing the Schumpeterian hypothesis give mixed results. Size is positively related to innovation.[18] This relationship is important for both product and process innovations.

While larger firms are more likely to be innovators, a lack of competition is not positively related to innovation. Indeed, intermediate levels of competition are more closely associated with innovation than the lowest level of competition – though the relationship is nonlinear. That is, as the number of competitors increases, the probability of innovating first increases and then decreases. This effect is particularly evident for the least novel innovations. Competition matters more in the diffusion process than for the introduction of the most novel world-first innovations.

Fifth, it is noteworthy that foreign-controlled firms are not significantly more likely to innovate generally. This result does, of course, depend upon the inclusion of the competency and size variables. Thus, differences in

---

[18] We also experimented with continuous variables by using a quadratic term to capture nonlinearities. Essentially the same results were obtained.

the raw innovation rates that exist between foreign and domestic firms are accounted for by differences in competencies.[19]

Finally, the scientific regime is related to the rate of innovation, but only marginally. It also is positively related to both product-only, product/process, and process-only innovations, and it is closely related to both more novel innovations and to those innovations that are just being introduced into Canada for the first time. While this points to the importance of scientific infrastructure across all types of innovation, it should be recognized that the industry effect that this variable measures is weak and the significance is sensitive to the formulation chosen. What is not sensitive to the formulation chosen is the strength and sign of the technological competency proxy at the firm level. Firm effects are therefore stronger than industry effects in this dimension.

---

[19] That these competencies account for most of the differences between foreign and domestic firms increases our confidence that these variables capture most of the important firm-specific effects in the regression.

FIFTEEN

Summary

Innovation originates from the firms and the institutions that make up the economy's actors. Using a comprehensive survey of Canadian firms in the manufacturing sector, this book examines the variety of actors that contribute to the innovation process.

Studies of economic phenomena must necessarily reduce complex processes to their essence. At times, this simplification can obscure important facets of the process. Rosenberg (1976) has noted that studies of innovation have often focused on discontinuous events, rather than on the continuous improvement process that he felt was important. Innovation studies have also tended to focus heavily on the role of one type of knowledge, arising from scientific R&D laboratories, and have paid less attention to the importance of engineering and other forms of knowledge. Moreover, with this focus on R&D, the importance of bringing products to market, mustering the human resource requirements, and managing the process have been given less attention.

In this study, we have profiled a range of innovation types and the actors that participate in the innovation process. This has enabled us to place the R&D-centric knowledge creation process in context. We have shown that there are competing systems at work and that some act as complements, while others act as substitutes to R&D. We have also shown that the various parts of the innovation system fulfil different functions. And together they constitute the Canadian innovation system in the manufacturing sector.

The study has asked whether there are a small number of archetypical participants in the innovation process, each of which fulfils a different but specific function, with a recognizable set of characteristics. It does so by investigating how innovations differ across size of firm, across industries, and across firms of different nationalities. It also focuses on differences in the production process across these dimensions. It finds that the innovation output of firms differs substantially across these dimensions. And it explains the different role that each participant plays.

Of course, within this heterogeneous world, there are overall patterns that are common to all participants and sectors. In the first part of this chapter, we provide an overview of these commonalities. In the second part of the chapter, we focus on important differences.

15.1 INNOVATION TYPES

Innovation is best thought of as a continuous process, one whose characteristics often change over the length of the product life cycle. The development of an entirely new product often initiates the first stage. Throughout the latter stages, as products become more standardized, process innovations become more important. At this stage, much progress is incremental, consisting more frequently of process innovations. Nevertheless, when taken together, the incremental changes often have a significant effect on technological progress.

Partially because of this view of the relationship between different types of innovation and the life cycle, some have placed considerable emphasis on the difference between product and process innovation. For example, Cohen and Klepper (1996a) argue that the two have different scale economies.

Our analysis has shown that the dichotomy between product innovation on the one hand and process innovation on the other is imperfect. A substantial proportion of Canadian innovators indicated that they introduced a product innovation with a change in manufacturing technology. Thus, many innovations are more complex than the simple established typology suggests.

This study has also shown that innovations vary considerably in terms of their novelty. The most original world-first innovations are understandably rarer than innovations that introduce to Canada new products or processes created abroad. Only one of six major innovations was a world-first; one of three was a Canada-first. Slightly more than half of innovations were new to the reporting firm, but had already been introduced elsewhere in Canada. The innovation process, then, consists of a relatively small number of world-shattering breakthroughs, and a considerable number of products that have had improvements made to them, ranging from the incremental to the substantial. Studies that focus on paradigm-shifting breakthroughs may be fascinating, but they are not relevant to the majority of innovation that is taking place.

This study has also demonstrated that there is considerable heterogeneity in the innovation regime across industries. Technological opportunity varies among industries and so does innovation intensity. A large proportion of innovations is created in a few *core* industries (chemical, petroleum refining, electronics, machinery, and instruments). The more mature industries in the *secondary* (transportation, primary metals, fabricated metals, nonmetallic metal products, plastics, and rubber) and in the tertiary *other* sector (clothing, food, textiles, leather, wood, paper, and other industries) receive their new materials and intermediary inputs, as well as new equipment and machinery, from the core sector. They typically produce more standard, homogeneous products that compete on price, or they produce differentiated consumer products competing on brand name. Firms in the core sector are twice as likely to report innovations as firms in the tertiary other sector. Almost half of Canadian core-sector firms introduced an innovation, while little more than 25% of firms in the least innovative industries of the tertiary other sector did so. This difference emphasizes the importance of diffusion from the core set of industries to other sectors.

The study has also shown that there is considerable specialization of function across different actors. While every third firm declared that it innovated over a three-year period, this overall average conceals large differences. The smallest firms (fewer than 20 employees) report innovations at about one-half the rate of the largest ones (more than 2,000 employees). But small firms rely on large firms as part of an innovation network. This size effect is pervasive across most industries.

There is also a considerable difference between firms that are internationally oriented and those that just serve domestic markets. The rate of innovation by multinational firms (whether foreign- or domestic-owned internationally oriented firms) was considerably greater than for purely domestic firms. Exporters are more intensive innovators than are nonexporters. This is strongly suggestive of a two-tiered economy – one that engages in innovation pushed by the highly competitive international market and the other less-innovative one that operates in a protected domestic sector.

15.2 THE IMPACT OF INNOVATION

Innovations can be aimed at exploiting scale economies, or they can be directed primarily at improving the flexibility of the production process, by reducing the length of production runs required to exploit economies of scale, by allowing for the quick changeover of products, or by facilitating the more rapid customization of products.

We find that the innovation process in Canada had its greatest impact on a firm's ability to respond flexibly to customers' needs. Innovations increased production flexibility and sped up the response to changing customer requirements in more than half of all firms accounting for two-thirds of total industry employment. This confirms other information indicating that the adoption of advanced manufacturing information systems has increased the flexibility of production processes – from design and engineering, to automated control and inspection of materials, to assembly, and to final product inspection (Baldwin and Sabourin, 1995).

This finding indicates that the innovation regime in Canada has a special focus. Canada is a small, open economy with a relatively limited market. Its foreign trade policy has been aimed at reducing barriers faced by Canadian firms in accessing the large North American markets of the United States and Mexico. Although barriers to these markets have been reduced in the 1990s, they have not been eliminated. The finding that innovation was focused primarily on flexibility suggests that the small size of the market is still a concern.

It should be recognized that innovation also served to reduce unit costs through the exploitation of product line and plant scale economies. But these impacts were listed less frequently than improvements in flexibility. It is thus flexibility, rather than volume costs, that innovation in Canada primarily affects. This difference in relative emphasis was even greater in larger than in smaller firms.

Our work also shows that innovation has tended to increase the demand for white-collar, relative to blue-collar, workers – that innovation has been skill enhancing. Although innovations reduced unit labour costs in many firms, they were more frequently associated with increases than decreases in employment. Firms reporting increases in the employment of production workers outnumbered by three to one those firms where innovation led to a decline in employment. But the results show a much stronger positive impact on the employment of nonproduction workers. The latter, on average, receive higher wages and are more likely to be white-collar workers. Innovation, therefore, has enhanced the demand for the latter, rather than the former.

Some innovations are introduced as a response to government regulations. The analysis of the impact of government regulations on innovation and on firm profitability suggests that broad negative generalizations about the effect of regulation on innovation are risky. Innovations introduced in response to government regulations increased product quality, improved working conditions, and had other positive effects on average more often than innovations not created in response to regulatory requirements. They were also more frequently associated with an increased share of domestic and foreign markets and with increased profit margins. This was particularly the case for innovations in the core sector.

Finally, innovations were invariably associated with the penetration of foreign markets. Significant proportions of the sales from innovations were exported. The more original, world-first, and Canada-first innovations had a higher proportion of export sales than did the imitative innovations. The importance of export orientation is confirmed by the finding that while there are large differences in the innovation intensity of foreign-controlled and domestic-controlled firms, once the export orientation of a firm is considered, these differences become statistically insignificant.

15.3 THE INNOVATION PROCESS

15.3.1 Sources of Innovation

Innovation activity depends crucially on the firm's capability to create and acquire knowledge that not only leads to inventions but also brings inventions successfully to the marketplace. This capability rests both on a firm's talent for internal problem solving and on its ability to forge productive external linkages via networks, strategic alliances, and user-producer relationships. The process by which firms acquire and generate knowledge is at the heart of innovation activity.

We have focused on how internal sources of knowledge are combined with external sources in order to produce innovation. In particular, we investigated the extent to which R&D is central to the internal process and facilitates the use of external sources. Since ideas and technology sources tapped by innovating firms are to an important degree determined by technological opportunity, we explored the extent to which complexity of innovation and type of industry are associated with the use of R&D units by innovators in the Canadian manufacturing sector, and with spillovers from external sources like universities. We have asked whether the R&D-centric model of innovation is adequate or, more particularly, where it is most appropriate.

Answers to questions on the type of R&D operation allow us to better understand the commitment that firms have to research and development – whether they conduct R&D on an ongoing basis or just occasionally; whether they conduct R&D in a separate research and development group, conduct it in other departments, or contract it out; and whether research networks are an important instrument through which a firm's internal research capabilities are enhanced.

The results discussed in this study show that the innovation process is fed from multiple sources, some internal to the firm, others external to the business. Ideas for new and improved products and processes are generated in the course of market transactions with clients and suppliers, with related and unrelated firms, and by other external sources. Ideas for new market opportunities are seized and adapted to a firm's advantage by the management, research, marketing, and engineering personnel of the firm.

While there is a broad array of information sources used for innovation, research is cited as the most important source of new ideas after management. Management is the principal source of innovative ideas primarily in smaller firms. Large firms are more likely to use R&D along with other sources. The results also show that other sources are used to complement R&D. The sales department, which links firms to their customers, is the most frequently reported internal source of new ideas that is combined with R&D. While the R&D-centric model is a relevant depiction of one mode of innovative activity, it needs to be modified to include an outward sales orientation.

This study has also addressed several questions relating to the process of knowledge creation. To what extent does imperfect appropriation of the results of R&D and innovation by private firms create spillovers that are a potentially important source of ideas and technology for other innovating firms? How important are the specialized public institutions that provide technical information and services – infrastructure that is seen to be an indispensable element of a national innovation system? Are specialized private firms that provide technical services as important a source of innovative ideas as the public sector? Are interfirm links between affiliated firms the most important source of information that is internalized, or do the links between firms that arise in the form of normal commercial relations (i.e., between suppliers and customers) also serve as an important conduit for the diffusion of ideas that drive innovation?

Interactions between firms are classified into one of three groups in this study – those associated with commercial transactions, those that fall in the traditional unpriced group, or those with infrastructure facilities like universities. Contrary to the emphasis that is usually placed on the problem of unpriced spillovers, the survey evidence shows that interfirm market and market-related transactions dominate the innovation process.

The main external sources of innovative ideas are customers, related firms, and suppliers, in that order. Knowledge diffusion from outside sources occurs through commercial partnerships arising either from trade or ownership and partnership connections. Arrangements established with related firms and strategic partners provide a privileged access to new technology that is less burdened by asymmetric information problems than are third-party contracts with independent licensors, research firms, and consultants. These commercial transactions and partnerships serve to internalize the externality problem associated with knowledge diffusion.

This evidence supports Mowery's (1983b) view that the traditional focus of much of the innovation literature on appropriability is misplaced because it implicitly assumes that knowledge transfer is costless. Since it is not, and since interfirm transactions are so important to knowledge creation, it is the research agendas like those of Von Hippel (1988) or Teece (1977, 1986) that are relevant. Studies of diffusion need to focus on the distribution of the benefits of cooperative R&D between firms and the contractual mechanisms for overcoming interfirm barriers to knowledge transmission.

Our study has also described three clusters of firm types that combine external links with internal capabilities. The first two groups of firms rely on R&D. One builds networks with market partners. The other relies more on its own extensive resources and develops a capacity to ingest outside sources of knowledge by combining internal expertise in R&D with spillovers of outside knowledge derived from research institutions. A third cluster, an alternative to the R&D-based model, consists of those who focus on internal engineering and production expertise and combine this with knowledge spillovers derived from universities. Universities are an important part of the innovation process, in particular when it comes to supporting applied research.

Although partners provide an important source of ideas, spillovers that entail more traditional externalities from competitors and public research facilities are also important, with almost half of firms receiving information from these sources. The most frequently mentioned source of spillovers comes from competitors. But some of these spillovers are also derived from trade fairs, trade conferences, public and private research institutes, universities, and trade or patent literature. R&D and production engineering absorbs and adapts information from these sources in a firm's innovation process.

Once the idea for an innovation has been developed, its introduction to the production process still presents a technical challenge. Production engineering is an important internal source of technological know-how, especially for process innovations.

Overall, the majority of firms (60%) combine both internal and external technology sources. The outside sources of technical expertise needed for process innovation come most frequently free of charge from suppliers of machinery and/or inputs and constitute an important complementary source of innovative ideas, especially in the tertiary other sector.

In summary, the knowledge production process relies heavily, but not exclusively, on R&D. Only a small minority (8%) of larger innovating firms (employing more than 20 persons) introduced an innovation without using R&D in some form or other. In addition to R&D, outside sources of information are important to most innovators. Classical spillovers can be found in about half of the cases – with competitors providing most of the stimulus here. But the largest proportion of outside knowledge comes through market partnerships between customers and suppliers, as well as between firms and their affiliates.

15.3.2 The R&D Process

Canada does not lead most OECD countries in terms of R&D spending as a function of gross output. This is not because firms ignore the R&D function. About two-thirds of Canadian manufacturing firms report that they conducted R&D. The proportion for larger firms increases to 90%.

In contrast, about one in three firms were found to be introducing innovations. An economy where some 30% of firms innovate in any three-year period may be one where all firms attempt to innovate and only 30% succeed in any three-year period, or it may be one where only 30% try and most of these succeed.

Although the R&D statistics show that over two-thirds of firms were conducting some form of R&D, most of this activity is done only on an occasional basis. Only about one-quarter of firms conduct R&D on an ongoing basis. It is this form of R&D that is most closely associated with innovation and that is conducted by firms who regard themselves as being more competitive with regards to their R&D. Of those firms conducting ongoing R&D, only about 40% perform their R&D in a separate R&D department. Therefore, less than 10% of firms regularly perform R&D and do so in a dedicated facility. The R&D process, then, consists of a small core of firms conducting R&D regularly in a dedicated facility and a large group who do so only occasionally and most often in other departments of the firm. The Canadian economy is characterized by a relatively small group of intense innovators and of specialized R&D performers.

Research networks are extremely important. Some 22% contract out R&D and another 14% have collaborative agreements with other firms. Indeed, the percentage of firms that have either of these outside links is just as high as those with a separate R&D department, thereby confirming observations of both Mowery and Rosenberg (1989) and Niosi (1995b) as to the importance of collaborative research. Universities are particularly important partners in this regard.

Chapter 4 on the sources of information indicated that other activities were combined with R&D facilities, sometimes as complements, sometimes as alternatives. In Chapter 14, which used a multivariate framework to analyze the factors that are associated with innovation, we found that being a continuous R&D performer, rather than doing no R&D, increased the probability of a firm's reporting an innovation to about 33% from 1%. This is an increase of about 30 percentage points. Performing R&D occasionally has only about a 10 percentage point effect. But placing a heavy emphasis on technology (the production and engineering side of the firm) increases the probability of being an innovator by 15 percentage points. Placing more emphasis on marketing increases the probability of success by about 20 percentage points. R&D is, therefore, an important factor behind success; but it is not the only factor that is associated with being a successful innovator.

15.3.3 Technology Acquisition

A focus on technology is important for innovation. Technological progress results when the information embodied in new and improved products and processes is diffused across firms, industries, and national economies. This diffusion is embodied mainly in products that flow from one firm to another. Much less frequent is the arm's-length sale of the innovation itself. Owing to difficulties associated with writing contracts that diffuse the information embedded in new products and processes, only a small proportion of firms (less than 2%) reported an arrangement where the disembodied innovation itself – that is, the idea, information, and technical know-how – was sold outright to other firms. For the small number of innovations that are sold outright, the intersectoral diffusion follows the flows of technology that others have found – from the core to the secondary to the tertiary other sector.

While only a few firms reported sales of disembodied innovations, the diffusion of technology does occur in other ways. About 60% of firms acquire some form of outside technical knowledge during their innovation process, and a good 40% of these firms obtain their technology via a licensing agreement.

A licensing agreement can be used either between arm's-length parties or between related members of a large firm. The evidence shows that these licensing arrangements differ dramatically between related and unrelated partners. Problems in negotiating the terms of these agreements in the face of rapidly changing technologies make them more appropriate for transactions between related partners who have a greater ability to adapt to changing circumstances. The evidence shows that related partners are more likely to transfer technical knowledge through contracts that allow for continuous exchange, while unrelated partners are more likely to rely on one-time transfers. This suggests that unrelated partners have more difficulty than foreign affiliates in obtaining licenses for the most recent technologies, which are less likely to be transferred by one-time transfers.

Technical transfers through licensing arrangements are used more frequently in situations where a firm's innovation does not originate in R&D. These transfers are often concluded by foreign affiliates to support the adaptation of innovations that have already been developed elsewhere in the world.

Joint ventures and strategic alliances also support the creation and transfer of knowledge needed for innovation, but in doing so they extend the boundaries of the firm as they internalize the externalities associated with knowledge creation. These contractual arrangements bring together the resources of several companies to create and transfer the technology that each firm needs for innovation. Joint ventures are utilized by about 10% of firms that successfully innovated. Contrary to the technology transfer process, which acts to a certain extent as a substitute for a firm's own research and development, joint ventures and research partnerships are directly associated with R&D activity. They are less frequent where the engineering department is the primary source of ideas for innovation. Thus, joint ventures and licensing of technology are substitutes, supporting competing models of innovation.

15.3.4 Intellectual Property Rights

Intellectual property rights (IPRs) are meant to protect the investments in knowledge creation and acquisition. The importance of IPRs may be determined by the extent to which they are owned and employed, and by whether firms rate them as effective. Both issues were examined here.

Several broad forms of protection for intellectual property are supported by the state – patents, copyrights, trademarks, industrial designs, and trade secrets. Only about one-quarter of the population of manufacturing enterprises make use of at least one of these forms of protection. Only about 7% specifically use patents. Since the use of all statutory mechanisms of intellectual property requires special expertise and is costly, their use increases with the size of the using entities.

Being innovative is a primary determinant of the use of intellectual property protection. There are substantial differences in the use of trademarks, patents, trade secrets, industrial designs, and copyrights between those who had just innovated in the three preceding years and those who had not. For example, the proportion of firms that were innovative over the three-year period preceding the survey and that possessed a patent was about 14%, but only 2% of firms that had not innovated over these three years did so. The size of this difference suggests that the non-innovating population possess relatively little of the knowledge-based capital that would be based on past activities. This, in turn, suggests that firms in the innovative group have not supplanted a previous set of innovators; rather, they have been the leaders in the development of knowledge and users of intellectual property for some time.

While being innovative is a prerequisite for the need for protection, not all forms of statutory protection are sought equally by innovative firms. When the effect of being innovative is separated from the effect of size, nationality, and industry, being innovative has its largest effect on the use of patents and trademarks. However, large and significant effects are also found on the use of industrial designs, trade secrets, and copyrights. Innovative firms, then, concentrate on patents but also use a wide range of other statutory forms of protection.

Previous work (Mansfield, 1986; Levin et al., 1987) has suggested that many firms do not find the IPR system very effective in preventing competitors from bringing copies of their innovations to market. This study confirms that these findings also apply to Canada. Firms tend to value alternate strategies more highly than the statutory forms of protection. Moreover, the population as a whole ranks such strategies as patent protection to be less than 'effective'. However, these rankings depend very much on the characteristics of a firm. If a firm is innovative, large, foreign owned, and in one of those core industries that tend to produce more innovations, the score given to statutory forms of protection like patents increases greatly. On average, these users of patents find them to be effective.

This points to a world where some firms choose to be innovators and develop a set of competencies that complement innovation – including a way to protect their intellectual property – and other firms choose not to lead the way with innovation. Firms that are generally innovative develop competencies and a strategic stance that engenders the use of protection – probably because a learning process is required before a firm understands how to exploit and protect its advantage. Invariably, users of intellectual property were more positive in their view of the effectiveness of the various forms of protection. These consistent differences suggest that intellectual property use – like any other strategy – involves acquired skills that only develop with practice. As firms innovate, they learn which strategies best protect their knowledge assets. The study also suggests that these skills, in that they are associated with size, are part of the growth experience and tend to increase as a firm successfully masters a range of strategies and grows. The evidence that the use of statutory IPRs, and particularly patents, increases with the size of firm is consistent with the contention that the costs of obtaining, maintaining, and enforcing an IPR are a heavier burden for small and medium-sized firms.

Our multivariate analysis shows that while there is a close connection between innovation and the appropriability climate or patent use, the causal relationship is much stronger going from innovation to the decision to use patents than from the use of patents to innovation. This extends the findings, based on survey evidence (Levin et al., 1987; Mansfield, 1990; Cohen et al., 2000), that patents are not seen by firms to be a very efficacious means of protecting innovations, even though they tend to be used once an innovation occurs.

15.3.5 Financing

The knowledge associated with innovation creates an asset that arises from specific activities like R&D. Knowledge investments require financing. Since financing of these soft assets associated with knowledge creation is difficult, governments tend to subsidize R&D spending.

Some, but not all, innovation-related investments (e.g., R&D) are encouraged by public subsidies. Setting R&D in context allows us to evaluate the extent to which subsidies for different activities facilitate innovation. The study shows that basic and applied research activities account for only about 17% of total innovation expenditures. Therefore, if subsidies are directed only at these areas, they will have a relatively small impact on total innovation costs.

Financing innovation offers particular challenges because the success of an innovation program is difficult to evaluate and the assets associated with knowledge-based investments do not provide the type of collateral that lenders generally require. Confirming these problems, we find that innovative firms are frequently forced to rely on internal sources of funds for investments that create innovations. More than two-thirds of all firms finance innovation entirely from internal sources.

In general, firms that are able to find outside funding (especially those that are Canadian controlled) rely more on the federal government than on other external sources. Financial institutions, provincial governments, and related firms follow by order of importance. Venture capital is generally unimportant as a source of funds, except in the core sector.

During the 1980s, tax credits supplanted R&D subsidies as the most important form of government support for R&D and innovation. Canada has one of the most generous programs of tax credits for R&D of all industrialized countries. Nevertheless, while almost two-thirds of firms performed R&D, only about one of seven of them claimed tax credits. Larger firms tended to make greater use of this program, as did firms in the more R&D-intensive sectors. This is not simply because R&D intensity differs across sectors. For example while 95% of electronics firms perform R&D, 55% of them claim tax credits; in contrast, while more than half of firms in leather and clothing and in printing and publishing industries perform R&D, less than 7% claim tax credits. Thus, the interindustry distribution of tax credit claims is not proportional to the use of R&D by sector. Firms in less R&D-intensive sectors either conduct R&D that is not eligible for subsidies or are less aware of the existence of government programs, or their lower R&D expenditures do not outweigh the costs of making claims under the tax-credit program. A similar argument can also be made about the small and medium-sized firms who make less use of the tax credit program than larger firms.

15.3.6 Complementary Strategies for Innovative Firms

Innovative firms are not those that serendipitously stumble across inventions. Innovators differ from non-innovators in that they adopt a purposive stance to find new products and to adopt new processes. The Canadian manufacturing sector is not a world where most firms are engaged in intensive innovative activity, where some are rewarded by chance and others are not. It is a world that divides into firms that heavily stress an innovative strategy and those that do not.

Strategic emphases differ considerably between innovative firms and non-innovative firms. Not surprisingly, innovators place greater weight on innovation-related strategies – R&D, technology, production processes, and aggressive marketing of new products in new markets. In addition to developing new products and penetrating new markets, they are more active in developing and refining technology.

The results of this study confirm other findings (Baldwin and Johnson, 1996b, 1999; Gellatly, 1999) that innovative firms also place greater emphasis on a broad range of strategies outside of scientific endeavours than do non-innovative firms. Innovative firms in the Canadian manufacturing sector are more aggressive in seeking out new markets for new products. Innovation also requires highly skilled workers. The greatest impediment to innovation that is perceived by firms in the manufacturing sector is a shortage of skills. In response, innovators are more likely to stress acquiring skills and developing them within their firm.

All of this indicates that innovative firms must master a set of complementary skills. While their competitive advantage lies in innovation-related areas such as spending on R&D, elsewhere (in marketing, in human resources, in finance), they give greater emphasis to these areas than do non-innovators. An innovator, then, develops a balanced approach that makes it a compleat firm.

This is confirmed in the multivariate analysis of the factors associated with being innovative. While an R&D emphasis is important, developing capabilities in a number of different areas is also generally a prerequisite for innovation. In particular, firms that give a stronger emphasis to technological capabilities and to marketing competencies are more likely to be innovators. The importance of skills in marketing and production confirms that innovators need to develop a capacity for processing information and to adapt to ever-changing market conditions.

15.4 HETEROGENEITY

The previous sections of this chapter have stressed several of the major findings – related to the nature of the inputs that the Canadian innovation process uses, and how it adapts to various problems that arise because of the specific problems associated with the knowledge creation process. We have shown that the various actors adopt measures to internalize the problems of diffusion, that they form networks to internalize research externalities, that they transfer technology via internal firm mechanisms, and that they develop a set of skills outside the scientific area, which are at times complements to R&D and at times substitutes. While they take advantage of state-supported intellectual property rights, they make use of their own strategies to protect themselves and feel that these strategies are at least equal to the formal rights associated with patents and trademarks.

But the study has also found that there is a substantial heterogeneity in the industry environment, between small and large firms and between foreign and domestic firms. In this section, we examine how our general conclusions vary across these types.

15.4.1 Differences in Innovation Regimes Across Industries

Others have found differences in the innovation regime across industries. Crépon, Duguet, and Kabla (1996) report that innovation intensity and type vary substantially across industries in France – much more so than across categories that depend upon size or competition, the factors that are so often stressed in the Schumpeterian literature. For Holland, Brouwer and Kleinknecht (1996) also stress that sector differences are related to differences in structure. Firms operating in sectors with a relatively large small-business presence tend to have a higher share of imitative innovations. In the United States, the innovation performance of small and large firms varies significantly from one industrial sector to another (Acs and Audretsch, 1988).

The Canadian innovation system, like that of both the United States and the United Kingdom, possesses distinct differences across the upstream and downstream sectors in the innovation chain – the core, secondary, and other sectors. This reflects the pattern of technological opportunity, which is decreasing from the core to the other sector. Firms in the core sector are more likely to have reported a major innovation. The average number of innovations per firm is higher in the core and secondary sectors than in the tertiary other sector.

There are substantial differences in the product/process focus across industry groups. Like Arvanitis and Hollenstein (1994), we have found that product and process innovations are affected by different factors. The core group of industries focuses relatively more on product than on process innovations. The secondary and tertiary other sectors typically compete more on price and downplay product innovation, relative to process innovation. This accords with the pattern described by Robson et al.

(1988) for the United States and the U.K. Nevertheless, it should be noted that the differences across sectors in the intensity of pure product innovation are greater than the differences in pure process innovation. Process technology is needed everywhere, regardless of whether product innovations are being produced for others or being ingested from elsewhere.

Nevertheless, this study found important differences not so much between those innovations that were just products, or just processes, but rather between these types of innovations and those innovations combining both product and process change. And here all of the same factors that shape whether a firm is innovative are at work – but they have a more powerful effect. For example, conducting R&D increases the probability of being just a product innovator by about 20 percentage points; it increases the probability of being both a product and a process innovator by almost 40 percentage points.

The proportion of innovations in the core sector that are world-first is about double that of the secondary and other sectors. The differences across the three sectors with regard to Canada-first innovations are rather small. But the pattern is reversed for imitative other innovations, which are more important in the sectors depending on technology developed in the upstream sectors. These differences have a dramatic effect on various parts of the innovation process – from the sources of innovation, to the types of firms required, to the production process, to the use of intellectual property rights.

The first notable sector difference lies in the importance of R&D. In general, the higher the industry is in the technology diffusion chain, the larger the firm, and the more original the innovation is, the more frequent and important the research and development activity of the firm as a source of innovation. On the other hand, firms in the more mature secondary and, above all, in the tertiary other sector receive their inspirations for innovation more frequently from production engineering or from management than from R&D.

R&D partnerships and agreements are most concentrated in the core sector and least in the tertiary other sector. In contrast, while joint ventures tend to be supported by, or are related to, collaborative research, they are not more concentrated in the most R&D-intensive core sector. Indeed, they are quite evenly distributed across sectors. The reason is that only the smaller firms in the core sector make very intensive use of joint ventures; large firms in the core sector do not. This demonstrates substitutability across instruments – this time across firm size classes.

While R&D is more important in the core sector, so too are external linkages. Both of the clusters involving R&D – the one that combines R&D with customers, the other with external R&D partners – can be found more frequently in the sectors that are higher up the innovation chain. This relationship is stronger in the core sector, because firms interact downstream with customers in other sectors and because of the greater use, in this sector, of related firms as sources of outside ideas. Other industries are less R&D intensive and tend to collaborate with their suppliers upstream in the core industries.

The importance of external spillovers from competitors and what we have referred to as the publicly available infrastructure (trade fairs, publications) increases with the distance from the core sector. In contrast, the R&D-oriented external sources (universities, industrial research labs) are used more frequently in the core sector and by those creating world-firsts. Spillovers from easily accessible public sources of knowledge are of relatively little importance in core-sector industries. The limited contribution of spillovers to innovation in these industries should be a warning against overestimating the contribution that these sources make to economic growth.

The finding that the intensity of recourse to market-related external sources of technology is inversely proportional, and that resort to outside research-oriented sources is directly proportional to the technological complexity of the sector, is compatible with the finding of Mowery and Rosenberg (1989), who argue that internal and external sources of knowledge are used in quite different situations. Their analysis of independent research laboratories in the United States indicated that these facilities specialize in routine research tasks, while innovations in technologically advanced industries require specific technical knowledge and often a close interaction with various corporate functions.

There are also significant differences in the sources of new technology across sectors. In keeping with the greater research orientation of the core sector, firms in this sector more frequently report experimental development and research departments as sources of technology solutions than they do production engineering in the case of product innovations. Production engineering is more important in the downstream sectors that ingest innovation into their own production processes in an innovative way.

Technology transfers through licensing arrangements are used as a substitute source of technical expertise to firms' own R&D. These transfers are more frequent in the core sector than elsewhere. They are the means by which subsidiaries of multinationals bring technology into Canada. But the tendency to use these arrangements is about the same across all firm size classes.

Differences in the originality of innovations and a greater emphasis on product innovations mean that the core sector places greater emphasis on statutory intellectual property rights. Thus, we confirm the findings of others (Taylor and Silbertson, 1973; Levin et al., 1987) that the industry environment affects the use that is made of intellectual property. Cross-industry differences in intellectual property usage in Canada are closely related to differences in innovativeness that have been described by Robson et al. (1988). Firms in core sector industries make greater use of almost all forms of statutory protection, particularly patents and trademarks.

Throughout the study, we have been careful to stress that the core sector should not be regarded as the most 'innovative' or the one that has the most difficult task to resolve before introducing innovations. Certainly, there is little evidence to suggest that firms in the core sector are any more likely to report most of the benefits that are associated with innovation. For example, firms operating in the tertiary other sector reported that innovation increased profitability just as often as did firms in the core and secondary sectors. And firms that have to find new ways to implement new equipment may face equally difficult problems as firms in the core sector.

In spite of innovation being just as profitable in the other sector, firms in this sector have more difficulties financing their innovations than do firms elsewhere. We found that the proportion of innovators relying exclusively on internal financing is highest in the tertiary other sector and lowest in the core sector. Less original innovations are internally financed more often than the more radical ones. This suggests that financial markets are less capable of evaluating innovators in the least innovative sectors. Innovative opportunities are most difficult to spot in these sectors, and the development of financing specialists in these sectors may therefore have lagged behind other sectors.

There are other substantial differences in the workings of the financial markets serving the innovation-creating and the innovation-using industries. Government programs are used more frequently in the upstream than in the downstream sectors. Firms in the core sector focus more of their innovation expenditures on R&D, which means that they would naturally benefit more from subsidy programs that support R&D. But they tend to make greater use of tax-credit programs than would be expected from their greater orientation to R&D. In addition to or because of the difference in the use of tax-credit programs, firms in this sector rely more on government for direct grants for funding of innovations. In contrast, firms in other sectors have to rely more on their own internal funding. They receive less outside funding, and, if they do receive it, it is less likely to come from government sources, and they make less use of tax credits.

15.4.2 Differences in Innovation Regimes Across Firm Size Classes

Differences in the innovation regimes of small firms have been the focus of several studies. Malerba (1993) outlines the differences in the innovation regime across firm size classes in the Italian innovation system. Two major players are identified – a small-firms network and a core R&D system. The small-firms network consists of a large population of small and medium-sized firms that interact intensively at the local level. The core R&D system is composed of large firms with industrial laboratories, public institutions, and universities. The small-firm network, which depends on capabilities developed through production experience, has shown itself capable of adopting technologies quickly. It innovates incrementally. These firms are characterized by a highly dynamic atomistic learning network, with advanced capabilities for absorbing, adapting, improving, and tailoring new technologies that are developed externally in order to meet specific market needs. Engineering skills, product know-how, and understanding customer requirements are the major sources of incremental innovations and product customizations.

A similar distinction has been drawn for Switzerland by Arvanitis and Hollenstein (1996). In Switzerland, small firms have been found to rely more on favourable demand conditions than on the technological opportunity in an industry – probably because their sources of financing are more limited and cash flow is more important. Small firms are less likely to use patents and rely more on other protection strategies – such as secrecy, lead time, and protection from complexity. Finally, technological opportunities in the industry tend to be more important for large than for small firms, while the latter rely more on existing public networks – trade journals, conferences, and suppliers. Small firms generate continuous improvements using external knowledge and protect themselves relatively more by secrecy and by first-mover advantages. Large firms operate more in international markets and depend more on their own internal capabilities for protection than on the formal intellectual property system or on external knowledge from science (whose successful absorption requires internal R&D capabilities).

The effect of innovation on employment in Germany depends both on the size of firm and on the type of innovation. Licht (1997) shows that employment growth is larger in small-firm innovators than in large firms. Baldwin, Diverty, and Sabourin (1995) find that when firms adopt new advanced technologies in Canada, employment increases in small firms but not in larger firms. Of course, some of these size differences may result from firms specializing in different types of innovation outputs. Small firms are more likely to be in the early stages of the life cycle and to focus on product, rather than process, innovation. Firms that are process innovators are more likely to have less employment growth than are pure product innovators – because process innovation is more likely to be labour saving.

The debate over the appropriate function of government policy for small firms, in general, and R&D subsidies, in particular, brings into focus the different roles that are played by large and small firms in the innovation process. This study demonstrates that in Canada, small firms are less likely than large firms to introduce new products and processes. Similar differences are observed between large and small firms with regard to the frequency of R&D activity. Small firms are less likely to engage in R&D, just as they are less likely to innovate.

While there are fewer innovators in the smaller size classes, those smaller firms that do innovate resemble larger innovators in several dimensions. There are few differences in the number of product innovations per innovative firm across size classes. The innovation cost structure is much the same. In addition, major product innovations in small innovative firms account for just as large a percentage of their sales as in larger innovative firms. There is, however, a tendency for smaller firms to produce fewer innovations that are extremely novel.

Large and small innovators do, however, rely on different sources of ideas for innovation. Large firms use R&D much more frequently than do small firms. Large firms are also more likely to rely on an external research network, through a relationship with a related firm, or to engage in collaborative research with other firms. On the other hand, small firms exhibit the same flexibility in their R&D that they show in many of their other operations. They are more likely to conduct R&D only occasionally, when the opportunity or need arises. They rely relatively less on the R&D

department and relatively more on the technical capabilities of their production departments than do large firms. They are much less likely to take advantage of R&D tax subsidies. There is, therefore, more of a difference in the way that R&D is conducted in small firms than whether it is conducted at all. Large firms have regularized the R&D process in order to shape their environment, whereas small firms use it to exploit opportunities in their environment when the need arises.

Many of these opportunities are brought to the attention of small firms by their customers or suppliers. Small firms also depend on networks – but these networks rely more heavily than do those of large firms on customers and their marketing departments for innovations. These differences suggest that large firms, as a whole, have mastered the need to create and acquire noncodifiable information, either by themselves or as part of a research network. Small firms, as a whole, utilize information sources that are more codifiable and easily transmitted through supplier relationships. The information networks of small firms are different than those of larger firms.

When the efficacy of the R&D process is compared across small and large firms, large firms conducting R&D are more likely to introduce innovations. This is true for firms conducting both ongoing and occasional R&D. Small firms, however, are more likely to conduct occasional R&D than are large ones, and occasional R&D has a lower success rate than ongoing R&D. For these two reasons, small firms employing fewer than 500 persons have a lower innovation rate than do large firms.

Small firms, then, can be divided into two groups. The first group consists of firms that resemble large firms in that they perform R&D and generate new products and processes primarily through their own efforts. The second group are those who rely upon management for internal sources of ideas and customers and suppliers for their external sources of ideas for innovation. Large firms, by way of contrast, tend to rely more heavily on R&D. While they too rely on networks for ideas, their networks focus more heavily on relationships with related firms.

Evidence shows that despite differences in R&D intensity, the success of small firms depends critically on their innovative capabilities – especially in the areas of R&D. Small firms also benefit from the R&D done in large firms because a larger proportion of their innovations come from interactions with their customers. All of this means that subsidies for R&D directly aid the most dynamic small firms that are conducting R&D and also aid other small firms indirectly because of the spillovers from large firms to small firms.

Despite the importance of R&D for innovation, there are other areas where small firms indicate that they are more likely to face impediments to innovation and, therefore, where public policy might develop special policies for small firms. Small firms perceive that externalities are relatively important in the area of information about technologies, market potential, and technical services. They also perceive that there are significant barriers to interfirm cooperation. This is a particularly serious problem since this is the method that they use most frequently for developing new ideas for innovation.

Small firms may also face barriers on the intellectual property front. They are less likely to make use of the intellectual property rights that are meant to facilitate the protection of intellectual ideas, and when they do so, to emphasize different forms of intellectual property protection. Over 60% of large firms protect themselves with any one of the statutory rights; less than 30% of those with fewer than 100 employees do so. Part, but not all, of this difference is accounted for by different tendencies to innovate. But even when these differences are taken into account, small innovative firms make less use of the statutory forms of protection than do large firms. Of those large firms reporting sales from a product innovation, some 80% possess one of the statutory rights; less than 40% of those with fewer than 100 employees who have recently introduced a major product innovation do so. When they do make use of intellectual property rights, small firms use trade secrets more frequently, relative to patents, than do large firms. According to evidence from the United States, this is primarily due to the complexity and the cost of patent filing (see Lerner, 1994, and Cohen et al., 2000). When other differences between small and large firms, such as innovativeness, nationality, and industry of activity, are considered, size has the greatest impact on patent and trademark use.

Financing also presents different problems to small firms. Small firms are more likely to have to finance their innovation investments exclusively from internal sources. Of those firms that receive outside funding, small firms are more likely to receive government funding. But smaller innovators are less likely to make use of government tax credits for R&D.

In summary, the effects of firm size on the propensity of a firm to innovate are pervasive. We have confirmed Malerba's (1993) finding for Italy that there is a considerable difference between the manner in which small and large firms innovate. There are major differences in the approach of small and large firms, in that small firms are less R&D focused and less export intensive. In Canada, there is a considerable difference in the rate of innovation by size class, whereas in Italy, there is very little difference in the innovation rates of small and large firms (Consiglio Nazionale delle Recherehe, 1993).

It is important to note that neither finding is incompatible with Acs and Audretsch's (1990) finding that the innovations per employee in small firms are equal to or larger than those for large firms. Small firms may be less likely to innovate, but their relative size is so much smaller that their innovation rate per employee is greater than for large firms.

While the firm-size effect is related to differences in the degree to which firms are engaged in innovative activities like R&D, it is not completely explained by it. In the multivariate analysis that examined the propensity to innovate and that accounted for whether firms were R&D performers, whether they were foreign-owned, or whether they had developed a wide range of competencies, the effect of firm size on the probability of being an innovator is always found to be positive and highly significant. For example, while performing continuous R&D adds some 30 percentage points to the probability of introducing an innovation, being in the largest size class rather than one of the smallest size classes (over 500 employees instead of fewer than 100 employees) adds 30 percentage points to the probability of reporting an innovation.

We have also shed light on a related issue in the Schumpeterian debate. We find that the degree of competition affects the probability of innovation. The degree of competition has the type of nonmonotonic U-shaped effect that Kamien and Schwartz (1985) reported. Moving from very oligopolistic conditions to mildly oligopolistic markets increases innovation, but moving to even more atomistic markets, then, reduces the probability of innovation. Despite this finding, the size of a firm has a much more important effect on the probability of innovating than does the competitive environment of an industry.

Size is a proxy for other differences between small and large firms that have not been measured in this study. There are many other differences between small and large firms than the measurable characteristics used in our multivariate analysis – R&D activity, marketing, and human resource skills. While we have extended the bounds of the analysis that examines the covariates associated with innovation by considering both R&D and other competencies, like human resource emphasis or marketing skills, we cannot purport to have fully covered all the competencies that matter.

It is also possible that size represents a different effect – that arising from recent growth – since size is invariably correlated with recent growth rates. Growing firms are more likely to generate higher cash flow, which this study and other research (Baldwin and Johnson, 1999b) have shown is required for financing investments in innovation. In other research, we have found that the probability of adopting new technologies is strongly related both to a measure of the initial firm size in an earlier period and to growth rates over the intervening period (Baldwin and Diverty, 1995). This suggests that the positive relationship between innovation and size described here is partially the result of this phenomenon.

15.4.3 Differences in Innovation Regimes Across Novelty Types

The development of such major inventions as steam engines, electric power, or the transistor often has dramatic, visible effects on the economic system. Other types of innovations are more incremental in nature and receive less attention, but their cumulative effect is often large and decisive for the success of major innovations. Indeed, measured productivity growth is so continuous that this suggests that the diffusion process is slow and more or less continuous.

In this study, we have followed Rosenberg's (1976) admonition to focus on both frontier innovations and more incremental innovations, finding that most innovation is of the imitative type. More importantly, we have outlined the differences in the regimes that govern the production process in each case.

Our survey has provided information on the most profitable innovation introduced by each firm. To assess different types of innovations, we contrasted world-first innovations with all other innovations. The latter group includes Canada-first and other innovations. Quantitatively, these non-world-first innovations are introduced more frequently than are world-firsts.

The main difference between world-first innovators and all other innovators lies in the extent to which world-firsts more frequently combine product and process innovation. Novelty is associated with more complexity, in that the production of the new or improved product also requires a change in the production process.

On the input side of the innovation process, world-first innovators tend to rely more frequently on research and development departments than on any of the other internal sources. They also are less likely to use external sources of innovative ideas. In short, they rely more on their own resources. In contrast, production engineering is an important internal source of technological know-how, especially for the less original, Canada-first, and other innovations. In this respect, we have confirmed the finding of Arvanitis and Hollenstein (1996) that incremental innovations are oriented towards the rapid application of existing technology.

The world-firsts are more likely to use outside groups that complement their research and development facilities – related firms, industrial research firms, and universities – than are non-world-first firms. Non-world-firsts are more likely to use the services of consultants, along with information garnered from publications.

In the area of the demand for labour, world-firsts are more likely to increase the demand for workers than are non-world-firsts, probably because the latter are at the stage of the product life cycle where process innovations are having a greater impact on unit costs and labour requirements. Both groups demand a more highly skilled workforce, but again, the world-firsts do so more frequently than their less original counterparts.

Public policy is directed at supporting the innovative process in a number of different ways. Policies are aimed at reducing impediments in several areas where markets are seen to have imperfections. The areas that give innovators the greatest difficulty are the lack of skilled personnel, the lack of market information, and government standards and regulations. In each of these cases, world-first innovators generally reported that they experience these problems more frequently than do non-world-first innovators. This suggests that these problems do not block innovation as much as they accompany more radical innovative efforts.

In summary, the Canadian innovation system produces a variety of innovations, from world leaders to the incremental changes that result from the general diffusion of knowledge about new production techniques. These different innovations have similar beneficial effects in terms of improving a firm's market share or profit margin. They are, however, the products of different innovation systems.

15.4.4 Differences in Innovation Regimes by Nationality

Since the Canadian manufacturing sector has more than half of its output under foreign control, its performance is affected by the multinationals operating in Canada. At issue is the extent to which multinationals operating in Canada have a truncated local capacity to innovate.

Foreign ownership in the manufacturing sector is highest in industries where multinationals can exploit their proprietary technological, marketing, and R&D assets. We find that even though they have a privileged access to their parents' R&D and technology, foreign subsidiaries perform R&D in Canada more often than do domestically owned firms.

This pattern is particularly significant in the technologically less intensive secondary, and above all, in the tertiary other sector. Foreign firms resemble one another across sectors far more than do domestic firms. Foreign affiliates are also more likely to complement their own research and development activity by participating in joint ventures and R&D partnerships.

The tendency of multinationals to be more likely to conduct R&D is found even when determinants of R&D activity, such as the size of the firm, the sector of activity, and various indicators of the competitive environment, are taken into consideration. Multinational firms not only exploit their proprietary advantages in Canada but also increasingly develop their own innovative initiatives and tap local sources of technology and scientific research. Their involvement in R&D partnerships and contacts with customers, suppliers, and unrelated firms indicates that there is no evidence to support the once-popular argument that foreign firms do not develop links in Canada and thus create a truncated corporate structure in Canadian manufacturing. Moreover, they were just as likely as domestic firms to develop collaborative relationships with Canadian universities.

Access to innovative ideas and technological expertise of their parent and sister companies confers a significant advantage on foreign affiliates over Canadian-owned firms. But even those multinationals who reported that this was an important source of innovative ideas indicated that their local R&D was an important source of innovation in their firm.

Overall, an increasing number of foreign subsidiaries pursue R&D as a part of a learning process, independent of, and often in competition with, their parent or sister companies. Learning relies on a close relationship with discerning customers and advanced research partners. In this respect, the Canadian situation is not different from developments observed elsewhere. Multinationals are increasingly moving their R&D abroad not only to transfer and exploit their own knowledge but also to absorb science and technology developed abroad (Niosi, 1995a).

Our analysis also suggests that there are two types of foreign subsidiaries in Canada. About two-thirds are relatively independent firms that do not rely heavily on their parent or sister firms for innovation; the remaining third do so. The technologically more independent foreign affiliates rely more on their own R&D and sales and marketing in Canada. They are also closer to their customers. The more dependent foreign affiliates are closer to the traditional image of a branch plant implementing ideas received from parent companies.

Foreign-owned firms innovate in all sectors more frequently than their Canadian counterparts. This is true for most size categories. They especially surpass domestic firms in industries most likely to be the recipients of new products or technologies developed in upstream core industries. Here foreign-owned firms do research more frequently than domestic firms do. The most significant laggards are the largest Canadian-owned firms belonging to the tertiary other sector.

Most of these differences are accounted for by competencies in R&D and technology. In the multivariate analysis that examines the characteristics of innovators, the difference between foreign and domestic firms shows up weakly only in the case of process innovations when factors like R&D activity, emphasis on technology, and size class differences are considered as explanatory variables. This, of course, does not mean that there are no differences between foreign- and domestic-owned firms in terms of innovation. It does tell us that these differences are related to differences in R&D intensity and in other competencies that separate small from large firms in general.

In keeping with their more frequent involvement in R&D, foreign-owned firms surpass Canadian firms in introducing the more original world-first and Canada-first innovations. Consequently, they also make more use of instruments of intellectual property than do Canadian-owned companies. They also have the advantage of arranging for the continuous transfer of technology from foreign affiliates.

Foreign-owned innovating firms are more export oriented than are domestic firms. Moreover, they export innovations more intensively than domestic firms do. Innovations introduced by foreign affiliates help them export and increase their share of foreign markets more often than they do for domestic firms. Innovations introduced by domestic firms led more often to increased share of the local market.

After controlling for other characteristics, this study finds that foreign-owned firms are not likely to use statutory intellectual property rights more frequently (with the exception of patents), but they perceive that almost all such property rights are more effective than domestic firms do. Nationality has the greatest impact on the firm's evaluation of the effectiveness of patents and trade secrets in preventing competitors from bringing to market copies of the innovation. In comparison with Canadian firms, foreign firms perceive almost all IP instruments as being more effective.

The evidence also suggests that domestic firms face more problems in financing. The proportion of Canadian firms that finance their innovations internally is higher than that for foreign firms. Owing to financial support from their parent and sister companies, foreign affiliates in the group that rely on some form of external financing depend much less on internal financing than do comparable Canadian–owned firms. This group of Canadian-owned firms relied more on government funding.

All of these comparisons to domestic firms as a whole provide evidence that multinational firms do not operate subsidiaries in Canada that are truncated relative to Canadian firms in general. They also show that the multinationals contribute significantly to innovation and technological progress in Canadian manufacturing. These results are similar to those of Tomlinson and Coombs (1998) for the U.K. and Dupont (1994) for France. But these latter comparisons do not deal explicitly with the issue of whether foreign subsidiaries are truncated relative to a standard that is required of competitive global corporations. For this purpose, we have also compared foreign subsidiaries to Canadian corporations that have an international orientation, referred to here as Canadian multinationals. These additional comparisons to Canadian multinationals show that the two groups of multinationals are quite similar, both with regard to the likelihood that they conduct R&D and that they introduce innovations. It is therefore the international orientation of a firm that is the most important factor differentiating firms by innovation regime.

APPENDIX

The Innovation and Advanced Technology Survey

A.1 Background

The usefulness of the information derived from a survey depends on the wording of the questions that are asked, the information guides that are provided to the respondents, and the professionalism of the interview teams.

In Chapter 2, we provide a general outline of the nature of the survey and its operations. Here we list the set of questions that were sent out to the sample of companies included in the survey and a brief outline of some of the definitions that were provided as a guide to the respondents.

A successful survey requires the development of questions that respondents willingly answer, that are clear, and that do not pose an onerous response burden. For that reason, preproduction tests were conducted with potential respondents. These meetings taught us that companies were willing to talk about their major innovations because they took pride in them, especially individuals who were closely associated with the innovation process. The title of the individual who could answer detailed questions about innovation varied firm by firm. In some cases, this would be the R&D manager; in other cases, it would be the vice president in charge of production engineering, or a vice president in charge of new product development. But after we had outlined the nature of the survey, our main company contact was able to direct us to the individual in the firm who could answer our questions about the innovation process at that particular firm – and these individuals had detailed firsthand knowledge about product and process innovation.

The survey contained eight sections: The first requests general background about the firm (including its strategic emphases); the second focuses on R&D performance, the third and fourth on innovation, and the fifth on the use of intellectual property; and the sixth, seventh, and eighth examine technology use. The preproduction tests taught us that it was difficult for one individual to answer all of the sections. As a result, questions on intellectual property often were sent to someone other than the innovation manager – for example, the corporate secretary. The section on the importance of general strategies being pursued by the firm was often filled in by someone in the corporate head office, rather than the innovation manager. The questions about technologies being used were sent to plant managers.

At the beginning of the survey, companies were contacted and the nature of the survey was explained to them. During this process, the names of individuals who could answer each section of the questionnaire were obtained from the main company contact. These individuals were then telephoned and their agreement to participate in the survey was obtained. Only then was the survey, customized with the name of the individual who had agreed to fill in the survey, mailed out. In large firms, each of these sections was often filled in by different persons in the firm. Therefore, while the survey instrument that follows is long, it did not pose an inordinate burden for any one individual. As a result, high response rates were achieved. For example, the large-firm response rate for the intellectual property section was over 85%; for small firms, it was over 92%. Other sections received equally high response rates.

All this required a complex operations platform that could contact different individuals and acquire their agreement to answer the relevant questions, send out customized questionnaires, and then pull all the responses back together again to provide the complete answers for a firm. A special computer assisted telephone interview (CATI) system was designed to perform this task.

The process also required an interview team who fully understood the nature of the survey that was being undertaken. The nature of the contact between this group and the survey respondents determines the quality of the responses received by a survey and the size of the response rate. The interviewers for this survey explained what was required in the initial telephone contact, were available after the survey was mailed out to answer questions about interpretations of questions, and conducted follow-ups over the phone to attain responses when questionnaires were not sent back. The latter were extremely important. The mail-out received an initial response of around 35%; but follow-ups moved that response rate up to over 80% for most sections.

Respondents' guides were prepared to helping the interview team and respondents. These guides included, inter alia, sets of definitions for terms used in the various questions. (These are also included in this appendix.) Equally important, the team who developed the survey worked closely with the interview team, briefing them on objectives and definitions, and making themselves available on short notice to answer questions from respondents and to resolve other difficulties that arose during the course of the survey.

A.2   The 1993 Survey of Innovation and Advanced Technology

1   General

1.1   Please indicate (√) the geographic region of citizenship/residence of the majority of your shareholders, if known, or that of the head office of your controlling firm otherwise.

Appendix

REGION	✓
Canada	
U.S.A.	
Europe	
Pacific Rim*	
Other (Please specify)	

* Pacific Rim is defined here as: Hong Kong, Indonesia, Japan, Malaysia, Singapore, South Korea, Taiwan, and Thailand.

1.2   Please indicate (✓) the regions/countries in which this firm has any of the following operations under its direct control.

REGION	Sales Office (✓)	R&D Unit (✓)	Production Unit (✓)	Assembly Unit (✓)
Canada				
U.S.A.				
Europe				
Pacific Rim				
Other (Please specify)				

1.3   What was the geographic distribution of your sales from Canadian production for 1991?

REGION	SALES %
Canada	
U.S.A.	
Europe	
Pacific Rim*	
Other (Please specify)	
Total	100

1.4   Please estimate the percentage distribution of your 1991 product sales and exports according to the classification below:

PRODUCTS	SALES %	EXPORTS %
Unchanged products during 1989–91		
Products with minor improvements during 1989–91		
Products resulting from major innovations introduced during 1989–91		
Total	100%	100%

1.5 Approximately what percentage of your production employees are covered by a collective agreement? [%]

1.6 Approximately what percentage of your output, by value, is custom designed and made? [%]

1.7 Please rate* the following regions by the extent to which their firms offer significant competition to your firm in the Canadian market.

* Not significant; 2: Somewhat significant; 3: Significant; 4: Very significant; 5: Extremely significant.

REGION	SCORE
Canada	
U.S.A.	
Europe	
Pacific Rim*	
Other (Please specify)	

1.8 Please indicate (√) how many firms (whether or not based in Canada) offer products directly competing with yours in Canada?

1 TO 5	6 TO 20	OVER 20	NONE

IF NONE, PLEASE GO TO 1.11

1.9 Please score* your competitive position relative to your main competitors in the Canadian market on each of the factors listed below:

* 1: Behind; 2: Somewhat behind; 3: About the same; 4: Somewhat ahead; 5: Ahead.

	FACTOR	SCORE (√)				
		1	2	3	4	5
PRODUCTS AND CUSTOMERS' NEEDS & SERVICES	Quality of products (goods/services)					
	Customer services					
	Range of products (goods/services)					
	Flexibility in responding to customers' needs					
	Frequency of introduction of new products (goods/services)					
MARKETING	Marketing					
PRODUCTION PROCESS	Frequency of introduction of new processes					
	Use of advanced manufacturing processes					
	Costs of production					
	Production management					
INNOVATION	Spending on innovation					
	Spending on research and development (R&D)					
	R&D management					
HUMAN RESOURCES	Spending on training					
	Labour climate					
	Skill levels of employees					
	Management of intellectual property (i.e., patents)					

1.10 Please indicate (√) the general level of the price of your products relative to that of your main competitors in the Canadian market.

- [ ] HIGHER
- [ ] ABOUT THE SAME
- [ ] LOWER

1.11 Please rate* the importance of the following factors in your firm's general development strategy.

    * 1: Not important; 2: Slightly important; 3: Important; 4: Very important; 5: Crucial.

FACTOR	SCORE (√)				
	1	2	3	4	5
MARKETS AND PRODUCTS (goods/services)					
Maintaining current production in present markets					
Introducing new products in present markets					
Introducing current products in new markets					
Introducing new products in new markets					
TECHNOLOGY					
Developing new technology					
Improving technology developed by others					
Using technology developed by others					
Improving own existing technology					
USE OF PRODUCTION INPUTS					
Using new materials					
Using existing materials more efficiently					
Cutting labour costs					
Reducing energy costs					
MANAGEMENT PRACTICES					
Improved management incentives via compensation schemes					
Innovative organizational structure					
Improved inventory control					
Improved process control					
Improved intellectual property management					
HUMAN RESOURCES STRATEGY					
Continuous staff training					
Innovative compensation package					
Staff motivation in other ways					
Others					

460 Appendix

2 Research and Development (R&D)

2.1   Please indicate (√) the frequency of R&D in your firm.

CHECK (√) AS MANY CATEGORIES AS APPLY

	√
R&D is performed on an ongoing basis	
R&D is performed on an occasional, or as needed, basis	

IF NO R&D IS PERFORMED, PLEASE GO TO THE NEXT SECTION OF THE QUESTIONNAIRE

2.2   Please indicate (√) the organization of R&D in your firm.

CHECK (√) AS MANY CATEGORIES AS APPLY

	√
There is a separate R&D department	
R&D work is carried out by other departments in the firm	
R&D work is contracted out to other companies or institutions	

2.3   Did your firm claim investment tax credits for R&D for any of tax years 1989–91?   YES ☐ NO ☐

2.4   During the period 1989–91, did you have any R&D collaboration agreements with other companies (domestic or foreign), R&D institutions, or universities?   YES ☐ NO ☐

IF NO PLEASE GO TO NEXT SECTION OR END

2.5   Please indicate (√) the types of collaboration partners and regions of your R&D collaboration agreements during the 1989–91 period.

COLLABORATION PARTNER TYPE	REGION OF COLLABORATION PARTNER				
	Canada	U.S.A.	Europe	Pacific Rim*	Other (Please specify)
Customers					
Suppliers					
Affiliated companies					
Competitors					
R&D institutions					
Universities or colleges					
Other					

* Pacific Rim is defined here as: Hong Kong, Indonesia, Japan, Malaysia, Singapore, South Korea, Taiwan, and Thailand.

3 Innovation

3.1 During the period 1989–91, did you introduce (or were you in the process of introducing) any PRODUCT or PROCESS innovations?

YES ☐   NO ☐

IF NO PLEASE GO TO NEXT SECTION OR END

3.2 Please indicate (√) the categories of your innovation activity for the period 1989–91:

	PRODUCT INNOVATIONS		PROCESS INNOVATIONS
STAGE	Without change in manufacturing technology	With simultaneous change in manufacturing technology	In manufacturing technology without product change
Introduced			
In progress			

3.3 Which, if any, of the following changes were associated with your innovations during the 1989–91 period?

	√
Increased plant specialization	
Increased scale of plant production	
Significant reorganization of work flows and/or functions	
Increased production flexibility	
Increased speed of response to customer requirements	
None of the above	

3.4 Please indicate (√) which of the following factors have particular significance to your firm as IMPEDIMENTS to your innovation program.

	√
Lack of skilled personnel	
Lack of information on technologies	
Lack of information on markets	
Deficiencies in the availability of external technical services	
Barriers to cooperation with other firms	
Barriers to cooperation with scientific and educational institutions	
Government standards and regulations	
Other	

4 Characteristics of Innovation

4.1 Please name and/or briefly describe your most important* innovation commercialized during the period 1989–91.

* The innovation which made the greatest contribution to the firm's profit.

4.2 What was the year of your firm's first commercial launch of this product innovation, or first application of this process innovation?

4.3 Please indicate (√) which of the following features describe the NOVELTY of this innovation:

FEATURE	√
PRODUCT INNOVATION INVOLVING:	
Use of new materials	
Use of new intermediate products	
New functional parts	
Fundamentally new functions	
Other	
PROCESS INNOVATION INVOLVING:	
New production techniques	
Greater degree of automation	
New organization (with regard to new technologies)	
Other	

4.4 How long did it take to commercialize this innovation from the time your firm first invested significant human or capital resources in it?

Months	Years

4.5 This innovation was:

	√
A world-first	
A Canadian-first	
Neither of the above	

PLEASE GO TO 4.7

Appendix 463

4.6 This innovation was introduced approximately how long after first introduction elsewhere?

Months	Years	Unknown ($\checkmark$)

4.7 Please indicate ($\checkmark$) the main sources of IDEAS AND INFORMATION for the generation and development of this innovation.

INTERNAL	$\checkmark$
Management	
Research and development	
Sales/marketing	
Production	
Other	

EXTERNAL	$\checkmark$
Suppliers	
Clients or customers	
Related firm (parent or subsidiary)	
Competitors	
Government regulations/standards	
Government industrial development & technology transfer agencies	
Patent offices or patent literature	
Public R&D institutions	
Private R&D institutions	
Consultants	
Software houses	
Universities/colleges	
Chambers of commerce	
Financial institutions	
Trade fairs/conferences/meetings	
Professional publications	
Other	

4.8 Please indicate ($\checkmark$) the principal sources of TECHNOLOGY used in the development of this innovation.

IF DEVELOPMENT OF THIS INNOVATION DID NOT REQUIRE NEW TECHNOLOGY PLEASE GO TO 4.12.

INTERNAL	√
Research	
Experimental development	
Production engineering	
Other	

EXTERNAL	√
A related firm	
An unrelated firm	
Government laboratories	
University laboratories	
Industrial research firms	
Consultants and service firms	
Joint ventures and strategic alliances	
Publications	
Trade fairs and conferences	
Customer firms	
Supplier firms	
Other	

4.9　Please indicate (√) whether obtaining the technology for this innovation involved:

A LICENSE OR OTHER TRANSFER AGREEMENT WHICH WAS:

Part of a continuous transfer, including access to future technology developed by the other party	
A one-time transfer of technology for a specific product or process	
A cross-licensing agreement	

OR WHICH SPECIFIED:

The right to manufacture	
The right to sell	
The right to use in manufacture	

OR WHICH SPECIFIED THE RIGHT TO USE:

– patents	
– industrial designs	
– trademarks	
– trade secrets	
– other	

IF NONE PLEASE GO TO 4.12

Appendix 465

4.10   Please indicate (√) whether any of these licensing or transfer agreements (written or unwritten) specified:

	√
The territory in which you may or may not manufacture the products or processes resulting from the innovation	
The territory in which you may or may not sell the products or processes resulting from the innovation	
The exclusive right to manufacture	
The exclusive right to sell	
The sources from which any inputs must be purchased	

IF NO TERRITORY RESTRICTIONS PLEASE GO TO 4.12

4.11   Please indicate (√) the regions to which these restrictions apply?

	May manufacture √	May not manufacture √	May sell √	May not sell √
Canada				
United States				
Europe				
Pacific Rim*				
Other				

* Pacific rim is defined as: Hong Kong, Indonesia, Japan, Malaysia, Singapore, South Korea, Taiwan, and Thailand.

4.12   Please indicate (√) which, if any, of the following methods you used to protect your innovation, and the geographic areas in which you used these methods.

METHODS	Canada	U.S.A.	Europe	Pacific Rim*	Other
Copyrights					
Patents					
Industrial designs					
Trademarks					
Trade secrets					
Integrated circuit designs					
Plant breeders' rights					
Other					

4.13 How much time did/will it take, from the time of your innovation, for others to copy it?

Months	Years	Unknown (√)

4.14 Please indicate (√) the methods, if any, by which your firm has transferred technology related to this innovation to other firms, and the regional locations of those firms.

METHODS OF TRANSFER	Canada √	U.S.A. √	Europe √	Pacific Rim √	Other √
License					
Sell					
Trade					
Joint venture/ strategic alliance					
Other					

4.15 Did your firm form a joint venture or strategic alliance with others to produce this innovation?   YES ☐ NO ☐

4.16 For this innovation, what was the approximate cost (in dollars or %) to the firm of each stage of the process set out below: (Please enter 0 where no expense was incurred.)

STAGE OF INNOVATION PROCESS	COST DISTRIBUTION	
	Cdn. $	%
Basic research		
Applied research		
Acquisition of technological knowledge (e.g., patents and trademarks, licenses, specialist consulting services, disclosure of know-how)		
Development (e.g., engineering, layout, design, prototype construction, pilot plant, acquisition of equipment, etc.)		
Manufacturing start-up (e.g., engineering, tooling, plant arrangement, construction, pilot plant, acquisition of equipment, etc.)		
Marketing start-up (for the process innovation or resulting products)		
TOTAL		100

Appendix

4.17   What is the percentage of the total cost of this innovation carried out by in the total cost of all innovations the firm during the period 1989–91? [%]

4.18   Please estimate the distribution of the sources of funds for the development of this innovation through to your first commercial launch or first application:

SOURCE OF FUNDS	Cdn. $	%
Internal		
PARENT OR AFFILIATED FIRM:		
Foreign		
Domestic		
Venture capital firms		
Other financial institutions		
Research consortia		
GOVERNMENTS:		
Federal		
Provincial		

OTHER		

TOTAL		100

4.19   If you have received government assistance for this innovation, please indicate ($\sqrt{}$) which of the following categories were used and the $ amount involved.

TYPE OF GOVERNMENT FUNDING	Federal		Provincial	
	$\sqrt{}$	Cdn. $	$\sqrt{}$	Cdn. $
Research support				
Technology development support				
Support for acquisition of machinery and equipment				
Training				
R&D investment tax credit				
Consulting/information				
Other				

4.20   Please estimate the percentage distribution of your sales of, or resulting from, this innovation by class of customer.

SECTOR	%
Households (i.e., consumers)	
Business/industry	
Government	

4.21 Please indicate (√) whether other firms in this or other industries have received this innovation via:

	√
Sale of intermediate products	
Sale of processes or capital goods	
Transfer of intellectual property	

IF NONE, PLEASE GO TO 4.23

4.22 To which (√) of the following industries or sectors have you sold or otherwise transferred this product or process innovation?

INDUSTRY	√
Agriculture, fishing and logging	
Mining and oil wells	
Manufacturing	
– Food, beverage, and tobacco	
– Rubber and plastic products	
– Textiles	
– Wood, furniture, and fixture	
– Paper and allied products	
– Printing and publishing	
– Primary metals	
– Fabricated metal products	
– Machinery	
– Aircraft and parts	
– Motor vehicle, parts and accessories	
– Telecommunication equipment	
– Electronic parts and components	
– Business machines	
– Nonmetallic mineral products	
– Refined petroleum and coal products	
– Pharmaceutical and medicine	
– Scientific and professional equipment	
– Other manufacturing industries	
Construction	
Utilities	
Trade	
Finance and insurance	
Other services	

4.23   Please indicate (√) whether this innovation had any of the following effects:

OVERALL EFFECTS

Improved profit margin	
Improved quality of products	
Improved technological capabilities	
Improved working conditions	
Reduced lead times	
Extending product range	

REDUCTION IN FACTORS OF PRODUCTION

Reduced labour requirements	
Reduced energy requirements	
Reduced capital requirements	
Reduced material requirements	
Reduced design costs	

IMPROVEMENT IN MARKET SHARE

Increased share in domestic market	
Increased share in foreign markets	

IMPROVEMENT IN FIRM'S INTERACTIONS WITH OUTSIDE PARTIES

Improved the firm's interactions with its customers	
Improved the firm's interactions with its suppliers	

IMPROVEMENT IN THE FIRM'S RESPONSE TO GOVERNMENT REGULATORY REQUIREMENTS

Environmental regulations	
Health and safety regulations	
Other	

OTHER	

4.24   Please indicate (√) the effect of this innovation on the number of workers in your firm.

WORKER GROUP	DECREASE	INCREASE	NO CHANGE
Production workers			
Nonproduction workers			
Overall			

4.25   Please indicate (√) how the skill requirements of your workers were changed as a result of this innovation?

Decreased	Unchanged	Increased

470 *Appendix*

4.26 Please provide estimates of sales and exports resulting from the innovation for 1991. In the case of process innovation, please provide the sales or exports of the products resulting from this innovation.

	Cdn. $
Sales resulting from this innovation	
Exports resulting from this innovation	

4.27 How many major innovations of the type described in this section did your firm make during the period 1989–91?

STAGE	PRODUCT INNOVATIONS		PROCESS INNOVATIONS
	Without change in manufacturing technology	With a simultaneous change in manufacturing technology	In manufacturing technology without product change
Introduced			
In progress			

5 *Intellectual Property*

5.1 Please indicate (√) the extent to which the following methods have been used by your firm to protect its intellectual property IN CANADA over the last three years (1989–91).

INTELLECTUAL PROPERTY	1989–1991				
	NONE	Number of usages (where relevant)			
		1 to 5	6–20	21–100	100+
Copyrights					
Patents					
Industrial designs					
Trade secrets					
Trademarks					
Integrated circuit designs (semiconductor chips)					
Plant breeders' rights (plant variety rights)					
Other					

5.2 How effective* are the following means of preventing your competitors from bringing to market copies of your new product or process technology?

* 1: NOT AT ALL EFFECTIVE; 2: SOMEWHAT EFFECTIVE;
3: EFFECTIVE; 4: VERY EFFECTIVE; 5: EXTREMELY EFFECTIVE

MEANS	SCALE
INTELLECTUAL PROPERTY RIGHTS	
RIGHTS ASSOCIATED WITH:	
Copyrights	
Patents	
Industrial designs	
Trade secrets	
Trademarks	
Integrated circuit designs	
Plant breeders' rights	
Other	
OTHER STRATEGIES:	
Complexity of product design	
Being first in the market	
Other	

5.3 During the last three years (1989–91), has your firm granted the right to use intellectual property to, or acquired the right to use intellectual property from, another firm?

YES   NO

IF NO PLEASE GO TO NEXT SECTION (OR END)

5.4 Please indicate (√) the type and direction of such intellectual property transfer:

INTELLECTUAL PROPERTY	GRANTED RIGHTS (S) TO		ACQUIRED RIGHT (S) FROM	
	Canadian Firms √	Foreign Firms √	Canadian Firms √	Foreign Firms √
Copyrights				
Patents				
Industrial designs				
Trade secrets licensing agreements				
Trademarks				
Integrated circuit designs				
Plant breeders' rights				
Other				

6 Advanced Technology Use

6.1 For EACH of the manufacturing technologies listed below, and currently used in your operations, please enter the approximate number of years in use: if NOT currently used, please indicate ($\checkmark$) which description best reflects plans for use:

MARK ONLY ONE COLUMN IN EACH ROW FOR EACH TECHNOLOGY

TECHNOLOGY	Used in operations — Approximate number of years in use	Not currently used — Plan to use within next 2 years $\checkmark$	Not currently used — No plans to use — No application $\checkmark$	Not currently used — No plans to use — Not cost effective $\checkmark$
FUNCTION: DESIGN AND ENGINEERING				
Computer-aided design (CAD) and/or computer-aided engineering (CAE)				
CAD output used to control manufacturing machines (CAD/CAM)				
Digital data representation of CAD output used in procurement activities				
FUNCTION: FABRICATION AND ASSEMBLY				
Flexible manufacturing cell(s) (FMC) or systems (FMS)				
Numerically controlled and computer numerically controlled (NC/CNC) machine(s)				
Materials working laser(s)				
Pick and places robot(s)				
Other robots				
FUNCTION: AUTOMATED MATERIAL HANDLING				
Automated storage and retrieval system (AS/RS)				
Automated guided vehicle systems (AGVS)				
FUNCTION: INSPECTION AND COMMUNICATIONS				
Automated sensor-based equipment used for inspection/testing of:				
– incoming or in-process materials				
– final product				
Local area network for technical data				
Local area network for factory use				
Intercompany computer network linking plant to subcontractors, suppliers, and/or customers				
Programmable controller(s)				
Computer(s) used for control on the factory floor				

6.2 For EACH item or class of software listed below, and currently used in your operations, please enter the approximate number of years in use; if NOT currently used, please indicate ($\sqrt{}$) which description best reflects plans for use:

MARK ONLY ONE COLUMN IN EACH ROW FOR EACH TECHNOLOGY

TECHNOLOGY	Used in operations Approximate number of years in use	Not currently used		
		Plan to use within next 2 years $\sqrt{}$	No application $\sqrt{}$	Not cost effective $\sqrt{}$
MANUFACTURING INFORMATION SYSTEMS				
Materials requirement planning (MRP)				
Manufacturing resource planning (MRP II)				
INTEGRATION AND CONTROL				
Computer integrated manufacturing (CIM)				
Supervisory control and data acquisition (SCADA)				
Artificial intelligence and/or expert systems				

7 Acquisition of Advanced Technology

FOR THE PURPOSE OF THIS SECTION OF THE QUESTIONNAIRE PLEASE REFER TO THE FUNCTIONAL GROUPING OF TECHNOLOGIES IN Q.6.1. YOU ARE ASKED TO ANSWER FOR EACH SUCH FUNCTIONAL GROUP.

IF NONE OF THE TECHNOLOGIES LISTED IN Q.6.1. ARE IN CURRENT USE IN YOUR OPERATIONS, PLEASE ANSWER ONLY QUESTIONS 7.14, 8.1, 8.2, AND 8.3.

7.1 Please indicate ($\sqrt{}$) the range that best reflects this plant's total investment in technologically advanced equipment and software for the period 1989–91. Please EXCLUDE education and training but

INCLUDE plant modifications, construction, integration, and equipment and software purchased or developed.

PLEASE ANSWER SEPARATELY FOR EACH FUNCTIONAL GROUP.

INVESTMENT	Design and Engineering	Fabrication and Assembly	Automated Materials Handling	Inspection and Communications
Less than $100,000*				
$100,000 to less than $1 million				
$1 million to less than $5 million				
$5 than million to less $10 million				
$10 million or more				
Not applicable				

* All $ are Canadian Currency.

7.2 For each functional technology group, please specify the percentage of total investment made up of technologically advanced equipment and software.

PLEASE ANSWER SEPARATELY FOR EACH FUNCTIONAL GROUP.

	Design and and Engineering	Fabrication and Assembly	Automated Materials Handling	Inspection and Communications
Percentage of total investment	%	%	%	%

7.3 Please indicate (√) which of the following best describes the impact of technologically advanced equipment and software on your education and training cost.

PLEASE ANSWER SEPARATELY FOR EACH FUNCTIONAL GROUP.

Appendix 475

IMPACT	Design and Engineering	Fabrication and Assembly	Automated Materials Handling	Inspection and Communications
Increased significantly				
Increased moderately				
Increased marginally				
No change				
Decreased				
Not applicable				

7.4 Please indicate ($\sqrt{}$) any factors that had particular significance over the last three years (1989–91) in HAMPERING OR DELAYING your acquisition of technologically advanced equipment and software from CANADIAN sources.

PLEASE ANSWER SEPARATELY FOR EACH FUNCTIONAL GROUP.

FACTORS	Design and Engineering	Fabrication and Assembly	Automated Materials Handling	Inspection and Communications
Overall cost				
Cost of technology acquisition				
Cost of education and training				
Worker uncertainty				
Time to develop software				
Cost to develop software				
Increased maintenance expense				
Need for market expansion				
Lack of financial justification				
Lack of technical support from vendors				
Other				
Not applicable				

7.5 Please indicate (√) any factors that had particular significance over the last three years (1989–91) in HAMPERING OR DELAYING your acquisition of technologically advanced equipment and software from FOREIGN sources.

PLEASE ANSWER SEPARATELY FOR EACH FUNCTIONAL GROUP.

FACTORS	Design and Engineering	Fabrication and Assembly	Automated Materials Handling	Inspection and Communications
Overall cost				
Cost of technology acquisition				
Cost of education and training				
Worker uncertainty				
Time to develop software				
Cost to develop software				
Increased maintenance expense				
Need for market expansion				
Lack of financial justification				
Lack of technical support from vendors				
Other				
Not applicable				

7.6 Please indicate (√) any factors that had particular significance over the last three years (1989–91) in HAMPERING OR DELAYING your acquisition of technologically advanced equipment and software.

PLEASE ANSWER SEPARATELY FOR EACH FUNCTIONAL GROUP

FACTORS	Design and Engineering	Fabrication and Assembly	Automated Materials Handling	Inspection and Communications
Overall cost				
Cost of technology acquisition				
Cost of education and training				
Worker uncertainty				
Time to develop software				
Cost to develop software				
Increased maintenance expense				
Need for market expansion				
Lack of financial justification				
Lack of technical support from vendors				
Other				
Not applicable				

Appendix

7.7 Please indicate (✓) any factors that have particular significance for your acquisition of technologically advanced equipment and software.

PLEASE ANSWER SEPARATELY FOR EACH FUNCTIONAL GROUP.

FACTORS	Design and Engineering	Fabrication and Assembly	Automated Materials Handling	Inspection and Communications
Lower price				
Internal familiarity with the technology				
Better technical support				
Lower maintenance expense				
Lower costs and shorter time of development of supporting software				
Ease of communication				
Faster delivery time				
Higher risk in dealing with unfamiliar sources				
Special arrangements				
Other				

7.8 How would you compare* your production technology with that of your most significant competitors in Canada and outside of Canada?

* 1: Much less advanced; 2: Less advanced; 3: About the same; 4: More advanced; 5: Much more advanced.

PLEASE ANSWER SEPARATELY FOR EACH FUNCTIONAL GROUP.

COMPETITORS	Design and Engineering	Fabrication and Assembly	Automated Materials Handling	Inspection and Communications
Other Canadian producers				
Producers abroad				

7.9   Please indicate (√) your principal INTERNAL sources of ideas for the adoption of technologically advanced equipment and software.

PLEASE ANSWER SEPARATELY FOR EACH FUNCTIONAL GROUP.

INTERNAL SOURCE	Design and Engineering	Fabrication and Assembly	Automated Materials Handling	Inspection and Communications
Research				
Experimental development				
Design work				
Production engineering				
Operating staff				
Management				
Corporate head office				
Other				

7.10   Please indicate (√) your principal EXTERNAL sources of ideas for the adoption of technologically advanced equipment and software.

PLEASE ANSWER SEPARATELY FOR EACH FUNCTIONAL GROUP.

EXTERNAL SOURCE	Design and Engineering	Fabrication and Assembly	Automated Materials Handling	Inspection and Communications
A related firm (with same parent firm)				
An unrelated firm				
Government laboratories				
University laboratories				
Provincial research organization				
Industrial research firms				
Research consortia				
Consultants and service firms				
Joint ventures and strategic alliances				
Publications				
Trade fairs, conferences				
Customer firms				
Supplier firms				
There was no significant external input				
Other				

Appendix 479

7.11 Please indicate (✓) the principal REGIONAL sources of your present technologically advanced equipment and software.

PLEASE ANSWER SEPARATELY FOR EACH FUNCTIONAL GROUP.

REGIONAL SOURCE	Design and Engineering	Fabrication and Assembly	Automated Materials Handling	Inspection and Communications
Canada				
United States				
Europe				
Pacific Rim*				
Other (please specify)				

* Pacific Rim is defined here as: Hong Kong, Indonesia, Japan, Malaysia, Singapore, South Korea, Taiwan, and Thailand.

7.12 Please indicate (✓) the average length of time between your becoming aware of the technologically advanced equipment and software that you eventually acquired and its implementation.

PLEASE ANSWER SEPARATELY FOR EACH FUNCTIONAL GROUP.

TIME PERIOD	Design and Engineering	Fabrication and Assembly	Automated Materials Handling	Inspection and Communications
Less than 1 year				
1–3 years				
3–5 years				
5–10 years				
more than 10 years				

7.13 Please indicate (✓) whether the adoption of technologically advanced equipment and software led to any of the following results.

PLEASE ANSWER SEPARATELY FOR EACH FUNCTIONAL GROUP

RESULTS	Design and Engineering	Fabrication and Assembly	Automated Materials Handling	Inspection and Communications
An improvement in productivity				

LOWER PRODUCTION COSTS BY REDUCING:				
Labour requirements				
Material consumption				
Energy consumption				
Product rejection rate				

Improvement in product quality				
Reduced setup time				
Greater product flexibility				
Improved working conditions				
Reduced environmental damage				
Reduced skill requirements				
Reduced capital investments				
Increased skill requirements				
Increased capital requirements				
Increased equipment utilization rate				
Lower inventory				
Other				

7.14 Please indicate ($\sqrt{}$) whether the adoption of technologically advanced equipment and software was associated with either of the following innovations.

PLEASE ANSWER SEPARATELY FOR EACH FUNCTIONAL GROUP

RESULT	Design and Engineering	Fabrication and Assembly	Automated Materials Handling	Inspection/ Communications Control
A new product				
A new process				
Neither				

Appendix

7.15 Please indicate (√) any plans to acquire technologically advanced equipment and software at this location over the next three years.

EXTENT OF PLANNED TECHNOLOGY ACQUISITION	Design and Engineering	Fabrication and Assembly	Automated Materials Handling	Inspection/ Communications Control
Total replacement (75% or more)				
Major upgrade (25% to less than 75%)				
Minor upgrade (less than 25%)				
Under consideration, but no firm plans				
None				

8 Acquisition of Advanced Technology: Impediments

8.1 Please indicate (√) which of the following factors have particular significance to your firm as IMPEDIMENTS to technology acquisition.

IMPEDIMENT	Source of technology	
	CANADIAN	FOREIGN
COST-RELATED PROBLEMS		
Cost of capital		
High cost of equipment		
Costs to develop software		
Increased maintenance expenses		
Cost of technology acquisition		
Lack of financial justification		
Tax regime: R&D investment tax credits		
Tax regime: capital cost allowances		
Government regulations/standards		
LABOUR-RELATED PROBLEMS		
Shortage of skills		
Training difficulties		
Labour contracts		
ORGANIZATIONAL/STRATEGIC PROBLEMS		
Difficulties in introducing important changes to the organization		
Management attitude		
Worker resistance		
OTHER PROBLEMS		
Lack of scientific and technical information		
Lack of technological services (e.g., technical and scientific consulting, tests, standards)		
Lack of technical support from vendors		
Other		

8.2 For each of the professional groups listed below, please indicate (√) first whether you have positions in your firm and, second, whether you are experiencing difficulty in filling any such positions.

PROFESSION	STAFF POSITIONS √	SHORTAGE √
Electrical engineers		
Aerospace engineers		
Engineering technologists and technicians		
Systems analysts and computer programmers		
Electronic data processing equipment operators		
Assemblers, printed circuit boards		
CAD draughtspersons		
CAD/CAM repair technicians		
CAD designers, printed circuit boards		
Computer hardware specialists		
Fibre optic cable splicers		
Laser beam welders		
Laser tube assemblers		
Machinists, numerically controlled machine tools		
Microcomputer specialists		
Numerical control operators		
Robotics technicians		
Other		

IF NONE, PLEASE GO TO 8.4

8.3 Which (√) of the following steps have you taken to deal with these shortages?

	√
Deferred acquisition of technology	
Subcontracted	
Given appropriate personnel training	
Improved wages, benefits	
Searched outside region	
Searched abroad	
Overtime	
Capital substitution	
Other	

8.4 During the period 1989–1991, have the technical and/or production employees of your plant received any training associated with your adoption of technologically advanced equipment and software?

YES   NO
☐   ☐
IF NO PLEASE GO TO 8.7

8.5 Please specify the nature of this training.

NATURE	Average Duration (days per trainee)	Number of trainees	Government assisted? YES ✓	NO ✓
Classroom (inside the firm)				
Classroom (outside the firm)				
On-the-job training				
Correspondence courses				

8.6 Please indicate (✓) the range of your cost for education and training related to the use of technologically advanced equipment and software for the period 1989–91.

Less than $10,000*	
$10,000–$50,000	
$50,000–$100,000	
$100,000–$250,000	
$250,000–$500,000	
$500,000–$1 million	
$1 million–$5 million	
Over $5 million	
Not applicable	

* All $ are Canadian currency.

8.7 If you have received government assistance for advanced technology acquisition in the last three years (1989–91) please indicate (✓) which of the following categories were used and, if possible, the $ amount involved.

TYPE OF GOVERNMENT FUNDING	Federal ✓	$*	Provincial ✓	$
Research support				
Technology development support				
Support for acquisition of machinery and equipment				
Training				
R&D investment tax credit				
Consulting information				
Other				

* All $ are Canadian currency.

A.3 Respondents' Guide: Survey of Innovation and advanced Technology

The purpose of this guide is to clarify the terms used in the questionnaire.

Section 1: Firm Characteristics

This section of the guide accompanies the first part of the Survey, which contains questions about your firm's activities and characteristics.

Definitions

Q 1.1   **Controlling firm** refers to a firm that directly or indirectly owns at least 50.1% of the voting stock of the company.

Pacific Rim is defined as Hong Kong, Indonesia, Japan, Malaysia, Singapore, South Korea, Taiwan, and Thailand.

Other includes all countries except Canada, the United States, those in Europe, and those defined above as part of the Pacific Rim.

Q 1.3   **Canadian production** refers to the goods and/or services produced in Canada by your firm.

Q 1.4   **Products with minor improvements** refers to those existing products whose technical characteristics have slightly been enhanced or upgraded. This can take two basic forms. A simple product may be improved (in terms of improved performance or lower cost) through the use of higher performance components or materials, or a complex product, which consists of a number of integrated technical subsystems, may be improved by partial changes to one of the subsystems.

Products with major innovations refers to newly marketed products whose intended use, performance characteristics, technical construction, design, or use of materials and components, is new or substantially changed. Such innovations can involve radically new technologies or can be based on combining existing technologies in new uses.

Q 1.5   **Collective agreement** refers to a contract between the employees and the corporation that is covered by the Canada Labour Act or comparable provincial legislation.

Q 1.6   **Custom designed** refers to products that are produced under customer specifications and that require alternative production configurations.

Q 1.7   **Significant competition** refers to other firms that market products similar to your own and that might be purchased by your customers.

Q 1.9   **Main competitors** refer to those that offer similar or differentiated products in the same market(s) as your firm.

Flexibility in responding to customers' needs refers to a company's ability to change marketing and production strategies to fit customer needs, i.e., modify product specifications, change distribution schedules, etc.

Research and development management refers to the organization, administration, and direction of research and development activities within the company.

Q 1.11 **General development strategy** refers to the firm's overall business strategy.

Innovative organizational structure refers to organizational structures that incorporate alternative features of control, authority, and responsibility.

Innovative compensation package refers to an alternative mix of compensation and/or benefits that is offered to employees. For example, you may have a performance-based pay and benefits package.

Section 2: Research and Development

This section of the guide is for the second part of the Survey, which contains questions about your firm's research and development activities.

Definitions

Q 2.1 **Research and development** is investigative work carried out:

1. to acquire new scientific and technological knowledge,
2. to devise and develop new products or processes, or
3. to apply newly acquired knowledge in making technically significant improvements to products or processes.

Research and Development normally does not include:

1. market research and sales promotion,
2. research in the social sciences,
3. operations research, except when required during the development phase of a product or a process,
4. quality control or routine testing of products and materials, and
5. activities necessary for commercial production of the new or improved product or process after development is completed.

Q 2.3 **Investment tax credit for research and development** refers to the tax credit offered to firms in Canada by the Canadian government for expenditures identified with research and development.

Q 2.5 **Affiliated companies** refer to companies under the same ultimate corporate control.

Research and development institutions refer to private or public organizations created and operated for the sole purpose of providing research and development.

Sections 3 and 4: Innovation

This section of the guide is for the third and fourth parts of the Survey, which contain questions about your firm's innovation process. The innovation section of the survey collects general data relating to product and process innovations in Canada.

Definitions

Q 3.1 **Innovation** is the practical use of an invention to produce new goods or services, to improve existing ones, or to improve the way in which they are produced or distributed. Changes that are purely aesthetic (such as changes in colour or decoration) or that simply involve minor design or presentation alterations to a product while leaving it technically unchanged in construction or performance are not considered as innovations.

Q 3.2 A **product innovation** is the commercial adoption of a new product. Technological change occurs when the design characteristics of a product change in a way that delivers new or improved services to consumers of the product.

A **process innovation** is the adoption of new or significantly improved production methods. These methods may involve changes in equipment or production organization or both. The methods may be intended to produce new or improved products, which cannot be produced using conventional plants or production methods, or to increase the production efficiency of existing products.

Q 3.3 **Increased plant specialization** occurs when a narrower range of products is produced.

Q 3.4 **Impediments to innovation** refer to such factors as conditions in input markets (labour, material, capital, technology) and other factors beyond the control of the firm, which slow or stop development of your innovations.

Q 4.2 **Commercial launch** refers to the activity involving the introduction of the product to the market, i.e., obtaining retail shelf space or being on industrial product lists.

Q 4.5 This refers to the innovation's entry to the market, i.e., whether it was a first in the world or a first in Canada.

Q 4.7 **Ideas and Information** refer to organized knowledge required for the development of the specific innovation, i.e., features of the new product/process, unsatisfied or potential needs, etc.

Q 4.8 **Joint ventures and strategic alliances** refer to cooperation agreements between two or more companies that are entered into for strategic reasons. They may involve the creation of a separate legal entity and may cover one or more projects.

Q 4.9 **Licensing** may involve a wide range of rights in the design, production, and distribution of goods and/or services. It may be restricted to a particular industry, sector, or geographic market, and it may be limited to a specific time frame.

Q 4.12 Brief descriptions of the methods of protecting intellectual property can be found in section 5.

Q 4.14 **Trade** refers to the activity of bartering technology with other companies.

Q 4.16 **Innovation Costs by Stage**: Innovation expenditures regarding the transfer of technological knowledge should cover 'acquisition', 'registration', and 'enforcement' costs.

Appendix

> **Basic research** is experimental or theoretical work undertaken primarily to acquire new knowledge of the underlying foundations of phenomena and observable facts, without any particular application or use in view.
>
> **Applied research** is also original investigation undertaken in order to acquire new knowledge. It is, however, directed primarily towards a specific practical aim or objective.

Q 4.18 Research consortia are institutions that are formed by various private and public organizations, with the objective to finance, undertake, and/or maintain research and development activities.

> **Affiliated firms** are firms under the same ultimate corporate control.

Q 4.21 **Capital goods** are final products that are used exclusively to increase the fixed assets of the organization and have an amortization period that runs longer than a year.

> **Intermediate products** are products that are used as inputs in the production of other products.

Q 4.24 **Production workers** are those directly associated with the production of goods and services.

Section 5: Intellectual Property Rights

This section of the guide is for the fifth part of the Survey, which contains questions about your firm's use of intellectual property rights.

Definitions

Q 5.1 Below are brief descriptions of the methods of protecting intellectual property.

> **Copyright:** A copyright is a form of protection, provided by a federal statute, given to authors and creators of original works, such as books, records, films, and works of art, against a variety of unauthorized uses.
>
> **Patent:** A Canadian patent is a document, issued by the federal government, that describes an innovation and creates a form of legal protection whereby the inventor or patent owner has the right to prevent others from making, using, and selling the invention in Canada.
>
> **Industrial Design:** The Industrial Design Act gives protection to designers of ornamental aspects of useful articles.
>
> **Trade Secret:** Canadian common law provides protection of trade secrets in respect of confidential, commercially valuable information. Obligations of trade secrecy can apply to such things as concepts, ideas, factual information, etc., and apply to persons who have acquired confidential information through their relationships with the firm.

Trade Secrets Licensing Agreement: A contract between firms in which trade secrets are licensed.

Trademark: A trademark is a visible sign, symbol, word, or picture that serves to distinguish the wares or services of an industrial or commercial enterprise.

Integrated Circuit Design: Registration of integrated circuit designs protects the original three-dimensional pattern or layout design embodied in an integrated circuit, also known as a semiconductor chip.

Plant Breeders' Rights: Plant breeders' rights legislation protects seeds and other propagating material and requires the use of a distinct generic name when selling the propagating material.

Section 6: Technology

This section of the guide accompanies the parts of the survey that contain questions about your firm's use of technologically advanced equipment and software. The section on technology is an expansion of earlier surveys done by Statistics Canada on the diffusion of advanced technologies, the source of technology and the impact of technology on resource use and product quality. The information collected will be used to analyze the extent of technology use in manufacturing and the factors affecting its adoption.

Definitions
Note: Technology definitions are presented below in the order in which the technologies are listed in Q 6.1.

Computer-Aided Design (CAD) and/or Computer-Aided Engineering: Use of computers for drawing and designing parts or products for analysis and testing of designed parts or products.

Computer-Aided Design (CAD) Computer-Aided Manufacturing (CAM): Use of CAD output for controlling machines used in manufacturing.

Digital Data Representation: Use of digital representation of CAD output for controlling machines used to manufacture the part or product.

Flexible Manufacturing Cells (FMC): Machines with fully integrated material-handling capabilities controlled by computers or programmable controllers, capable of single path acceptance of raw material and delivery of finished product.

Flexible Manufacturing Systems (FMS): Two or more machines with fully integrated material-handling capabilities controlled by computers or programmable controllers, capable of single or multiple acceptance of raw material and multiple path delivery of finished product.

NC CNC Machines: A single machine either numerically controlled (NC) or computer numerically controlled (CNC) with or without automated material-handling capabilities. NC machines are controlled by numerical commands, punched on paper or plastic mylar tape, while CNC machines are controlled electronically through a computer residing in the machine.

Materials Working Laser(s): Laser technology used for welding, cutting, treating, scribing, and marking.

Robot: A reprogrammable, multifunctional manipulator designed to move materials, parts, tools, or specialized devices through variable programmed motions for the performance of a variety of tasks.

Pick and Place Robot: A simple robot, with one, two, or three degrees of freedom, which transfers items from place to place by means of point-to-point moves. Little or no trajectory control is available.

Automated Storage and Retrieval System (AS/RS): Computer-controlled equipment providing for the automatic handling and storage of materials, parts, subassemblies, or finished products.

Automated Guided Vehicle Systems (AGVS): Vehicles equipped with automatic guidance devices programmed to follow a path that interfaces with work stations for automated or manual loading and unloading of materials, tools, parts, or products.

Technical Data Network: Use of local area network (LAN) technology to exchange technical data within design and engineering departments.

Factory Network: Use of local area network (LAN) technology to exchange information between different points on the factory floor.

Programmable Controller: A solid state industrial control device that has programmable memory for storage of instructions, which performs functions equivalent to a relay panel or wired solid state logic control system.

Industrial Computers Used for Control on the Factory Floor include computers on the factory floor that may be dedicated to control, but which are capable of being reprogrammed for other functions; excluded are computers imbedded within machines, or computers used solely for data acquisition or monitoring.

Q 7.1   **Total investment** refers to the actual amount paid (undepreciated book value) for the specified technology.
Q 7.4   Technology from **Canadian** sources refers to Canadian production.
Q 7.5   Technology from **foreign** sources refers to foreign production.
Q 7.8   **Significant competitors** refers to other firms that market products similar to your own or that might be purchased by your customers.
Q 7.9   **Technology adoption** is the process of selecting and acquiring a technology, which has already been fully developed, for the purpose of integrating it into existing or new products and processes.

References

Abernathy, W. J. and J. M. Utterbach. 1978. 'Patterns of Industrial Innovation'. *Technology Review* 80: 41–47.

Acs, Z. J., and D. B. Audretsch. 1988. 'Innovation in Large and Small Firms: An Empirical Analysis'. *American Economic Review* 78(4): 678–90.

Acs, Z. J., and D. B. Audretsch. 1990. *Innovation and Small Firms*. Cambridge, MA: MIT Press.

Acs, Z. J., and D. B. Audretsch. 1991. 'R&D, Firm Size and Innovative Activity'. In *Innovation and Technical Change: An International Comparison*. Edited by Z. J. Acs and D. B. Audretsch. Ann Arbor: University of Michigan Press. Pp. 39–59.

Acs, Z. J., and D. B. Audretsch (eds.). 1991. *Innovation and Technological Change: An International Comparison*. Ann Arbor: University of Michigan Press.

Aitken, H. G. 1961. *American Capital and Canadian Resources*. Cambridge, MA: Harvard University Press.

Äkerblom, M., M. Virtaharju, and A. Leppäahti. 1996. 'A Comparison of R&D Surveys, Innovation Surveys and Patent Statistics Based on Finnish Data'. *Innovation, Patents and Technological Strategies*. Paris: OECD. Pp. 57–70.

Amemiya, T. 1978. 'The Estimation of a Simultaneous Equation Generalized Probit Model'. *Econometrica* 46: 1193–205.

Amey, L. R. 1964. 'Diversified Manufacturing Businesses'. *Journal of the Royal Statistical Society*, Series A 127: 251–90.

Arminger, G., J. Wittenberg and A. Schepers. 1996. *Mecosa* (A program for the analysis of general mean- and covariance structures with non-metric variables: Users guide). Friedrichsdorf, Germany: Additive Gmbh.

Arrow, K. J. 1962. 'Economic Welfare and the Allocation of Resources for Invention'. In *The Rate and Direction of Inventive Activity*. National Bureau Committee for Economic Research. Princeton, NJ: Princeton University Press. Pp. 609–24.

Arundel, A. 2001. 'The Relative Effectiveness of Patents and Secrecy for Appropriation'. *Research Policy* 30: 611–24.

Arundel, A., G. van de Paal, and L. Soete. 1995. *Innovation Strategies of Europe's Largest Industrial Firms.* Maastricht: MERIT.

Arvanitis, S., and H. Hollenstein. 1994. 'Demand and Supply Factors in Explaining the Innovative Activity of Swiss Manufacturing Firms'. *Economics of Innovation and New Technology* 3: 15–30.

Arvanitis, S., and H. Hollenstein. 1996. 'Industrial Innovation in Switzerland: A Model-Based Analysis with Survey Data'. In *Determinants of Innovation: The Message from New Indicators.* Edited by A. Kleinknecht. London: Macmillan Press. Pp. 13–62.

Audretsch, D. B. 1995. *Innovation and Industry Evolution.* Cambridge, MA: MIT Press.

Baldwin, J. R. 1996. 'Innovation: The Key to Success in Small Firms'. In *Evolutionary Economics and the New International Political Economy.* Edited by J. de la Mothe and G. Paquette. London: Francis Pinter. Pp. 238–56.

Baldwin, J. R. 1997a. *Innovation and Intellectual Property.* Catalogue No. 88-515-XPE. Ottawa: Statistics Canada.

Baldwin, J. R. 1997b. *The Importance of Research and Development for Innovation in Small and Large Canadian Manufacturing Firms.* Research Paper Series No. 107. Ottawa: Statistics Canada.

Baldwin, J. R. 1998. *Innovation and Training in New Firms.* Research Paper No. 123. Analytical Studies Research Paper Series. Ottawa: Statistics Canada.

Baldwin, J. R. 1999. *Innovation, Training and Success.* Research Paper No. 137. Analytical Studies Branch. Ottawa: Statistics Canada.

Baldwin, J. R., D. Beckstead, and R. E. Caves. 2001. *Changes in the Diversification of Canadian Manufacturing Firms.* Research Paper No. 151. Analytical Studies Research Paper Series. Ottawa: Statistics Canada.

Baldwin, J. R., and R. E. Caves. 1991. 'Foreign Multinational Enterprises and Merger Activity in Canada'. In *Corporate Globalization Through Mergers and Acquisitions.* Edited by L. Waverman. Calgary: University of Calgary Press. Pp. 89–122.

Baldwin, J. R., W. Chandler, C. Le, and T. Papailiadis. 1994. *Strategies for Success: A Profile of Growing Small and Medium-Sized Enterprises in Canada.* Catalogue 61-523-RPE. Ottawa: Statistics Canada.

Baldwin, J. R., and M. Da Pont. 1996. *Innovation in Canadian Manufacturing Enterprises.* Catalogue 88-514-XPB. Ottawa: Statistics Canada.

Baldwin, J. R., and B. Diverty. 1995. *Advanced Technology Use in Manufacturing Establishments.* Research Paper No. 85. Analytical Studies Branch. Ottawa: Statistics Canada.

Baldwin, J. R., B. Diverty, and D. Sabourin. 1995. 'Technology Use and Industrial Transformation: Empirical Perspectives'. In *Technology, Information, and Public Policy.* Edited by T. Courchene. John Deutsch Institute for the Study of Economic Policy. Kingston, Ontario: Queens University. Pp. 95–130.

Baldwin, J. R., and G. Gellatly. 1999a. 'Developing High-Tech Classification Schemes: A Competency-Based Approach'. In *New Technology-Based Firms in the 1990s.* Vol. 6. Edited by R. Oakey, W. During, and S. Mukhtar. Oxford: Elsevier Science Ltd. Pp. 185–99.

Baldwin, J. R., and G. Gellatly. 1999b. 'A Firm-Based Approach to Industry Classification: Identifying the Knowledge-Based Economy'. In *Doing Business in the Knowledge-Based Economy.* Edited by L. Lefebvre, E. Lefebvre, and P. Mohnen. Holland: Kluwer Academic Publishers. Pp. 199–238.

Baldwin, J. R., G. Gellatly, J. Johnson, and V. Peters. 1998. *Innovation in Dynamic Service Industries.* Catalogue No. 88-516-XPB. Ottawa: Statistics Canada.

Baldwin, J. R., and P. K. Gorecki. 1994. 'Concentration and Mobility Statistics'. *Journal of Industrial Economics* 42: 93–104.

Baldwin, J. R., T. Gray, and J. Johnson. 1996. 'Advanced Technology Use and Training in Canadian Manufacturing'. *Canadian Business Economics* 5: 51–70.

Baldwin, J. R., T. Gray, and J. Johnson. 1997. 'Technology-Induced Wage Premia in Canadian Manufacturing Plants During the 1980s'. Research Paper No. 92. Analytical Studies Branch. Ottawa: Statistics Canada.

Baldwin, J. R., T. Gray, J. Johnson, J. Proctor, M. Rafiquzzaman, and D. Sabourin. 1997. *Failing Concerns: Business Bankruptcy in Canada.* Catalogue No. 61-525-XPE. Ottawa: Statistics Canada.

Baldwin, J. R., and P. Hanel. 2000. *Multinationals and the Canadian Innovation Process.* Research Paper No. 151. Analytical Studies Branch. Ottawa: Statistics Canada.

Baldwin, J. R., and J. Johnson. 1996a. 'Human Capital Development and Innovation: A Sectoral Analysis'. In *The Implications of Knowledge-Based Growth for Micro-Economic Policies.* Edited by P. Howitt. Calgary: University of Calgary Press. Pp. 83–110.

Baldwin, J. R., and J. Johnson. 1996b. 'Business Strategies in More- and Less-Innovative Firms in Canada'. *Research Policy* 25: 785–804.

Baldwin, J. R., and J. Johnson. 1998b. 'Innovator Typologies, Competencies, and Performance'. In *Microfoundations of Economic Growth: A Schumpeterian Perspective.* Edited by C. Green and C. McCann. Ann Arbor: University of Michigan Press. Pp. 227–53.

Baldwin, J. R., and J. Johnson. 1999a. 'Entry, Innovation and Firm Growth'. In *Are Small Firms Important? Their Role and Impact.* Edited by Z. J. Acs. Dordrecht: Kluwer. Pp. 51–71.

Baldwin, J. R., and J. Johnson. 1999b. *The Defining Characteristics of Entrants in Science-Based Industries.* Catalogue No. 88-517-XPB. Ottawa: Statistics Canada.

Baldwin, J. R., and Z. Lin. 2002. 'Impediments to Advanced Technology Adoption for Canadian Manufacturing'. *Research Policy* 31: 1–18.

Baldwin, J. R., and V. Peters. 2001. *Training as a Human Resource Strategy: The Response to Staff Shortages and Technological Change.* Research Paper No. 154. Analytical Studies Branch. Ottawa: Statistics Canada.

Baldwin, J. R., and G. Picot. 1995. 'Employment Generation by Small Producers in the Job-Turnover Process'. *Small Business Economics* 7: 1–14.

Baldwin, J. R., and M. Rafiquzzaman. 1994. *Structural Change in the Canadian Manufacturing Sector: 1970 to 1990.* Research Paper No. 61. Analytical Studies Branch. Ottawa: Statistics Canada.

Baldwin, J. R., and M. Rafiquzzaman. 1995. *Restructuring in the Canadian Manufacturing Sector: 1970 to 1990. Industry and Regional Dimension of Job*

Turnover. Research Paper No. 78. Analytical Studies Branch. Ottawa: Statistics Canada.

Baldwin, J. R., and M. Rafiquzzaman. 1999. 'The Effect of Technology and Trade on Wage Differentials Between Non-production and Production Workers in Canadian Manufacturing'. In *Innovation, Industry Evolution and Employment.* Edited by D. B. Audretsch and R. Thurik. Cambridge: Cambridge University Press. Pp. 57–85.

Baldwin, J. R., E. Rama, and D. Sabourin. 1999. *Growth in Advanced Technology Use in Canadian Manufacturing During the 1990s.* Research Paper No. 105. Analytical Studies Branch. Ottawa: Statistics Canada.

Baldwin, J. R., and D. Sabourin. 1995. *Technology Adoption in Canadian Manufacturing.* Catalogue No. 88-512-XPB. Ottawa: Statistics Canada.

Baldwin, J. R., and D. Sabourin. 1997. 'Factors Affecting Technology Adoption: A Comparison of Canada and the U. S.'. *Canadian Economic Observer* (August): 3.1–3.17.

Baldwin, J. R., and D. Sabourin. 2001. *Impact of the Adoption of Advanced Information and Communication Technologies on Firm Performance in the Canadian Manufacturing sector.* Research Paper No. 174. Analytical Studies Branch. Ottawa: Statistics Canada.

Baldwin, J. R., D. Sabourin, and M. Rafiquzzaman. 1996. *Benefits and Problems Associated with Technology Adoption in Canadian Manufacturing.* Catalogue No. 88-514-XPE. Ottawa: Statistics Canada.

Baldwin, J. R., D. Sabourin, and D. West. 1999. *Advanced Technology in the Canadian Food-Processing Industry.* Catalogue No. 88-518-XPE. Ottawa: Statistics Canada.

Baldwin, W., and J. T. Scott. 1987. *Market Structure and Technological Change.* Chur, Switzerland: Harwood.

Bartel, A. P., and F. R. Lichtenberg. 1987. 'The Comparative Advantage of Educated Workers in Implementing a New Technology'. *Review of Economics and Statistics* 69: 1–11.

Bartlett, C. A., and S. Ghoshal. 1989. *Managing Across Borders.* Boston, MA: Harvard Business School Press.

Berman, E., J. Bound, and Z. Griliches. 1993. *Changes in the Demand for Skilled Labor Within US Manufacturing Industries.* Working Paper No. 4255. Cambridge, MA: National Bureau of Economic Research.

Bernstein, J. I. 1997. 'Interindustry R&D Spillovers for Electrical and Electronic Products: The Canadian Case'. *Economic Systems Research* 9(1): 111–25.

Birkenshaw, J. 1995. *Business Development Initiatives of Multinational Subsidiaries in Canada.* Occasional Paper No. 2. Ottawa: Industry Canada.

Bosworth. D., and P. Stoneman. 1997. 'Information and Technology Transfer in Europe'. In *Innovation Measurement and Policies.* Edited by A. Arundel and R. Garrelfs. Luxembourg: Office for Official Publications of the European Communities. pp. 114–20.

Bosworth, D., and T. Westaway. 1984. 'The Influence of Demand and Supply Factors in the Quantity and Quality of Inventive Activity'. *Applied Economics* 16: 131–146.

Britton, J. N. H. 1980. 'Industrial Dependence and Technological Underdevelop-

ment: Canadian Consequences of Foreign Direct Investment'. *Regional Studies* 14: 181–99.
Britton, J. N. H., and J. M. Gilmour. 1978. *The Weakest Link*. Background Study. Ottawa: Science Council of Canada.
Brouwer, E., and A. Kleinknecht. 1996. 'Determinants of Innovation: A Microeconometric Analysis of Three Alternative Innovation Output Indicators'. In *Determinants of Innovation: The Message from New Indicators*. Edited by A. Kleinknecht. London: Macmillan. Pp. 99–124.
Brower, E., A. Kleinknecht, and J. O. N. Reijnen. 1993. 'Employment, Growth and Innovation at the Firm Level: An Empirical Study'. *Journal of Evolutionary Economics* 3: 153–59.
Bussy, J. C., I. Kabla, and T. Lehoucq. 1994. *The Role of Patents as an Appropriation Mechanism and the Other Function of Patents*. Paris: Insee.
Caves, R. E. 1971. 'International Corporations: The Industrial Economics of Foreign Investment'. *Economica* 38: 1–27.
Caves, R. E. 1975. *Diversification, Foreign Investment and Scale in North American Manufacturing Industries*. Ottawa: Economic Council of Canada.
Caves, R. E. 1982. *Multinational Enterprise and Economic Analysis*. Cambridge, MA: Cambridge University Press.
Caves, R. E., M. E. Porter, and A. M. Spence with J. T. Scott. 1980. *Competition in the Open Economy: A Model Applied to Canada*. Cambridge, MA: Harvard University Press.
Caves, R. E., and M. Uekasa. 1976. 'Industrial Organization in Japan'. In *Asia's New Giant*. Edited by H. Patrick and H. Rosovsky. Washington, DC: Brookings Institution. Pp. 459–523.
Chakrabati, A. K., and M. R. Halperin. 1990. 'Technical Performance and Firm Size: Analysis of Patents and Publications of U.S. Firms'. *Small Business Economics* 2(3): 183–90.
Chandler, A. D. 1977. *The Visible Hand*. Cambridge, MA: Harvard University Press.
Chesnais, F. 1993. 'The French National System of Innovation'. In *National Innovation Systems: A Comparative Analysis*. Edited by Richard R. Nelson. Oxford: Oxford University Press.
Cockburn, I., and Z. Griliches. 1988. 'Industry Effects and Appropriability Measures in the Stock Market's Valuation of R&D and Patents'. NBER Working Paper No. 2645. Published in abridged form in *American Economic Review* 78(2): 419–23.
Coe, D. T., and E. Helpman. 1994. *International Monetary Fund, North-South R&D Spillovers*. Washington, DC: International Monetary Fund.
Cohen, W. M. 1996. 'Empirical Studies of Innovative Activity'. In *The Handbook of the Economics of Technological Change*. Edited by P. Stoneman. Oxford: Basil Blackwell. Pp. 182–264.
Cohen, W. M., and S. Klepper. 1992. 'The Tradeoff Between Firm Size and Diversity in the Pursuit of Technological Progress'. *Small Business Economics* 4: 1–14.
Cohen, W. M., and S. Klepper. 1996a. 'A Reprise of Size and R&D'. *Economic Journal* 106: 925–52.

Cohen, W. M., and S. Klepper. 1996b. 'Firm Size and the Nature of Innovation Within Industries: The Case of Process and Product R&D'. *Review of Economics and Statistics* 78: 232–43.
Cohen, W. M., and R. C. Levin. 1989. 'Empirical Studies of Innovation and Market Structure'. In *Handbook of Industrial Organization*. Vol. 2. Edited by R. Schmalensee and R. D. Willig. Amsterdam: North-Holland. Pp. 1059–107.
Cohen, W. M., R. C. Levin, and D. C. Mowery. 1987. 'Firm Size and R&D Intensity: A Reexamination'. *Journal of Industrial Economics* 35: 543–65.
Cohen, W. M., and D. A. Levinthal. 1989. 'Innovation and Learning: Two Faces of R&D'. *Economic Journal* 99: 569–96.
Cohen, W. M., R. R. Nelson, and J. Walsh. 1997. 'Appropriability Conditions: Why Firms Patent and Why They Do Not'. Carnegie Mellon Working Paper, June 24.
Cohen, W. M., R. Nelson, and J. Walsh. 2000. *Protecting Their Intellectuall Assets: Appropriability Conditions and Why U.S. Manufacturing Firms Patent (or Not)*. Working Paper No. 7552. Cambridge, MA: National Bureau of Economic Research.
Consiglio Nazionale delle Recherche. 1993. *Patterns of Innovation in Italian Industry*. Empirical Studies and the Community Innovation Survey. Luxembourg: European Commission.
Consumer and Corporate Affairs Canada. 1990. *Intellectual Property Rights and Canada's Commercial Interests: A Summary Report*. Ottawa: Ministry of Supply and Services.
Crépon, B., E. Duguet, and I. Kabla. 1996. 'Schumpeterian Conjectures: A Moderate Support from Various Innovation Measures'. In *Determinants of Innovation: The Message from New Indicators*. Edited by A. Kleinknecht. London: Macmillan. Pp. 63–98.
Crépon, B., E. Duguet, and J. Mairesse. 1998. 'Research, Innovation, and Productivity: An Econometric Analysis at the Firm Level'. *Economics of Innovation and New Technology* 7(2): 115–58.
Dasgupta, P., and J. Stiglitz. 1980a. 'Uncertainty, Industrial Structure and the Speed of R&D'. *Bell Journal of Economics* 11(1): 1–28.
Dasgupta P., and J. Stiglitz. 1980b. 'Industrial Structure and the Nature of Innovative Activity'. *Economic Journal* 90: 266–93.
De Melto, D. P., K. E. McMullen, and R. M. Wills. 1980. *Preliminary Report: Innovation and Technical Change in Five Canadian Industries*. Discussion Paper No. 176. Ottawa: Economic Council of Canada.
Doern, B. G. 1995. *Institutional Aspects of R&D Tax Incentives: The SR&ED Tax Credit*. Occasional Paper, No. 6. Ottawa: Industry Canada.
Dunning, J. H. 1958. *American Investment in British Manufacturing Industry*. London: George Allen and Unwin.
Dunning, J. H. 1992. 'Multinational Enterprises and the Globalization of Innovatory Capacity'. In *Technology Management and International Business: Internationalisation of R&D and Technology*. Edited by O. Granstrand, L. Hakanson, and S. Sjolander. Chichester: John Wiley and Sons. Pp. 19–52.
Dunning, J. H. 1993. *Multinational Enterprises and the Global Economy*. New York: Addison-Wesley.

Dupont, M.-J. 1994. 'Les filiales étrangères en France: Des atouts majeurs pour innover'. *Les Chiffres Clés; L'innovation technologiques.* Paris: Ministère de l'Industrie des Postes et Télécommunications et du Commerce Extérieur.

Eastman, H. C., and S. Stykolt. 1967. *The Tariff and Competition in Canada.* Toronto: Macmillan.

Eden, L (ed). 1993. *Multinationals in Canada.* Calgary: University of Calgary Press.

Edmunds, S. E., and S. L. Khoury. 1986. 'Exports: A Necessary Ingredient in the Growth of Small Business Firms'. *Journal of Small Business Management* 24: 54–65.

Edquist, C. 1997. 'A Systems Approach to Innovation Indicator Development'. In *Innovation Measurement and Policies.* Edited by A. Arundel and R. Garrelfs. Luxembourg: Office for Official Publications of the European Communities. Pp. 74–79.

Enos, J. 1962. *Petroleum Profits and Progress: A History of Process Innovation.* Cambridge, MA: MIT Press.

Ergas, H. 1987. 'The Importance of Technology Policy'. In *Economic Policy and Technological Performance.* Edited by P. Dasgupta and P. Stoneman. Cambridge: Cambridge University Press. Pp. 51–96.

Etemad, H., and L. Séguin-Dulude. 1987. 'Patenting Patterns in 25 Large Multinational Enterprises'. *Technovation* 7: 1–15.

European Commission. 1994. *The Community Innovation Survey: Status and Perspectives.* Luxembourg: Office for the Official Publications of the European Community.

European Commission. 1997. *Innovation Measurement and Policy.* Luxembourg: Office for the Official Publications of the European Community.

Evangelista, R., T. Sandven, G. Sirilli, and K. Smith. 1997. 'Measuring the Cost of Innovation in European Industry'. In *Innovation Measurement and Policies.* Edited by A. Arundel and R. Garrelfs. Luxembourg: Office for Official Publications of the European Communities. Pp. 109–13.

Evans, D., and B. Jovanovic. 1989. 'An Estimated Model of Entrepreneurial Choice Under Liquidity Constraints'. *Journal of Political Economy* 97: 808–27.

Felder, J., G. Licht, E. Nerlinger, and H. Stahl. 1996. 'Factors Determining R&D and Innovation Expenditure in German Manufacturing Industries'. In *Determinants of Innovation: The Message from New Indicators.* Edited by A. Kleinknecht. London: Macmillan. Pp. 125–54.

Fellner, W. 1951. 'The Influence of Market Structure on Technological Progress'. *Quarterly Journal of Economics* 65(3): 556–77.

Finance Canada. 1998. *The Federal System of Income Tax Incentives for Scientific Research and Experimental Development: Evaluation Report.* Ottawa: Department of Finance.

Firestone, O. J. 1971. *Economic Implications of Patents.* Ottawa: University of Ottawa Press.

Freeman, C. 1971. *The Role of Small Firms in Innovation in the United Kingdom Since 1945.* Research Report No. 1. Committee of Inquiry on Small Firms. London: Her Majesty's Stationery Office.

Freeman, C. 1982. *The Economics of Industrial Innovation.* 2d ed. London: Francis Pinter.
Freeman, C. 1991. 'Networks of Innovators: A Synthesis of Research Issues'. *Research Policy* 20(5): 363–79.
Gellatly, G. 1999. *Differences in Innovator and Non-innovator Profiles: Small Establishments in Business Services.* Research Paper No. 143. Analytical Studies Branch. Ottawa: Statistics Canada.
Gellatly. G., and V. Peters. 1999. *Understanding the Innovation Process: Innovation in Dynamic Service Industries.* Research Paper No. 127. Analytical Studies Branch. Ottawa: Statistics Canada.
Geroski P. A. 1990. 'Innovation, Technological Opportunity, and Market Structure'. *Oxford Economic Papers*, n.s. 42(3): 586–602.
Geroski P. A. 1995. 'Do Spillovers Undermine the Incentive to Innovate?' In *Economic Approaches to Innovation.* Edited by S. Dowrick. Aldershot, UK: Edward Elgar.
Geroski, P. A., S. Machin, and J. Van Reenen. 1993. 'The Profitability of Innovating Firms'. *RAND Journal of Economics* 24: 198–211.
Geroski, P. A., J. Van Reenen, and C. F. Walters. 1997. 'How Persistently Do Firms Innovate?' *Research Policy* 26: 33–48.
Ghoshal, S., and C. A. Bartlett. 1990. 'The Multinational Corporation as an Interorganizational Network'. *Academy of Management Review* 15(4): 603–25.
Globerman, S. 1979. 'Foreign Direct Investment and "Spillover" Efficiency Benefits in Canadian Manufacturing Industries'. *Canadian Journal of Economics* 12(1): 42–56.
Globerman, S., J. C. Ries, and I. Vetinsky. 1994. 'The Economic Performance of Foreign Affiliates in Canada'. *Canadian Journal of Economics* 27(1): 143–56.
Gorecki, P. K. 1975. 'An Inter-industry Analysis of Diversification in the UK Manufacturing Sector'. *Journal of Industrial Economics* 24: 131–46.
Gort, M. 1962. *Diversification and Integration in American Industry.* Princeton, NJ: Princeton University Press.
Gort, M., and S. Klepper. 1982. 'Time Paths in the Diffusion of Product Innovations'. *Economic Journal* 92: 630–53.
Granstrand, O. H. L. 1993. 'Internationalization of R&D: A Survey of Some Recent Research'. *Research Policy* 22: 413–30.
Grant, R. M. 1977. 'The Determinants of the Inter-Industry Pattern of Diversification by UK Manufacturing Industries'. *Bulletin of Economic Research* 29: 84–95.
Grégoire, P. E. 1992. *Compte rendue de l'enquête sur la perception des mesures fiscales en R-D.* Québec: Ministère de l'Industrie, Commerce, et Technologie.
Griliches, Z. 1979. 'Issues in Assessing the Contribution of Research and Development to Productivity Growth'. *Bell Journal of Economics* 10(1): 92–116.
Griliches, Z. 1990. 'Patent Statistics and Economic Indicators'. *Journal of Economic Literature.* 28: 1661–707.
Gu, W., and L. Whewell. 1999. *University Research and the Commercializtion of Intellectual Property in Canada.* Research Paper No. 21. Ottawa: Industry Canada.

Hall, B. H. 1992. 'Investment and Research and Development at the Firm Level: Does Source of Financing Matter?' Working Paper No. 4096. Cambridge, MA: National Bureau of Economic Research.

Hall, B. H., A. N. Link, and J. T. Scott. 2000. *Universities as Research Partners*. Working Paper No. W7643. Cambridge, MA: National Bureau of Economic Research.

Hanel, P. 1976. *The Relationship Existing Between the R&D Activity of Canadian Manufacturing Industries and Their Performance in the International Market*. Technological Innovation Studies Program. Ottawa: Industry, Trade and Commerce, Office of Science and Technology.

Hanel, P. 1989. 'Technology and Canadian Exports of Machinery for Wood and Paper Industry'. *Canadian Journal of Administrative Studies* 6(1): 9–17.

Hanel, P. 1994. 'Interindustry Flows of Technology: An Analysis of the Canadian Patent Matrix and Input-Output Matrix for 1978–1989'. *Technovation* 14(8): 529–48.

Hanel, P. 2000. 'R&D, Interindustry and International Technology Spillovers and Total Factor Productivity Growth of Manufacturing Industries in Canada: 1974–1989'. *Economic Systems Research* 12(3): 345–61.

Hanel, P., and K. Palda. 1982. 'Les entreprises innovatrices et leur performance à l'exportation'. *L'Actualité Économique* 58: 380–97.

Hanel, P., and K. Palda. 1992. 'Appropriability and Public Support of R&D in Canada'. *Prometheus* 10(2): 204–26.

Hanel, P., and A. St.-Pierre. 2002. 'Effects of R&D Spillovers on the Profitability of Firms'. *Review of Industrial Organization* 20(4): 305–22.

Harabi, N. 1997. 'Channels of R&D Spillovers: An Empirical Investigation of Swiss Firms'. *Technovation* 17: 627–35.

Himmelberg, C. P., and B. C. Petersen. 1994. 'R&D and Internal Finance: A Panel Study of Small Firms in High-Tech Industries'. *Review of Economics and Statistics* 76: 38–51.

Holbrook, J. A. D., and R. J. Squires. 1996. 'Firm-Level Analysis of Determinants of Canadian Industrial R&D Performance'. *Science and Public Policy* 23(6): 369–74.

Hollander, S. 1965. *The Sources of Increased Efficiency: A Study of Dupont Rayon Plants*. Cambridge: Cambridge University Press.

Hughes, A., and D. J. Storey. 1994. *Finance and the Small Firm*. London: Routledge.

Jaffe, A. B. 1988. 'Demand and Supply Influences in R&D Intensity and Productivity Growth'. *Review of Economics and Statistics* 70: 431–37.

Jaffe, A. B., and K. Palmer. 1994. *Environmental Regulation and Innovation: A Panel Data Study*. Resources for the Future Discussion Paper: No. 95-03. Washington, DC: Resources for the Future.

Jensen, M. C., and W. Meckling. 1976. 'Theory of the Firm: Managerial Behaviour, Agency Costs and Ownership Structure'. *Journal of Financial Economics* 3(4): 305–60.

Johnson, J., J. R. Baldwin and C. Hinchley. 1997. *Successful Entrants: Creating the Capacity for Survival and Growth*. Catalogue No. 61-524-XPE. Ottawa: Statistics Canada.

Kamien, M. I., and N. L. Schwartz. 1982. *Market Structure and Innovation*. Cambridge, MA: Cambridge University Press.
Kamien, M. I., and N. L. Schwartz. 1985. 'Market Structure and Innovation: A Survey'. *Journal of Economic Literature* 13: 1–37.
Kleinknecht, A. 1987. 'Measuring R&D in Small Firms: How Much are We Missing?' *Journal of Industrial Economics* 36(2): 253–56.
Kleinknecht, A. 1989. 'Firm Size and Innovation: Observations in Dutch Manufacturing Industry'. *Small Business Economics* 1(1) 215–22.
Kleinknecht, A. (ed.). 1996a. *Determinants of Innovation: The Message from the New Indicators*. London: Macmillan Press.
Kleinknecht, A. 1996b. 'New Indicators and Determinants of Innovation: An Introduction'. In *Determinants of Innovation: The Message from New Indicators*. Edited by A. Kleinknecht. London: Macmillan Press. Pp. 1–12.
Kleinknecht, A., and D. Bain. 1993. *New Concepts in Innovation Output Measurement*. New York: St. Martin's Press.
Kleinknecht, A., T. P. Poot, and J. O. N. Reijnen. 1991. 'Technical Performance and Firm Size: Survey Results from the Netherlands'. In *Innovation and Technological Change: An International Comparison*. Edited by Z. J. Acs and D. B. Audretsch. Ann Arbor: University of Michigan Press. Pp. 71–83.
Kleinknecht, A., J. O. N. Reijnen, and W. Smits. 1992. 'Collecting Literature-Based Innovation Output Indicators. The Experience in the Netherlands'. In *New Concepts in Innovation Output*. Edited by A. Kleinknecht and D. Bain. London: Macmillan. Pp. 42–84.
Kleinknecht, A., and B. Verspagen. 1990. 'Demand and Innovation: Schmookler Re-examined'. *Research Policy* 19: 387–94.
Klepper, S. 1996. 'Entry, Exit, Growth and Innovation over the Product Life Cycle'. *American Economic Review* 86: 562–83.
Klepper, S., and J. H. Millar. 1995. 'Entry, Exit and Shakeouts in the United States in New Manufactured Products'. *International Journal of Industrial Organization* 13: 567–91.
Kochan, T. A. 1988. 'Looking to the year 2000: Challenges for Industrial Relations and Human Resource Management'. In *Perspective 2000*. Ottawa: Economic Council of Canada.
Kraft, K. 1989. 'Market Structure, Firm Characteristics and Innovative Activity'. *Journal of Industrial Economics* 37: 329–36.
Kraft, K. 1990. 'Are Product and Process Innovations Independent of Each Other?' *Applied Economics* 22: 1029–38.
Krugman, P. 1994. *Peddling Prosperity: Economic Sense and Nonsense in the Age of Diminished Expectations*. New York: W. W. Norton & Company Inc.
Kumar V. 1995. *The Role of R&D Consortia in Technology Development*. Occasional Paper No. 2. Ottawa: Industry Canada.
Kumar, V., and U. Kumar. 1991. *Technological Innovation in Canadian Manufacturing Industry: An Investigation of the Speed and Cost of Innovation*. Ottawa: School of Business, Carleton University.
Leo, H. 1996. 'Determinants of Product and Process Innovation'. *Économies et Sociétés*, Dynamique technologique et organisation, Série W 3: 61–77.

Lerner, J. 1994. 'Patenting in the Shadow of Competitors'. *Journal of Law and Economics* 38: 463–96.

Levin, R. C. 1982. 'The Semi-conductor Industry'. In *Government and Technical Progress: A Cross-Industry Analysis.* Edited by R. R. Nelson. New York: Pergammon Press. Pp. 9–100.

Levin, R. C., W. M. Cohen, and D. C. Mowery. 1985. 'R&D Appropriability, Opportunity, and Market Structure: New Evidence on Some Schumpeterian Hypotheses'. *American Economic Review: Papers and Proceedings* 75: 20–24.

Levin, R. C., A. Klevorick, R. Nelson, and S. G. Winter. 1987. 'Appropriating the Returns from Industrial Research and Development'. *Brookings Papers on Economic Activity* 3: 783–820.

Levin, R. C., and P. C. Reiss. 1984. 'Tests of a Schumpeterian Model of R&D and Market Structure'. In *R&D, Patents and Productivity.* Edited by Z. Griliches. Chicago: Chicago University Press. Pp. 175–208.

Levin, R. C., and P. C. Reiss. 1988. 'Cost-Reducing and Demand-Creating R&D Spillovers'. *RAND Journal of Economics* 19(4): 538–56.

Licht, G. 1997. 'The Impact of Innovation on Employment in Europe'. In *Innovation Measurement and Policies.* Edited by A. Arundel and R. Garrelfs. Luxembourg: Office for Official Publications of the European Communities. Pp. 135–41.

Link, A. N., and B. Bozeman. 1991. 'Innovation Behavior in Small-Sized Firms'. *Small Business Economics* 3: 179–84.

Lundvall, B. A. (ed.). 1992. *National Systems of Innovation: Towards a Theory of Innovation and Interactive Learning.* London: Francis Pinter.

Lunn, J. 1986. 'An Empirical Analysis of Process and Product Patenting: A Simultaneous Equation Framework'. *The Journal of Industrial Economics* 34: 319–30.

Lunn, J. 1987. 'An Empirical Analysis of Firm Process and Product Patenting'. *Applied Economics* 19: 743–51.

Main, O. W. 1955. *The Canadian Nickel Industry.* Toronto: University of Toronto Press.

Mairesse J., and P. Mohnen. 1995. 'Research & Development and Productivity: A Survey of the Econometric Literature'. Montréal: Université du Québec à Montréal. Manuscript.

Malerba, F. 1993. 'The National System of Innovation: Italy'. In *National Innovation Systems: A Comparative Analysis.* Edited by R. R. Nelson. Oxford: Oxford University Press.

Mansfield, E. 1986. 'Patents and Innovation: An Empirical Study'. *Management Science* 32: 173–81.

Mansfield, E. 1988. 'The Speed and Cost of Industrial Innovation in Japan and the U.S.: External vs. Internal Technology'. *Management Science* 34(10): 1157–68.

Mansfield, E. 1990. 'Patents and Innovation: An Empirical Study'. *Management Science* 32: 173–81.

Mansfield, E., A. Romeo, M. Schwartz, D. Teece, S. Wagner, and P. Brach. 1982. 'Basic Research and Productivity Increase in Manufacturing'. In *Technology Transfer, Productivity and Economic Policy.* Edited by E. Mansfield, A. Romeo,

M. Schwartz, D. Teece, S. Wagner, P. Brach, et al. New York: W. W. Norton & Company.

Mantoux, P. 1961. *The Industrial Revolution in the Eighteenth Century*. New York: Harper and Row.

Matzner, E. F., R. Schettjat, and M. Wagner. 1990. 'Labour Market Effects of New Technology'. *Futures* 20: 687–709.

McDonald, S. 1986. 'Theoretically Sound: Practically Useless? Government Grants for Industrial R&D in Australia'. *Research Policy* 16: 269–84.

McFetridge, D. G. 1993. 'The Canadian System of Innovation'. In *National Innovation Systems*. Edited by R. R. Nelson. Oxford: Oxford University Press. Pp. 299–323.

McFetridge, D. G. 1995. *Science and Technology: Perspectives for Public Policy*. Occasional Paper No. 9. Ottawa: Industry Canada.

McGuiness, N. W., and B. Little. 1981. 'The Impact of R&D Spending on the Foreign Sales of New Canadian Industrial Products'. *Research Policy* 10: 78–98.

Mohnen, P. 1992. *The Relationship Between R&D and Productivity Growth in Canada and Other Industrialized Countries*. Ottawa: Economic Council of Canada.

Mowery, D. C. 1983a. 'The Relationship Between Intrafirm and Contractual Forms of Industrial Research in American Manufacturing, 1900–1940'. *Explorations in Economic History* 20: 351–74.

Mowery, D. C. 1983b. 'Economic Theory and Government Technology Policy'. *Policy Sciences* 16: 27–43.

Mowery, D. C., and N. Rosenberg. 1989. *Technology and the Pursuit of Economic Growth*. Cambridge: Cambridge University Press.

Myers, S. C. 1984. 'The Capital Structure Puzzle'. *Journal of Finance* 39: 575–92.

Napolitano, G. 1991. 'Industrial Research and Sources of Innovation: A Cross Industry Analysis of Italian Manufacturing Firms'. *Research Policy* 20: 171–78.

Nelson, R. R. 1959. 'The Economics of Invention: A Survey of the Literature'. *The Journal of Business* 32(2): 101–37.

Nelson, R. R. 1987. *Understanding Technical Changes as An Evolutionary Process*. Amsterdam: North-Holland.

Nelson, R. R. 1993. *National Systems of Innovation: A Comparative Analysis*. Oxford: Oxford University Press.

Nelson, R. R., and N. Rosenberg. 1993. 'Technical Innovation and National Systems'. In *National Innovation Systems: A Comparative Analysis*. Edited by R. R. Nelson. Oxford: Oxford University Press. Pp. 3–21.

Nelson, R. R., and S. G. Winter. 1982. *An Evolutionary Theory of Economic Change*. Cambridge, MA: Harvard University Press.

Niosi, J. 1995a. 'Technical Alliances in Canadian High-Technology'. In *Flexible Innovation, Technological Alliances in Canadian Industry*. Edited by J. Niosi. Montréal: McGill and Queen's University Press.

Niosi, J. 1995b. *Vers l'innovation flexible*. Montréal: Les presses de l'Université de Montréal.

Organization for Economic Cooperation and Development. 1992. *Frascati Manual, 1993: Proposed Standards for Surveys of Research and Experimental Development*. Paris: OECD.
Organization for Economic Cooperation and Development. 1993a. *The Measurement of Scientific and Technical Activities: Standard Practice for Surveys of Research and Development – Frascati Manual, 1993*. Paris: OECD.
Organization for Economic Cooperation and Development. 1993b. *Comparison of Innovation Survey Findings*. EC/OECD Joint Seminar on Innovation Surveys, OECD-DSTI, Paris, April 29, 1993.
Organization for Economic Cooperation and Development. 1994. *Main Science and Technology Indicators, 1994*. Paris: OECD.
Organization for Economic Cooperation and Development. 1997. *Proposed Guidelines for Collecting and Interpreting Technological Innovation: Oslo Manual*. Paris: OECD.
Pakes, A., and Z. Griliches. 1984. 'Patents and R&D at the Firm Level: A First Look'. In *R&D, Patents, and Productivity*. Edited by Z. Griliches. National Bureau of Economic Research. Chicago: University of Chicago Press. Pp. 55–72.
Palda, K. 1993. *Innovation Policy and Canada's Competitiveness*. Vancouver: The Fraser Institute.
Palys, E. 1992. *Qualitative and Quantitative Perspectives*. Toronto: Harcourt Brace Jovanovich.
Patel, P., and K. Pavitt. 1991. 'The Limited Importance of Large Firms in Canadian Technological Activities'. In *Foreign Investment, Technology and Economic Growth*. Investment Canada Research Series. Edited by D. G. McFetridge. Calgary: University of Calgary Press. Pp. 71–88.
Pavitt, K. 1984. 'Sectoral Patterns of Technical Change: Towards a Taxonomy and a Theory'. *Research Policy* 13: 343–73.
Pavitt, K. 1988. 'Uses and Abuses of Patent Statistics'. In *Handbook of Quantitative Studies of Science and Technology*. Edited by A. F. J. van Raan. Amsterdam: Elsevier. Pp. 508–36.
Pavitt, K. 1998. 'Technologies, Products and Organization in the Innovating Firm: What Adam Smith Tells Us and Joseph Schumpeter Doesn't'. *Industrial and Corporate Change* 7(3): 433–52.
Pavitt K., M. Robson, and J. Townsend. 1987. 'The Size Distribution of Innovating Firms in the U.K.: 1945–83'. *The Journal of Industrial Economics* 35: 297–316.
Phillips, A. 1966. 'Patents, Competition, and Technical Progress'. *American Economic Review* 56: 301–10.
Porter, M. E., and C. van-der-Linde. 1995. 'Toward a New Conception of the Environment-Competitiveness Relationship'. *Journal of Economic Perspectives* 9: 97–118.
Raynauld, A. 1972. 'The Ownership and Performance of Firms'. In *The Multinational Firm and the Nation State*. Edited by G. Paquette. Toronto: Ryerson Press. Pp. 96–101.
Reshef, Y. 1993. 'Employees, Unions and Technological Changes'. *Journal of Labor Research* 44: 111–29.

Robinson, J. C. 1995. 'The Impact of Environmental and Occupational Health Regulation on Productivity Growth in U.S. Manufacturing'. *Yale Journal on Regulation* 12(2): 387–434.

Robson, M., J. Townsend, and K. Pavitt. 1988. 'Sectoral Patterns of Production and Use of Innovations in the UK: 1945–1983'. *Research Policy* 17: 1–14.

Rohmer, P. 1994. 'The Origins of Endogenous Growth'. *The Journal of Economic Perspectives* 8: 3–22.

Rosenberg, N. 1976. *Perspectives on Technology*. Cambridge: Cambridge University Press.

Rosenberg, N. 1982. *Inside the Black Box*. Cambridge: Cambridge University Press.

Rosenberg, N. 1990. 'Why Do Firms Do Research (with Their Own Money)?' *Research Policy* 19: 165–74.

Rosenberg, N., and R. R. Nelson. 1994. 'American Universities and Technical Advance in Industry'. *Research Policy* 23: 323–48.

Rosenbloom, R. S., and W. J. Abernathy. 1982. 'The Climate for Innovation in Industry'. *Research Policy* 11: 209–25.

Rothwell, R., and W. Zegveld. 1982. *Innovation and the Small and Medium-Sized Firm*. London: Francis Pinter.

Safarian, A. E. 1973. *Foreign Ownership of Canadian Industry*. Toronto: University of Toronto Press.

Sakurai N., G. Papaconstantou and E. Ioannidis. 1995. *The Impact of R&D and Technology Diffusion on Productivity Growth: Evidence from 10 OECD Countries in 1970 and 1980*. Macroeconomic and Sectoral Evidence. Paris: OECD Expert Workshop on Technology, Productivity, and Employment.

Salter, W. 1966. *Productivity and Technical Change*. 2d ed. Cambridge: Cambridge University Press.

Saunders, R. S. 1980. 'The Determinants of Productivity in Canadian Manufacturing Industries'. *Journal of Industrial Economics* 29: 167–84.

Schankerman, M. 1991. 'How Valuable is Patent Protection? Estimates by Technology Field Using Patent Renewal Data'. Working Paper No. 3780. Cambridge, MA: National Bureau of Economic Research.

Scherer, F. M. 1966. 'Firm Size, Market Structure, Opportunity, and the Output of Patented Inventions'. *American Economic Review* 55: 1097–125.

Scherer, F. M. 1980. *Industrial Market Structure and Economic Performance*. 2d ed. Chicago: Rand McNally.

Scherer F. M. 1982a. 'Inter-industry Technology Flows and Productivity Growth'. *Review of Economics and Statistics* 64: 627–34.

Scherer F. M. 1982b. 'Inter-industry Technology Flows in the United States'. *Research Policy* 11: 227–45.

Scherer, F. M. 1982c. 'Demand-Pull and Technological Invention: Schmookler Revisited'. *Journal of Industrial Economics* 30: 225–37.

Scherer, F. M. 1983. 'The Propensity to Patent'. *International Journal of Industrial Organization* 1: 107–28.

Scherer, F. M. 1984. *Innovation and Growth: Schumpeterian Perspectives*. Cambridge, MA: MIT Press.

Scherer, F. M. 1992. 'Schumpeter and Plausible Capitalism'. *Journal of Economic Literature* 30: 1416–34.

Schmookler, J. 1966. *Invention and Economic Growth*. Cambridge, MA: Harvard University Press.

Schumpeter, J. A. 1942. *Capitalism, Socialism, and Democracy*. New York: Harper and Row.

Séguin-Dulude, L. 1982. 'Les flux technologiques interindustriels: Une analyse exploratoire du potentiel canadien'. *L'Actualité Économique* 58(3): 259–81.

Séguin-Dulude, L., and C. Desranleau. 1989. *The Individual Canadian Inventor*. Catalogue 88–510. Ottawa: Statistics Canada.

Serapio, M. G. 1997. 'Internationalization of Industrial R&D: Theoretical Perspectives and Empirical Evidences'. Presented at The Internationalization of Corporate R&D: An International Seminar, Montréal, UQAM-CIRST, August.

Soete, L. G. 1979. 'Firm Size and Inventive Activity: The Evidence Reconsidered'. *European Economic Review* 12: 319–40.

Soete, L. 1981. 'A General Test of Technological Gap Theory'. *Weltwirtschaftliches Archiv* 117: 638–66.

Soete L. 1987. 'The Impact of Technological Innovation on International Trade Patterns: the Evidence Reconsidered'. *Research Policy* 16: 101–30.

Statistics Canada. 1990. *The Statistics Canada Business Register*. Ottawa: Systems Development Division.

Statistics Canada. 1991. *Industrial Research and Development*. Catalogue No. 88-202. Ottawa: Statistics Canada.

Sterlacchini, A. 1994. 'Technological Opportunities, Intraindustry Spillovers and Firm R&D Intensity'. *Economic Innovation and New Technology* 3: 123–37.

Stiglitz, J. E. 1972. 'Some Aspects of the Pure Theory of Corporate Finance: Bankruptcies and Take-overs'. *Bell Journal of Economics and Management Science* 3(2): 458–82.

Stuckey, J. A. 1983. *Vertical Integration and Joint Ventures in the Aluminum Industry*. Cambridge, MA: Harvard University Press.

Tassey, G. 1991. 'The Functions of Technology Infrastructure in a Competitive Economy'. *Research Policy* 20: 345–61.

Taylor, C. T., and Z. A. Silbertson. 1973. *The Economic Impact of the Patent System: A Study of the British Experience*. Cambridge: Cambridge University Press.

Teece, D. J. 1977. 'Technology Transfer by Multinational Firms: The Resource Cost of Transferring Technological Know-how'. *The Economic Journal* 87 (346): 242–61.

Teece, D. J. 1986. 'Profiting from Technological Innovation: Implications for Integration, Collaboration, Licensing and Public Policy'. *Research Policy* 15: 285–305.

Tisdell, C. 1995. 'Evolutionary Economics and Research and Development'. In *Economic Approaches to Innovation*. Edited by S. Dowrick. Aldershot, UK: Edward Elgar.

Tomlinson, M., and R. Coombs. 1998. *Comparing the Innovative Behaviour of 'British' and 'Foreign' Firms Operating in the UK*. CRIC Discussion Paper No. 23. University of Manchester.
Townsend, J., F. Henwood, G. Thomas, K. Pavitt and S. Wyatt. 1981. *Science and Technology Indicators for the UK: Innovations in Britain Since 1945*, SPRU Occasional Paper Series No. 16. University of Sussex.
Utterback, J. M., M. Meyer, E. Roberts, and G. Reitberger. 1988. 'Technology and Industrial Innovation in Sweden: A Study of Technology Based Firms Formed Between 1965 and 1980'. *Research Policy* 17: 25–26.
Vernon, R. 1966. 'International Investment and International Trade in the Product Life Cycle'. *Quarterly Journal of Economics* 80: 190–207.
Von Hippel E. 1982. 'Appropriability of Innovation Benefit as a Predictor of the Source of Innovation'. *Research Policy* 11: 95–115.
Von Hippel, E. 1988. *The Sources of Innovation*. New York: Oxford University Press.
Walsh, V. 1991. 'Demand As a Stimulus to Innovation'. Paper given to Demand, Public Markets and Innovation in Biotechnology Workshop. Manchester School of Management, U.K: UMIST.
Warda J. 1990. *International Competitiveness of Canadian R&D Tax Incentives: An Update*. Ottawa: The Conference Board of Canada.
Wilkinson, B. W. 1968. *Canada's International Trade: An Analysis of Recent Trends and Patterns*. Montréal: The Private Planning Association of Canada.
Williamson, O. E. 1975. *Markets and Hierarchies: Analysis and Antitrust Implications*. New York: The Free Press.
Williamson, O. E. 1985. *The Economic Institutions of Capitalism*. New York: The Free Press.
Wolff, E. N., and M. I. Nadiri. 1993. 'Spillover Effects, Linkage Structure, and Research and Development'. *Structural Change and Economic Dynamics* 4: 315–63.
Woodward, J. 1958. *Management and Technology*. London: HMSO.
Wynarczyk, P., R. Watson, D. Storey, H. Short and K. Keasey. 1993. *Managerial Labour Markets in Small- and Medium-Sized Enterprises*. London: Routledge.
Zimmerman, K. F. 1987. 'Trade and Dynamic Efficiency'. *Kyklos* 40: 73–87.

Index

appropriability, 23, 28, 209–10, 219, 225, 233, 254, 330, 357, 406–8, 423, 433, 438
arm's-length transactions, 74, 167, 349–52, 436
asymetric information (*see* innovation: financing: asymetric information)

benchmarking, 382

Canadian multinationals (*see* multinational firm: Canadian-owned)
capital
  abundance of, 315
  cost of, 140, 315, 328–9
  goods, 297, 372–41
  market, 25, 328–9, 330–2, 348
  venture, 330–2, 439
collaboration
  with competitors, 37, 105–8, 120–1
  with customers and suppliers, 105–8
  foreign-owned firms, 277, 286–7
  with R&D partners, 118, 120–1, 123–5, 174, 365
  with related firms (affiliates), 105–8
  with universities and colleges, 108, 120
competencies (*see* firm-specific competencies)
competition 416, 421
  conditions of, 149
  oligopolistic, 185, 368
competitors, 77, 149
copyright
  effectiveness of, 229–31, 257–9
  use of, 225–57
cross-licensing, 353

diffusion
  ideas of, 435–6
  innovation of, 371–4
domestic multinationals (*see* multinational firm: Canadian-owned)

effects of innovation activity on
  capital requirement, 133–8
  economies of scale, 133–7, 430
  employment, 132, 155
  energy requirement, 133–7
  interaction with customers, 133–7
  interaction with suppliers, 133–7
  labour costs, 155
  labour requirements, 133–7
  lead times, 133–7
  material requirements, 133–7
  organization of production, 131
  plant specialization, 133–7
  product diversity, 133–7
  product range, 133–7
  production, 131
  production costs, 133–7
  production flexibility, 134–7, 154
  reorganization of workflow, 133–7
  response to consumer needs, 430
  share in foreign market, 154–5
  workers' skills, 131–2, 154–5
effects of the most profitable innovation on
  capital requirements, 140–1
  design requirements, 140–1
  domestic market, 138–40, 155
  employment, 144–52, 154–5, 212–14, 217, 315–16, 430, 446
  environmental standards and regulations, 142–4, 155
  energy savings, 140–1

effects of the most profitable innovation on (*cont.*)
  foreign market share, 138–40, 155
  interaction with customers, 138–9
  interaction with suppliers, 138–9
  labour cost, 140–1
  lead times, 138–9
  material costs, 140–1
  product diversity, 197–8
  product quality, 138–9, 154, 196, 431
  product range, 138–9, 154
  production cost, 140–1, 154, 430
  profitability, 138–9, 154–5
  technical improvements, 140–1
  workers' skills, 154, 316
  working conditions, 154, 197–8
endogeneity of innovation and intellectual property use, 411
engineering, 65–6, 200–1
export intensity, 165, 312–13, 431, 449
exports, 153–5, 198–9
  by Canadian-owned firms, 312–13
  by foreign-owned firms, 312–13
externalities, 64, 433
  internalization of, 10, 65, 75, 106, 349, 365, 436
external sources of ideas for innovation, 166–70, 182, 199–202, 430–4
  competitors, 77, 79, 93, 120, 201–3, 417
  customers, 75, 120, 201–3, 417, 425
  demand related, 78–9, 201–3
  industry differences, 78–9
  market partners, 66–7
  market-related partners, 74–5
  by ownership, 294–5
  patent system, 77, 79, 200
  public R&D institutions, 77, 79, 94, 202
  publications, 77
  related firms, 75–6, 79, 83–4, 121, 201–3, 287–8, 290–3
  spillovers, 67, 75, 79, 83–5, 201–3, 433
  suppliers, 75, 79, 81–2, 83–4, 120, 201–3
  technological infrastructure, 66, 85, 202
  trade fairs, 77, 79
  universities and colleges, 77, 79, 94, 294–5
external sources of technology, 89–90, 205–7, 294, 443–4
  consultants, 89–90
  customers, 89–90
  internal and external, 90, 94
  related firms, 206–7, 287
  relationship with internal ideas, 90, 94
  R&D institutions, 206–7
  suppliers, 89–90, 94, 206–7
  trade fairs, 90, 94
  universities and colleges, 89–90, 206–7
  unrelated firms, 206–7

financing innovation (*see* innovation: financing)
firm
  multiplant, 65
  operations of, truncation, 24, 273–6
  ownership of, 24
  related (parent and sister companies), 287–8, 436
firm size
  cost of innovation, 323–5
  effects of innovation, 150, 446
  impediments to innovation, 179–81
  innovation incidence, 21–2, 39, 45–7, 50, 58, 121–2, 134, 147, 156–64, 182–4, 445–50
  innovation intensity, 189–91
  joint venture, 360
  large (IP), 39, 159
  licensing, 358
  ownership, 271–2
  probability of innovation, effect on, 405, 413, 418–19, 423
  R&D, 171–8, 182–4, 446–8
  small (NIP), 39, 348
  sources of funds, 328–330, 448–450
  sources of innovation, 166–70, 446–7
  tax credit claims, 337, 348, 448
  use of intellectual property, 234–7, 261, 448
firm-specific competencies (also strategies), effect on employment, 148
foreign direct investment,
  alternatives to, 266–7
  licensing, 266
  new theories of, 267–8
  transfer of technology, 294
foreign-owned firm
  dependent, 293, 319, 452
  employment, 313–15
  exports, 312–13, 453
  financing innovation, 453
  independent, 293, 319, 452
  number of innovations introduced, 306–8
  originality of innovation, 303–6, 320
  property rights, 308–10, 453
  type of innovation, 302–3
foreign ownership
  effect on probability of innovation, 405, 415
  by industry sector, 271
foreign subsidiaries (*see also* foreign-owned firm; multinational firm)
  dependent, 293, 319, 452, 452–3
  independent, 293, 319, 452, 452–3
Frascati manual, 15, 97, 100, 115

Index

government support to
  innovation, 25, 184, 217–18, 333–4, 438–9, 451
  R&D, 181–4, 333–48, 438–9, 446–7

health and safety regulations, 142–3, 155
human resource strategy, 148, 389–92, 440

impediments to innovation
  Canadian-owned firms, 297–9
  effect on employment, 149
  foreign-owned firms, 297–9
  improved management, 148
  lack of market information, 210–12
  lack of skilled labour, 210–12
  by originality, 209–12
  public policy to reduce, 451
  by size of firm, 179–81, 184, 448
  standards and regulations, 211
industrial design
  effectiveness of, 229–33, 240–3, 253–4
  industry differences, 249–53
  originality of innovation, 248–9
  type of innovation, 246–8
  use of, 224–9, 237–40
industry
  effect on innovation, 417
  taxonomy, 8, 48, 441–2
infra-technologies (*see* technological infrastructure)
innovation
  appropriation of benefits of, 210
  Canada-first (*see* innovation: originality of)
  by Canadian and foreign-owned firms, 270–2, 299–302, 405–7, 451–3
  combined product-process (complex), 192–4
  competencies (strategies) for, 27, 402
  competitive conditions, effect of, 409, 416
  cost of
    composition, 24–5, 323–7, 345, 438–9
    manufacturing start-up, 326–7
    marketing, 326–7
    R&D, 326–7
    technology acquisition, 326–7
  definitions of, 31, 33, 121, 187
  determinants (*see* innovation: probabilistic model of innovation determinants)
  diffusion of, 350, 371–4
  effects of (*see* effects of innovation activity; effects of the most profitable innovation)
  expenditures on, 98–103
  features of, 37, 57, 194–6

  financing, 438–9
    asymetric information, 328
    domestic vs. foreign-owned firms, 332–3, 452–4
    external sources, 328–33, 439
    firm size by, 6–7, 21, 46, 156, 418, 429
    foreign-owned firms, 332–3
    by government, 333–45, 439, 444
    internal sources, 328–34, 345, 439, 444
    tax credits, 335–45 (*see also* tax credits)
    venture capital, 330–2, 439
  by foreign-owned firms, 299–302, 405–7, 451–3
  government funding of, 333–5, 451
  government support, theory of, 322–3, 333–4
  imitative, 164, 189, 450
  incidence of, 45, 163, 188–91, 412–34
  incremental, 186
  industry differences, 6, 8, 48, 61, 70, 148–9, 190–1, 441–5
  inputs, 14–16, 37–8
  intensity of, 17, 43, 55, 163, 183, 435
  interindustry flows of, 372–4
  linear model of, 66, 98–100
  major, 34, 160
  measure of, 160–2
  minor, 34, 160
  importance of, 53
  originality of, 13, 22, 34–6, 52–5, 61, 125–7, 164, 187, 192–4, 421–4, 428–9, 450–1
  outputs, 12–14
  ownership of firm, nationality of, 402, 429
  private returns to, 333
  probabilistic model of innovation determinants, 397–426
    conclusion, 423
    dependent variables, 401
    explanatory variables, 402
    novelty (originality), 421
    patent use, 417
    probability of, 400–7
    regression results, 412
    type of innovation, 418
  process, 34, 57, 61, 191–4, 418–21, 441–2
  product, 34, 57, 191–4, 418–21, 441–2
    vs. process, 7, 428
    vs. process, cost composition of, 325–8
  in response to government regulations, 132, 154, 430–1
  sectoral flows of, 48–51, 371–4, 435
    Canada, 373–4
    U.S. and U.K., 373
  skilled workers, use of, 389–92
  social returns to, 333–4

innovation (*cont.*)
  sources of (*see* external sources of ideas for innovation; internal sources of ideas for innovation)
  strategies of, 3, 27, 380–96, 404–10
  surveys of (*see* surveys of innovation)
  transfer of, 371–5
  type of, 7, 51, 59, 428
  users of, 371–4
  world-first (*see* innovation: originality of)
innovation regimes, 430
  firm size differences, 445–50
  by nationality of ownership, 451
  by originality, 185–218
innovators
  characteristics of, 3, 27, 439–40
  competencies and strategies of, 415
  competitive strategies of, 380–3, 410
  human resource strategies of, 389, 410, 416
  management strategies of, 392–4, 410
  marketing strategies of, 387–9, 410, 415, 421
  production strategies of, 385–7, 410, 416
  strategic capabilities and competencies of, 380–3, 440
  technological strategies of, 383–5, 410, 415–21, 444
integrated circuit design, 226–8
intellectual property rights
  characteristics of innovations, 244–9
  comparison Canada-U.S., 233
  effectiveness of, 9, 23, 37, 229–33, 253–60, 262–4, 406
  firm size, 226, 234–7, 240–3, 257–9, 261, 422, 448
  industry differences, 249, 256, 263
  innovative and non-innovating firms, 237–40
  originality of innovation, 248–9, 262
  other strategies, 233–4, 416–17
  products vs. processes, 246–9
  regression model of, 255–60
  use of, 9, 23, 37, 207–9, 219–21, 226–8, 237–40, 251–3, 260–1
    in Canada, 226–9
    by domestic firms, 257
    by foreign affiliates, 257
  in technology transfer agreements, 357
intermediate products, 372–3
internal sources of ideas for innovation, 70–2, 182, 199–201, 430–4
  industry differences in, 72
  management, 83, 91
  R&D, 66–8, 83–4, 91–2, 167–70, 199–200
  by ownership, 287–9
  sales/marketing department, 79, 93

internal sources of technology, 169, 204–5, 294, 443
  experimental development, 87, 204–5
  process versus product innovation, 81
  production engineering, 87, 94–5, 200, 204–5
  R&D, 87, 204–5

joint ventures, 364–371, 436
  collaboration, 364–71
  incidence of, 365–6
  model of, 368
  R&D collaboration, 366–8, 376

knowledge
  appropriation of, 10, 106, 406, 417
  codifiable, 167, 180–2
  economic characteristics of, 8, 63–5
  externalities (*see* knowledge: spillovers)
  generation of, 63–5, 86
  internalization of, 365
  market failure, 9
  processing, 65
  proprietary, 299
  spillovers, 9, 64, 75–6, 86, 433, 443
  tacit, 167, 180–2, 225
  transfer, 65, 349–50

labour
  demand for, 213–14
  skilled, 212–14, 217, 389–92
    lack of, 149, 210–11
  labour market, 212–14
large firms (IPS), 39, 159
learning, 71
license agreement, 351, 357, 375–6, 436

management strategy, 148
manufacturing start-up, 326–7
  cost (*see* innovation: cost of)
market
  imperfections of, 9, 209–10
  share, 130, 138–40, 142–5, 154–5, 197–9, 212
marketing strategy, 148, 387–9, 395, 402, 404
microfirms, 160–2, 171–2
multinational firm, 23–4, 39, 44, 451–4,
  Canadian innovation system in, 268–9
  Canadian-owned, 24, 269, 273, 276, 288, 301–2, 320–1
  collaboration, R&D, 284–7
  hub and spoke model of, 268–9
  incidence and organization of R&D, 273
  parent company of, 267, 452
  R&D, 452
  theory of, 266–8

neoclassical model of innovation, 5–6
non-innovators, 12
  competencies of, 16
  intellectual property rights, 37, 241, 261
  R&D characteristics, 121, 128–9
  strategies of, 27, 378–80, 383–96, 439–40
  vs. innovators, 401, 422

Oslo manual 13, 17 (*see also* surveys of innovation: The European Community Innovation Survey)
ownership, 7–8, 24, 44, 74–5, 93, 147, 180, 256–60, 265–321, 332–3, 402–3, 405–6, 429, 451–4
  vs. trade orientation, 316–17

patent
  alternate strategies to, 379, 395
  appropriability, 406–8
  cross-licensing, 353–4
  definition of R&D, 97, 117
  effect on innovation, 415–21, 423
  effectiveness, 229–33, 257–9
  filings, international comparison of, 221–4
  incentives for innovation, 9, 406–8
  individual inventors, 162
  propensity to, 157
  races, 9
  right to use, 356–7
  sectoral flows of, 373–4
  spillovers, 9, 93
  statistics, 221–4
  use of, 23, 37, 79, 207–9, 220–64, 436–8
    by firm ownership, 308–11
    by industry, 398
  world-first innovation, 218
patent office as source of information, 75, 77, 202
patent system, 77
plant breeders' right, 225–7, 230, 235, 241, 244–9
product life cycle, 30, 43–4, 185–6, 199, 212, 217, 451
production engineering, 67, 71, 169, 186, 294, 435, 443
production strategy, 148, 385–7
profitability, 20, 110, 130–2, 138–41, 154–5
public good, 64, 167, 209–10
  labor market, 209
  universities, 120
public policy
  on innovation, 217–18
  small firms for, 184
publications, professional, 77–8

R&D (research and development)
  basic and applied research, 100
  collaboration, 74, 105–8, 112–15, 123–5, 365
    joint ventures, 365–8
  conductor of, 103, 170–2
  consortium, 365
  contracted out, 105–6, 174, 203–4
  definitions, 96–7, 115
  department, 37, 66, 104, 114, 203–5, 434–5
  efficacy, large and small firms, 176–9
  experimental development, 169
  firm size, 156–9
  foreign competition, 284–7
  foreign-owned firms, 110, 128, 269
  foreign vs. domestic firms, 275, 284–6, 320
  impediment to innovation, 180
  incidence of, firm, foreign and Canadian-owned, 269, 413
  industry differences, 109, 118–19
  innovating firms, 177–9, 182–3, 200, 203–4
  innovation performance, 340–5, 347
  institutions, 120
    public, 77–8
  intensity, 67, 128
  joint ventures, 364–71, 376, 433
  management, 105–6
  networks (*see* R&D: collaboration; R&D: partnerships)
  non-innovating firms, 121, 128, 177–9, 183
  occasional, 37, 100, 110–12, 434–5
  ongoing (continuous), 37, 110–12, 126–7, 434–5
  organization, 104, 112, 172–4, 203, 418
    in other departments, 104–5, 203–4
  oversupply of, 9
  partnerships, 93, 435, 442 (*see also* R&D: collaboration)
  performer of, 103, 170–2
  role in innovation, 3, 19, 36–7, 98, 121, 171–5, 422, 425, 434–5, 446–8
  spillovers, 77, 179
  strategic alliances, 364–71, 376, 433
  subsidies (grants), 334
  tax credits, 117,
    claims for, 334–45, 347–8
    claims for, by industry, 337–9, 347
    use by originality of innovation, 339
    use by small vs. large firms, 338, 348, 439
R&D-centric model, 67, 72, 92, 431

Schumpeterian hypothesis, 279
Scientific Research and Experimental Development Tax Credit Program, 102, 334–5

skilled labour (*see* labour: skilled)
sources of ideas for innovation
  external (*see* external sources of ideas for innovation)
  internal (*see* internal sources of ideas for innovation)
sources of technology
  external (*see* external sources of technology)
  internal (*see* internal sources of technology)
spillovers of ideas, 49, 77, 93, 433
standards and regulations, 22, 142–4
  impediment to innovation, 210–11, 297–9
strategic alliance, 365–8
  model of, 368–71
strategies (*see* innovators: strategic capabilities and competencies of)
surveys of innovation
  The (European) Community Innovation Survey, 13, 31, 45
  OECD – Oslo manual, 13, 17
  questionnaire for, 457–89
  Survey of Innovation and Advanced Technology 1993, 1, 17, 34, 38, 87, 99, 148, 187, 455–6

tax credits, 25, 102–3, 109, 116–18, 127, 334–48, 438–9, 445, 448
technological infrastructure, 60, 66, 75–7, 83–7, 93, 202–3, 399–402, 426, 432–3, 443
technological opportunities, 7–8, 18, 48, 59, 67, 71, 76, 320, 363, 408–9, 415
technology
  acquisition of, 15, 87, 326–7, 329, 363, 375, 435–6
  adoption of, 137, 154, 266, 389–94, 430
  by foreign-owned firms, 266, 270
  information, lack of, 180
  labour saving, 212–14
  proprietary, 266, 272, 317–19, 451
technology strategy, 148
  foreign-owned firms, 297
technology transfer, 11, 26, 83, 167–8, 294–7, 350–3, 370–1, 375–7, 436, 441
  agreements, characteristics of, 353–4
  continuous, 297, 351, 363–4, 436
  cost, 352
  forms of, 350–3
  intrafirm, 168
  joint ventures, 26, 40, 364–71
  model of, 360–4
  one-time, 351, 363–4, 436
  originality of the innovation, 354
  rights and restrictions, 354–60, 363–4
trade fairs, 75–87, 90, 93–4, 202, 433, 443
trade secrets
  effectiveness of, 229–33, 257–9
  use of, 224, 226–55, 255–7, 416–17
trademarks
  effectiveness of, 229–33, 257–9
  use of, 224, 226–53, 255–7
truncated structure, 24, 268–77, 319–21, 451–4

unionization, 147
universities and colleges, 11, 15, 19, 22, 26, 75–8, 81–4, 90, 94, 118–21, 124–8, 205–7, 286, 295, 320, 408, 431–2, 435

WESTMINSTER COLLEGE
LIBRARY